# Springer Complexity

Springer Complexity is an interdisciplinary program publishing the best research and academic-level teaching on both fundamental and applied aspects of complex systems - cutting across all traditional disciplines of the natural and life sciences, engineering, economics, medicine, neuroscience, social and computer science.

Complex Systems are systems that comprise many interacting parts with the ability to generate a new quality of macroscopic collective behavior the manifestations of which are the spontaneous formation of distinctive temporal, spatial or functional structures. Models of such systems can be successfully mapped onto quite diverse "real-life" situations like the climate, the coherent emission of light from lasers, chemical reaction-diffusion systems, biological cellular networks, the dynamics of stock markets and of the internet, earthquake statistics and prediction, freeway traffic, the human brain, or the formation of opinions in social systems, to name just some of the popular applications.

Although their scope and methodologies overlap somewhat, one can distinguish the following main concepts and tools: self-organization, nonlinear dynamics, synergetics, turbulence, dynamical systems, catastrophes, instabilities, stochastic processes, chaos, graphs and networks, cellular automata, adaptive systems, genetic algorithms and computational intelligence.

The two major book publication platforms of the Springer Complexity program are the monograph series "Understanding Complex Systems" focusing on the various applications of complexity, and the "Springer Series in Synergetics", which is devoted to the quantitative theoretical and methodological foundations. In addition to the books in these two core series, the program also incorporates individual titles ranging from textbooks to major reference works.

## Editorial and Programme Advisory Board

# Understanding Complex Systems



Future scientific and technological developments in many fields will necessarily depend upon coming to grips with complex systems. Such systems are complex in both their composition - typically many different kinds of components interacting simultaneously and nonlinearly with each other and their environments on multiple levels - and in the rich diversity of behavior of which they are capable.

The Springer Series in Understanding Complex Systems series (UCS) promotes new strategies and paradigms for understanding and realizing applications of complex systems research in a wide variety of fields and endeavors. UCS is explicitly transdisciplinary. It has three main goals: First, to elaborate the concepts, methods and tools of complex systems at all levels of description and in all scientific fields, especially newly emerging areas within the life, social, behavioral, economic, neuroand cognitive sciences (and derivatives thereof); second, to encourage novel applications of these ideas in various fields of engineering and computation such as robotics, nano-technology and informatics; third, to provide a single forum within which commonalities and differences in the workings of complex systems may be discerned, hence leading to deeper insight and understanding.

UCS will publish monographs, lecture notes and selected edited contributions aimed at communicating new findings to a large multidisciplinary audience.

Octavian Iordache

# Polystochastic Models for Complexity

Springer

**Author**

Dr. Octavian Iordache

Polystochastic
3205 Blvd Pitfield
St-Laurent, Montreal Quebec
Canada H4S 1H3
E-mail: polystochastic@bellnet.ca

ISBN 978-3-642-10653-8 e-ISBN 978-3-642-10654-5

DOI 10.1007/978-3-642-10654-5

Understanding Complex Systems ISSN 1860-0832

Library of Congress Control Number: 2010920829



*Typeset & Cover Design:* Scientific Publishing Services Pvt. Ltd., Chennai, India.

Printed on acid-free paper

9 8 7 6 5 4 3 2 1

springer.com

# Preface

This book is devoted to a domain of highest industrial and scientific interest, the complexity. The complexity understanding and management will be a main source of efficiency and prosperity for the next decades.

Complex systems are assemblies of multiple subsystems and are characterized by emergent behavior that results by nonlinear interactions among the subsystems at multiple levels of organization. Evolvability that is the ability to evolve is the method to confront and surpass the successive boundaries of complexity. Evolvability is not biological but should be considered here in the sense that the corresponding systems have, at different levels, characteristics that are naturally associated to the living systems. The significance of the complexity and the phenomena of emergence are highlighted in the first chapter of the book. The implication of concepts as level of reality, circularity and closure for evolvable systems is evaluated.

The second chapter of the book exposes the methodology to analyze and manage complex systems. The polystochastic models, PSMs, are the considered mathematical tools. PSMs characterize systems emerging when several stochastic processes occurring at different conditioning levels, are capable to interact with each other, resulting in qualitatively new processes and systems. Innovative are the higher categories approach and the introduction of a partial differential model for multiple levels modeling. This imposes making use of appropriate notions of time, space, probabilities and entropy.

Category theory is the formalism capable to outline the general framework, shared by the functional organization of biological organisms, of cognitive systems, by the operational structure of evolvable technologies and devices and after all by the scientific and engineering methods.

The modeling hierarchy, which is modeling at several abstraction levels, is deep-rooted in the higher categories study. It is the high complexity that imposes to develop higher dimensional modeling. The difficulty to study complexity for a system is related to the lack of a higher dimensional theory for that system. Models of models, that is, meta-models allowing studying processes of processes, then meta-meta-models, are presented in the second chapter of the book.

The next three chapters illustrate the applicability of PSMs for the main domains of the scientific knowledge-sciences of matter, biosciences and cognitive sciences. Since complex systems demands a transdisciplinary approach the book is inherently transdisciplinary. The book follows principles from different disciplines, highlights them at abstract level and shows how to correlate these to the detailed models of data for real systems.

The third chapter of the book is devoted to physical and chemical systems. Presented case studies are flow-sheet synthesis, cyclic operations of separation, drug delivery systems and entropy production. Evolvability issues are emphasized.

Studies of biomimetic systems represent the main objective of the fourth chapter.

The included analyses refer to bioinspired calculation methods, to the role of artificial genetic codes, neural networks and neural codes for evolutionary calculus and for evolvable circuits as example of biomimetic devices.

The fifth chapter, taking its inspiration from systems sciences and cognitive sciences outlines the potentiality of the PSMs for engineering design, case base reasoning methods, failure analysis, and multi-agent manufacturing systems.

Perspectives and integrative points of view are discussed in the sixth chapter of the book with reference to the classification of sciences, cybernetics and its extensions, and to categorification as the new wave of transdisciplinarity, coming after complexity.

An appendix introduces the necessary elements of category theory.

This monograph is a development of the research program presented in the previous books devoted to PSM and applications (Iordache 1987, 2009). The conventional demand for optimal and adaptive technologies and devices is challenged today by the request to build up systems that are, at different degrees, adaptive, cognitive, intelligent and lastly evolvable and autonomous in their environment. PSMs offer an answer to this change of requirement from adaptable to evolvable, from a low dimensional to a higher dimensional insight.

The domain of PSM is relatively new and not well established yet. Some of the models and methods presented here are still tentative, and some implications are unexpected and partially verified. The book is a personal view, which is in line with the existing state of the art but contains perspectives that may be considered to some extent as controversial. In spite of inherent difficulties, we are confronted with one of the key field of major practical interest and a promising area of investigation for the general science of complex systems and processes.

# Contents

# List of Figures

# List of Tables

# Abbreviations

AI-artificial intelligence
AL-artificial life
ANN-arificial neural network
BDI-beliefs desires intentions
BGPI-beliefs goals plans intentions
CAD-computer aided design
CAPE-computer aided process engineering
CBR-case based reasoning
DDS-drug delivery systems
DNA-deoxyribonucleic acid
DOE-design of experiment
EC-evolvable circuit
ECF-electrochemical filament
EDOE-evolvable design of experiment
FA-failure analysis
GA-genetic algorithm
GDT-general design theory
GF(m)-Galois field
GP-genetic programming
IRDS- information resource dictionary system
ISO-international standardization organization
MAS-multi agent system
MML-meta modeling language
MOF-meta object facility
NN-neural network
OA-orthogonal array
OMG-object management group
PCB-printed circuit board
PSA-pressure swing adsorption
PSM-polystochastic model
RDTE-research development test evaluation

RNA-ribonucleic acid
RSCC-random system with complete connections
SASO-self adaptive self-organizing
SKUP-S-states, K-conditions, U-operators, P-possibilities
SMB-simulated moving bed
UML-unified modeling language
WE-wave equation

# Chapter 1
# Introduction

**Abstract.** Complexity and emergence are introduced here in relation with the self-organization of systems in levels of reality.

Evolvability defined as the ability to evolve is the projected way to confront and surpass the successive levels of complexity. Polystochastic models allow refocusing from adaptable to evolvable, from a low dimensional to a higher dimensional insight.

Significant concepts for evolvability as level of reality, circularity, semantic closure, functional circle, circular schema and integrative closure are presented.

The correlation with organic computing or autonomic computing research areas is highlighted.

## 1.1 Complexity

What is a complex system, and what does it means for a system to be complex and emergent?

The study of complex systems or more generally the science of complexity has been a hot research topic for the last decades.

Usually a complex system is described as a structure or process involving non-linear interactions among many parts and levels, which displays emergent properties. In other words this means that the aggregate system activity is not derivable from the linear summations of the activity of individual components and that novel structures, patterns or properties arise, from simple interactions among parts.

Complex systems are ones in which patterns can be seen and understood, but interplay of individual parts cannot be reduced to the study of individual parts considered in isolation from one another. A survey of the literature indicates that there is no standard agreed upon definition of a complex or emergent system. Some of the definitions may even seem contradictory but they may make sense when applied to particular types of systems and from which perspective one choices to observe (Adami 2002, Boschetti et al. 2005, Cariani 1992, Goldstein 1999, Kauffman 1995). This suggests considering several domains of complexity and a hierarchy of levels for complexity. The complexity for an inorganic system should be different from the complexity of a biological, cognitive or intelligent system.

An example of physical complex system is the global climate, including all components of the atmosphere and oceans and taking into account the effects of extraterrestrial processes as solar radiation, tides, and meteorites. An illustration of complex biological system is the human brain composed of millions of nerve cells. Their collective interaction allows recognizing visual, acoustic or olfactory patterns, speaking and performing different mental activities. An example of complex social system is the human society with its participants, natural resources and capital goods, financial and political systems. For the logical and mathematical realm, examples of high complex calculus systems may consists of large scale distributed software systems or hierarchies of layered computing subsystems organized and running together to achieve particular objectives.

What is remarkable is that systems that have apparently little in common-material systems as an array of polymers in a test tube, biological systems as a group of receptors on a cell's surface, knowledge or cognitive systems as a group of ants in a swarm or human agents in a company -often share remarkably similar structures and means of organization. This explains and justifies the need for a science of complexity.

Features such as non-linearity, hierarchy of levels, time-scales, connectivity, non-equilibrium, unpredictability, interconnectivity, collective behavior, self-organization, self-production, self-reference, and multi-agency are associated with complexity studies. Complexity is correlated to non-linearity, which is a necessary but not sufficient condition of complexity, as well as to interconnectivity, self-organization, self-similarity and collective behavior (Mainzer 1996).

The understanding of complexity changes with the domains of application. Some surveys consider that complexity has not an absolute meaning, and it is only a relative notion depending on the level of observation or abstraction. However, it is commonly stated and accepted that some objects and processes are more complex than others.

We must to take into account this facet of complexity as a relative concept which depends both on the task at hand and on the tools available to achieve this task.

For industrial systems, despite the fact that numerous physical or chemical processes are identified as complex, more of the conventional ones may be operated in regimes were complexity properties are neglected. For several centuries, physical and chemical sciences made great steps by experimenting and constructing simplified models of complex phenomena, deriving properties from the models, and verifying those properties by new experiments. This approach worked because the complexities ignored in that models were not the essential properties of the phenomena. It does not work when the complexity becomes the essential characteristic. In an increasing number of cases the complexity is not transient or atypical, but it is an intrinsic property of that systems.

Given this situation the challenge for engineers and scientists is not only to identify complexity domains but also to show how to overtake the successive complexity barriers. The next defy in science and technology is to surmount the complexity, finding the ways from complexity to a new simplicity. The $21^{st}$ century concerns as energy, nutrition, health, ecology, finance and security pertain without doubt to high complexity sphere and need new methodologies.

## 1.2 Emergence

As for complexity there is not a consensus on a standard definition of emergence (Bonabeau and Desalles 1997, Boschetti et al. 2005, De Wolf and Holvoet 2004).

However emergence is the property that distinguishes and makes complex systems interesting to study. Emergence is the phenomenon that differentiates complex systems from complicated or multi-component ones. A complicated device is a simple addition of separable preexistent components, whereas a complex system is made out of a large number of components interconnected by a network of relations in such a way that a modification somewhere in the system modifies the whole system.

Emergence is what parts of a system do together that they would not do by themselves. An example is the collective behavior, that a system does by virtue of its relationship to its environment but it would not do by itself. Important facets of emergence concept are: novelty, micro-macro effects, coherence, nonlinearity, interacting parts, local control, robustness, and flexibility.

There are some general trends in the definition and study of emergence.

We will focus mainly towards the emergence of levels. The notion of emergence involves the existence of levels of organization, of description, of behavior and so on.

A direction of studies tries to relate the emergence to semantics and meaning (Pattee 1995, 1997). The emergence is considered as linked to the concept of evolution and its bioinspired models as the genetic algorithm. In addition, the Darwinian-type of evolution was considered the engine for emergence in biology.

According to this route of research, a higher-level of investigation is completely defined by considering not only how the entities involved interact but also what meaning is associated to the interactions by an external observer or by the entities themselves. Correlated to the above studies is the definition of emergence relative to a model (Cariani 1989, 1992). This definition does not consider emergence to be an intrinsic absolute property of a phenomenon, but that it can only be defined by considering the phenomenon with respect to an observer which could be a formal model, for instance. This represents the source of a functional theory of emergence giving an account of how new necessary functions of the observer-measurements, computations and controls, can come into being.

Another class of definitions for emergence emphasizes the role of multiple levels of organization and their interaction. The starting point is that the majority of complex systems exhibit hierarchical self-organization in levels under selective constraints. Self-organization will occur when individual independent parts in a complex system interact in a jointly cooperative manner that is also individually appropriate, such as to generate a higher level organization. Complex systems can be observed at different levels of investigation. For example we can observe an industrial installation at the level of molecules or at the level of devices interactions.

The number of reality and observation levels is inherently finite.

Among the significant versions of the modern theory of levels are that developed by Hartmann (Hartmman 1940, 1952), Piaget (Piaget 1977), Poli (Poli 2001, 2007), and Nicolescu (Nicolescu 2002).

Hartmann considers four levels of reality and associated objects: material or inanimate, biological or animate, mind-related or psychological, and intelligent or spiritual and emphasizes the finite number of sublevels to be taken into account at any basic level.

Poli distinguishes at least three ontological strata of the real world: the material, the psychological and the social stratum.

Fig. 1.1 shows the Hartmann's four-levels of the real world. Hartmann ranked these four levels in the hierarchy: material $<$ biological $<$ cognitive or psychological $<$ intelligent or spiritual. Intelligence may be assimilated to higher cognitive potentiality.

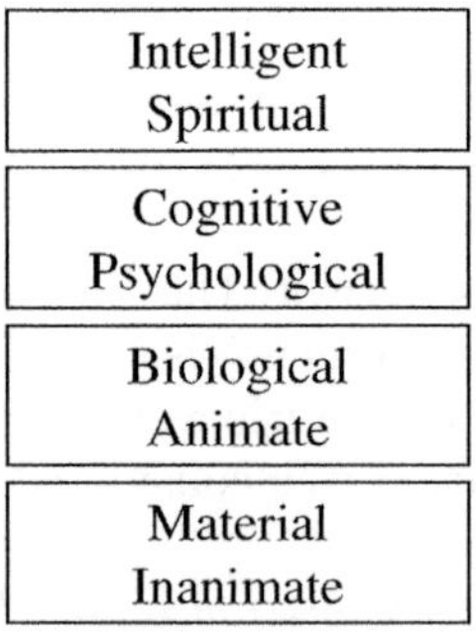


**Fig. 1.1** Four levels world-hierarchical framework

The complexity levels follow the hierarchy of reality levels.

Hartmann elaborated several laws for his hierarchy of levels.

- Law of recurrence: Lower categories recur in the higher levels as subaspects of higher categories, but never reversely.

- Law of modification: The categorial elements modify during their recurrence in the higher levels: they are shaped by the characteristics of the higher levels.
- Law of the *novum*: The higher category is composed of a diversity of lower elements. It contains a specific *novum* which is not included in the lower levels.
- Law of distance between levels: The different levels do not develop continuously, but in jumps. The levels can be clearly differentiated. This supposes that there exists a kind of autonomy of a level implying existence of properties, relations, behavioural laws concerning entities at a given level independently from other levels.

The law of the *novum* ascertains Hartmann as one of the modern discoverers of emergence—originally called by him "categorial *novum*" that is a new category.

A property is considered as emergent if it cannot be explained from the properties and interactions of the lower level entities. However, it is mandatory to explain from were novelty comes.

Following the hierarchy of levels shown in Fig.1.1, the hierarchy of emergence types is resulting. There are four levels of emergence, each level corresponding to a reality level. On the basic level there are atoms and organic molecules. On the next level there is the emergence of life, on the next level the emergence of conscious states and then, the emergence of product of human mind, such as the scientific theories (Hartmann 1940, Popper 1987). The existence of material systems allowed the $1^{st}$ order emergence of biosystems, followed by the $2^{nd}$ order emergence of cognitive systems, and this in turn by the $3^{rd}$ order emergence of intelligent systems.

Cariani (1992) differentiates physical emergence, from biological emergence, psychological emergence and social emergence. Physical emergence is related to the appearance of new physical structures, as for instance the Bénard cells in natural convection. Biological emergence is related to the increase in morphological complexity and to the appearance of new functions in biological evolution. The immune system is an example. The psychological that is cognitive emergence is correlated to the appearance of new ideas, of explanations for instance. Another type of emergence is the one encountered in social evolution, which corresponds to the appearance of new social structures and cultural or scientific innovations. Stigmergy example refers to social insects and the stock markets organization is an example for human society.

This kind of approach requires clarifying the relation between higher levels and lower levels in hierarchy. This study refers to the micro-macro, local to global or the two-ways mechanism of emergences and also to the concept of downward causation called also immergence (Pattee 2000). An emergent feature is supposed to have some kind of causal power on lower level entities. Downward form of causality, operates from upper to lower level, and complements the upward causation. Lower entities exercise an upward causation on the emergent features.

Obviously this approach implies a two-levels and two-way causal relation. As an illustrative example we may consider several component transport processes organizing in a compliant installation. The component transport processes affect how the installation develops, upward, and the development of the installation affect the behavior and the interaction of the component processes, downward.

As a working approach, the emergence is considered as a dynamic, nonlinear process that leads to novel properties, structures, patterns at the macro-level or global level of a system from the interaction of the part at the micro-level or local level. Such novelty cannot be understood by reductionism but may be studied by looking at each of the parts in the context of the system as a whole. Detecting and breaking complexity frontier allows in fact exploiting properties of emergence, as non-linearity of interactions and coherence. This may be accomplished by distributed systems constructed as a group of interacting autonomous entities that are designed to cooperate, to have an emergent globally coherent behavior. Due to non-linear reinforcement, local interactions may result in a larger effect in the form of a novelty at the macro-level.

Several studies associate the emergence to predictability and to complexity decrease. A feature use to be considered emergent if it can provide better predictability on the system behavior, compared to the lower level entities. The predictability needs an information theoretic approach and this naturally involves an agent or observer. The measure of emergence the system provides should be an intrinsic property of the whole system including the agents. Entropy and algorithmic complexity represents the appropriate candidates for such quantifiable aspects of complexity and emergence.

## 1.3 Evolvability

The ability to evolve that is the evolvability means that the system has life-like capabilities and it is viable. Evolvability is the way to confront and surpass the successive boundaries of complexity.

Despite the lack of general designation of life there are some features that necessarily characterize all living systems.

To clarify the implications of evolvability we start from the fact that there are essential differences between adaptability and evolvability. Adaptability refers to optimization or adjustment on the time scale of the existence of a system as for instance some industrial product or organisms. In a hierarchy like that shown in Fig.1.1 adaptability may refer to just two levels as for example animate and its inanimate environment. It is a low dimensional perspective. Evolvability is not the same as adaptability or evolutionary versatility although it might be argued that it subsumes such candidates.

Evolvability requires capacity for change to march into new life cycles, for instance new type of products, new niches, new organisms and new levels. In a hierarchy as shown in Fig.1.1, the evolvability refers to the four levels, to a transformation from inanimate towards animate then cognitive and then intelligent systems. This is a higher dimensional perspective for evolvability. If the functioning of cognitive and intelligent systems is based or not on mechanisms similar to that shown by ordinary Darwinian selection for the biological realm is an open problem (Edelman 1987, Piaget and Garcia 1989).

Although survival is still to the fittest, as in Darwinian scenario, the fittest are those that evolve most ably with the dynamic environment. An example, is to consider that if some individuals are equally fit at a given level but the successors of one at the next level are likely to be more fit that the successors of others, than this individual is considered as the more evolvable (Turney 1999). This is a kind of higher-level equivalence of Darwinian selection.

Several characteristics of the evolvable systems have been outlined in the study of living systems and in the study of real and artificial life, AL (Farmer and Belin 1992, Bedau et al. 2000).

Farmer and Belin (1992) selected eight criteria to define the AL:

- Life is a pattern in space time
- Self-reproduction
- Information storage of a self-representation
- Metabolism
- Functional interaction with the environment
- Interdependence of parts
- Stability under perturbations
- Capacity to evolve that is evolvability

Several other lists of properties have been suggested in the literature to discern the inert material systems from the living systems. These refer to features as: self-organization, growth, development, functionality, adaptability, agency, reproduction, or inheritance.

Usually a system is considered alive or showing viability behavior if it meets criteria as for instance: has a high level of organization and complexity, survive in a dynamic environment, responds to stimuli, is capable of reversing entropy and is capable of open-ended evolution.

This means:

- To have a separate symbolic description, genotype, that makes use of a code
- That genotypes describe the structure of the entity, with phenotype as the low part level
- That parts are self-organizing such that, when put together in different orders, they link up and interact with each other in ways that cannot be easily predicted
- That reproduction makes provision of random mutations of the genotype code

Obviously there is an assembly of acknowledged features that necessarily characterize living systems.

In the next section we focus on the circularity and closure as have been outlined by the study of all evolvable systems, both natural as artificial.

## 1.4 Closure and Circularity

### *1.4.1 General Concepts*

The notion of closure plays a prominent role in systems theory where it is used to identify or define the system in distinction from its environment and to explain the autonomy of the systems. Closure and circularity are critical for emergence and evolvability (Emmeche 2000).

Significant is the relation between self-adaptivity, cognitivity, intelligence and different notions of closure as encountered in systems theory: organizational closure (Maturana and Varela 1980, Varela 1989), closure to efficient cause (Rosen 1991), semantic closure (Pattee 1995), and operational closure (Luhmann 1995).

Closure does not mean that the system is not in contact with its environment. Rather the term closure refers to the closed loop which connects the structures and the functions of individual life-like entities.

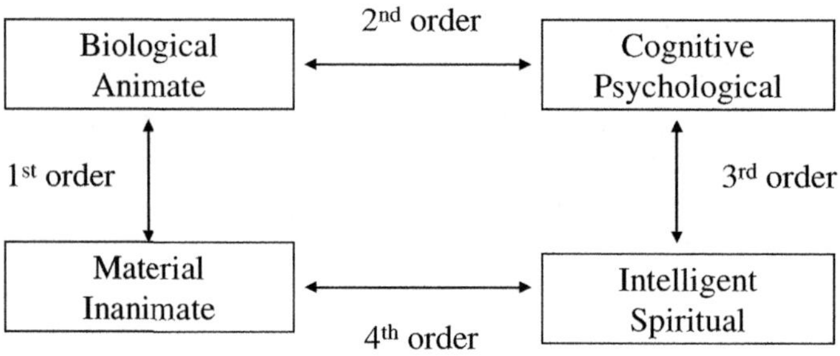


**Fig. 1.2** Four realms world-network frame

Hartmann's foresight of four-level hierarchy may be developed toward a network or an integrative closure of the four realms (Fig. 1.2).

The Hartmann's hierarchy shown in Fig.1.1 is retrieved if the network is pursued clockwise but the network perspective is richer. The hypothetical circularity starts with the material realm and follows with biological, cognitive and intelligence realms.

There are four evolvability levels, each level corresponding to the attainment of a new reality level. Fig.1.2 outlines the different orders of emergence. The $1^{st}$ order emergence refers to biosystems, the $2^{nd}$ order emergence refers to cognitive systems, and the $3^{rd}$ order emergence refers to intelligent systems.

The living systems are not limited to the grounding material or biological level but would include also cognitive, intelligent and social systems, that is, the higher levels in Hartmann's hierarchy and higher order emergence.

The fully accomplished evolvability for technologies corresponds to the embedding in the basic material level that is, to the $4^{th}$ order emergence and integrative closure as shown in Fig.1.2. After one cycle in the network, the material repeated embodiment of cognitive, mathematical and computing capacities may support the emergence of another type of material realm and material science.

For Fig.1.2 it was assumed that different realms in the cycle interact with the neighboring others. The network should take into account the direction of interconnecting arrows. There exist prevailing interactions and directions as well as interdicted ones. In some cases diagonal interactions of the four realms shown in Fig. 1.2 may be considered too. It is possible to suppress or to neglect one level, to connect level n-1 directly to the level n+1.

At the level of biological systems, the closure outlines the qualitative difference between the material non-living systems and the living systems in the framework of autopoiesis. For the cognitive systems, in a constructivist perspective, the closure refers to the ability to employ new distinctions and generate meaning in the system. In the context of psychological or social system the concept of closure directs to organizational and intentional aspects. Finally the intelligent systems realm may have an interaction with the material realm, a kind of embodiment closing in this way the cycle.

Further ontological analysis of the systems requires construction of several sub-realms for each realm of the main structure. Material realm includes

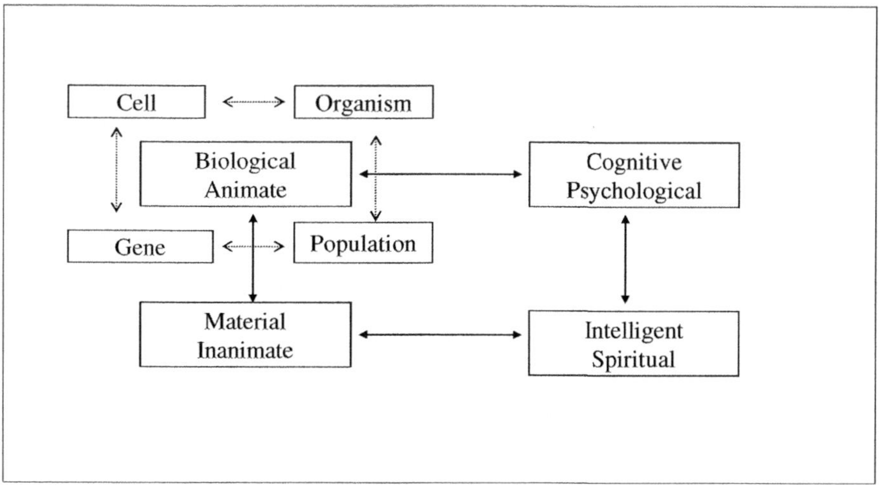


**Fig. 1.3** Four realms and sub-realms for biosystems

physical or chemical sub-realms; cognitive realm may include representations, goals, beliefs, desires and so on.

For the example shown in Fig. 1.3 the biosystems realm is represented by four sub-realms: molecules as genes, cells, organisms and population that is, the ecosystem (Poli 2002). The splitting in four sub-realms appears as a natural one and it parallels and in some sense recapitulates the basic splitting of reality in four realms.

The gaining of an integrated understanding of the different scales is a difficult task for disciplinary research. For the biosystems realm the sub-realms includes cellular and sub-cellular spatial and temporal organizations and multi-cellular systems integrating gene regulation networks with cell-cell signaling and bio-mechanical interactions. The biosystems underlies larger scale physiological functions which emerge from sets of cells, tissues and organs in complex interaction within a given environment. At the highest level, the understanding and control of ecosystems involves deeply integrated interactions among organisms in a given biotope.

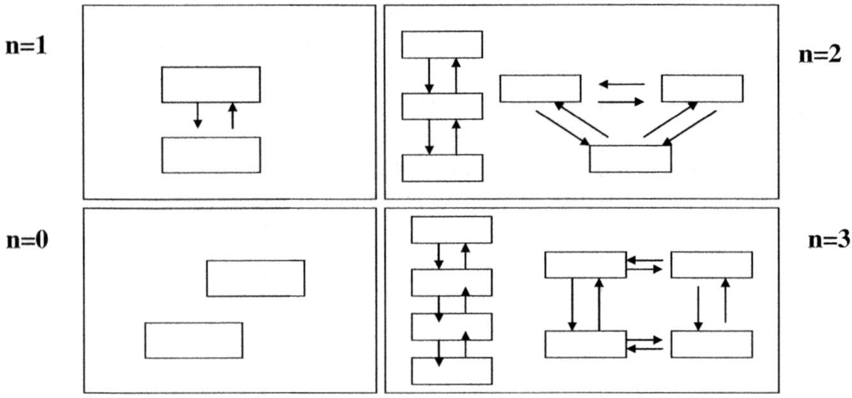


**Fig. 1.4** Multiple-realms frameworks

Fig.1.4 summarizes regular types of ontological frameworks.

The number of interacting modules, levels or realms varies from 1 to 4. The n=0 structures are those of the realms studied separately.

The n=1 structures outline interactions between two realms. For n=2 and n=3 the hierarchical frameworks and network frames as the cyclic ones should be taken into account.

The connection between the theory of levels and causality allows inferring that every levels or realm may trigger its own causal chain. Taking into account both upward and downward causation, that is emergence and immergence, we need to distinguish between material, biological, cognitive and intelligent causality chains.

## 1.4.2 Closure Paradigms

### 1.4.2.1 Semantic Closure

Some particular concepts of closure or circularity, specifically the semantic closure, the functional circle, the circular schema and the integrative closure will be considered more in detail.

In a significant investigation applicable to both real life and artificial devices, Pattee pointed out that the evolution requires complementary description of the material and symbolic aspects of biological events (Pattee 1995, 2000, Rocha 2001). Life and evolvability involve a semantically closed self-referencing organization between symbolic records and dynamics. Symbols, as discrete functional switching-states, are seen in all evolvable systems in the form of genetic codes, and at the core of all neural systems in the form of informational mechanisms that switch behavior. Symbolic information such as that contained in genotype has no intrinsic meaning outside the context of an entire symbol system in addition to a material organization that interprets the symbol leading to specific function such as construction, classification, control and communication. Self-reference that has evolvability potential is an autonomous closure between the dynamics, that is, physical laws of the material aspects and the constraints, that is, syntactic rules of the symbolic aspects of a physical organization. Pattee refers to this condition as semantic or semiotic closure and concludes that it requires separating and complementing the symbolic description (genotype, design, software, and logical systems) from the material embodiment (phenotype, machine, computer, and physical systems). Semantic closure concept allows a synthetic perspective for the logical and physical aspects. The symbolic description must be capable to generate the material embodiment. Symbolic descriptions that generate local dynamics that promotes their own stability will survive and therefore serve as seeds of evolvable autonomous systems. Finally, the material embodiment must be capable of re-generating cyclically the symbolic description with the possibility of mutation. Cariani (1989, 2001) evaluated the semantic closure principle relation with the design of adaptive and evolutionary devices with emergent semantic functions. Self-modification and self-construction were recognized as important to the symbol-matter problem and as requirements for semantically adaptive devices or for evolutionary ones.

### 1.4.2.2 Functional Circle

The circularity and closure are correlated also to the "*Umwelt*" concept that was introduced by von Uexküll in theoretical biology to describe how cognitive organisms perceive and interpret their environments. The *Umwelt* concept is basic for biosemiotics and the theory of levels (Hoffmeyer 1997). The *Umwelt* was defined as the part of the environment that an organism selects with its specific sense organs according to its needs (von Uexküll 1973).

*Umwelt* theory asserts that a complex system doesn't responds to its environment but rather to its perception of the environment. A complex system actively creates its *Umwelt*, through repeated interactions with the environment. It simultaneously observes the world and changes it, the phenomenon which von Uexküll called a functional circle (Fig. 1.5).

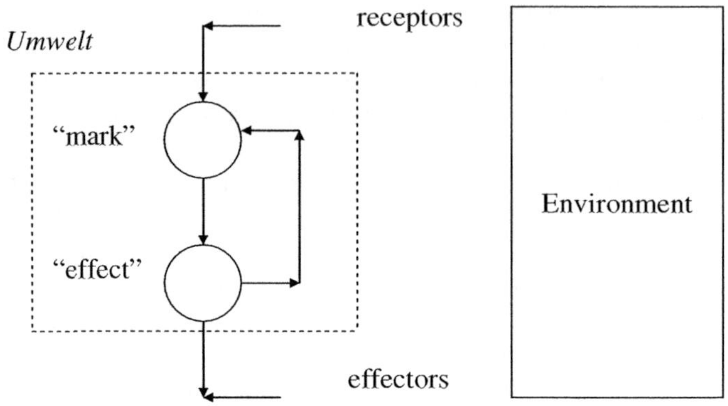


**Fig. 1.5** Functional circle

The functional circle includes receptors and effectors. The sensory experience is based on interactions and these have specific purposes. The elementary unit of evolvable systems includes the functional circle of the following four parts: the environmental object, the receptors, the command generator and the effectors. The command generator includes a "mark" organ and an "effect" organ.

The *Umwelt* concept offers suggestions for the study of animats and of artificially evolvable circuits (Iordache 2009). It should be emphasized that *Umwelt* theory takes into account three of the levels of the Hartman hierarchy namely the material, biologic and cognitive level (Cariani 2001). This is illustrated in Fig. 1.5. The material realm corresponds to the environment. The biological level includes the sensors and the effectors while the cognitive level corresponds to the cycle including "mark" and "effect" organs.

#### 1.4.2.3 Circular Schema

Circular reactions have been emphasized in the study of action schema done by Piaget (Piaget 1970, 1971). Piaget called his general theoretical framework "genetic epistemology" because he was primarily interested in how knowledge develops in living organisms. Cognitive structures are patterns of physical or mental actions that underlie specific acts of intelligence and correspond to the stages of development. According to Piaget, there are four primary cognitive development stages: sensory-motor, pre-operations, concrete operations, and formal operations.

The Piaget's action schema, which constitutes the foundation of his learning theory, is a cycle including three elements: a recognized situation, an activity that has been associated with this situation, and an expected result. The recognition of a situation involves assimilation, that is to say, the situation must manifest certain characteristics which the organism has abstracted in the course of prior experience. The recognition then triggers the associated activity. If the expected result does not occur, the organism's equilibrium is disturbed and an accommodation may occur, which may eventually lead to the formation of a new action scheme. Accommodation does not take place unless something unexpected happens.

Assimilation integrates new information in pre-existing structures while accommodation change and build new structure to understand information. The equilibration through assimilation and accommodation, takes into account three of the levels of the Hartman hierarchy, the material, the biologic level, more close to assimilation, and the cognitive level, more close to accommodation concept.

Piaget general equilibration theory offers a standpoint to consider the three level chains of interactions, namely biologic, cognitive and intelligent.

Piaget theory of emergence in the context of social sciences emphasizes on the concepts of feedback and circularity (Piaget 1980). For Piaget there are three types of processes describing the behaviour of populations:

- Composition process, which defines the properties of the population as a whole, that is, the global behaviour
- Emergence, the process through which the whole generates new properties with respect to individuals
- Relation processes, the system of interaction modifying the individuals, and therefore explaining the properties of the whole

It can be observed that semantic closure, functional circle and circular reaction concepts have basic similarity despite the fact that they may refer to different levels of reality. They describe cycles or in other words loops of interaction between two or three successive levels or realms. Pattee focuses on two levels frameworks, the material versus biologic, or biologic versus cognitive, while von Uexküll and in part Piaget focuses on three level frameworks, material, biologic and cognitive level. It should be emphasized that some of the Piaget schemas embraces a four-level perspective including intelligence level (Piaget and Garcia 1989).

#### 1.4.2.4 Integrative Closure

What is less understood and under active investigation for a real word perspective as that shown in Fig.1.2 is the link between intelligence or logical levels and the material level involving the four realms as a whole and allowing the integrative closure, the full evolvability and autonomy. Integrative closure imposes a re-examination of the link between physics and information and implies the $4^{th}$ order emergence based on the material embodiment of the logical or calculus capabilities.

The autonomic computing (Horn 2001, Kephart and Chess 2003), cyber-physical systems (Lee 2007), organic computing (Müller-Schloer et al. 2004), and other paradigms were identified among the fundamental concepts and paradigms that may be beneficial for integrative closure understanding and achievement.

Autonomic computing addresses complexity and evolution problems in software systems. Autonomic computing refers to computing elements disseminated throughout the system which beyond the material, biologic or cognitive units, embeds computational power. The autonomic computing systems are supposed to share certain feature with living systems.

Actually, the large-scale deployment of computational systems will not be possible without making those systems autonomous and thereby endowing them with properties of living systems as natural robustness, reliability, resilience and homeostasis.

A cyber-physical system is a system featuring the combination and coordination between, the system's computational and physical elements. An antecedent generation of cyber-physical systems is often referred to as embedded systems. In embedded systems the emphasis tends to be more on the computational elements, and less on an intense link between the computational and physical elements, as in cyber-physical systems studies.

Organic computing is the research field developing around the principle that problems of organization in different domains as complex materials, molecular biology, neurology, computer science, manufacturing, ecology and sociology can be studied scientifically in a unified way (Würtz 2008). Instead of looking to ensure statically and a priori the correct execution of programs or designs, organic computing intends to modify these incrementally so that they achieve the prescribed tasks. This approach is tightly coupled with concepts and theories like self-organization, emergence, evolvability and constructivism.

Inspired by the organisms observed in nature, organic computing research extends the autonomic computing objectives and focus on emergence and technological embodiment aspects. Organic computing is broad in scope, in that it touches upon the full range of bioinspired computing, without reference to any particular type of environment, and addresses not only the problems underlying cognitive and intelligent control and cognitive robotics but also the cognitive middleware, sensor networking, artificial immune systems and models of computing derived from chemistry, biology, cognition and mathematics.

## References

Adami, C.: What is complexity? Bioessays 24(12), 1085–1094 (2002)

Bedau, M.A., McCaskill, J.S., Packard, N.H., Rasmussen, S., Adami, C., Green, D.G., Ikegami, T., Kaneko, K., Ray, T.S.: Open problems in artificial life. Artificial Life 6, 363–376 (2000)

Bonabeau, E., Desalles, J.L.: Detection and Emergence. Intellectica 2(25), 85–94 (1997)

Boschetti, F., Prokopenko, M., Macreadie, I., Grisogono, A.: Defining and detecting emergence in complex networks. In: Khosla, R., Howlett, R.J., Jain, L.C. (eds.) KES 2005. LNCS (LNAI), vol. 3684, pp. 573–580. Springer, Heidelberg (2005)

Cariani, P.: On the Design of Devices with Emergent Semantic Functions. Ph D Dissertation, Binghamton University (1989)

Cariani, P.: Emergence and Artificial Life. In: Langton, C.G., Taylor, J., Farmer, D., Rasmussen, S. (eds.) Artificial Life II, vol. X, pp. 775–798. Addison Wesley, Redwood City (1992)

Cariani, P.: Symbols and dynamics in the brain. Biosystems 60, 59–83 (2001)

De Wolf, T., Holvoet, T.: Emergence and self-organization: a statement of similarities and differences. In: Proceedings of the Second International Workshop on Engineering and Self-Organizing Applications, New York, USA, pp. 96–110 (2004)

Edelman, G.M.: Neural Darwinism: the theory of neuronal group selection. Basic Books, New York (1987)

Emmeche, C.: Closure, Emergence, Semiosis and Life: The Same Idea? In: Chandler, J.L.R., Van de Vijveir, G. (eds.) Closure: Emergent Organizations and their Dynamics. Annals of the New York Academy of Science, vol. 901 (2000)

Farmer, D., Belin, A.: Artificial Life: The Coming Evolution. In: Langton, C.G., Taylor, J., Farmer, D., Rasmussen, S. (eds.) Artificial Life II. Addison Wesley, Redwood City (1992)

Goldstein, J.: Emergence as a construct: History and issues. Emergence 1(1), 49–72 (1999)

Hartman, N.: Der Aufbau der realen Welt.Grundriss der allgemeinen Kategorienlehre. Walter de Gruyter, Berlin (1940)

Hartmann, N.: The new ways of ontology. Greenwood Press, Westport (1952)

Hoffmeyer, J.: Signs of Meaning in the Universe. Indiana University Press, Bloomington (1997)

Horn, P.: Autonomic computing: IBM's perspective on the state of information technology. Technical Report, IBM Research (2001)

Iordache, O.: Polystochastic Models in Chemical Engineering. VNU Science Press, Utrecht (1987)

Iordache, O.: Evolvable Designs of Experiments Applications for Circuits. J. Wiley VCH, Weinheim (2009)

Kauffman, S.: At Home in the Universe: The Search for Laws of Self-Organization and Complexity. Oxford University Press, Oxford (1995)

Kephart, J.O., Chess, D.M.: The vision of autonomic computing. IEEE Computer 36(1), 41–50 (2003)

Lee, E.A.: Computing foundations and practice for cyber-physical systems: A preliminary report. Tech Report Univ of California Berkeley/EECS-2007-72 (2007)

Luhmann, N.: Social Systems. Stanford University Press, California (1995)

Mainzer, K.: Thinking in complexity, the complex dynamics of matter, mind and mankind. Springer, Berlin (1996)

Maturana, H., Varela, E.: Autopoiesis and Cognition. The Realization of the Living. Reidel, Dordrecht (1980)

Müller-Schloer, C., von der Malsburg, C., Wurtz, R.P.: Organic Computing. Informatik Spektrum (27), 332–336 (2004)

Nicolescu, B.: Manifesto of Transdisciplinarity. SUNY Press, New York (2002)

Pattee, H.H.: Evolving self-reference: matter, symbols, and semantic closure. Communication and Cognition –Artificial Intelligence 12(1-2), 9–28 (1995)

Pattee, H.H.: The physics of symbols and the evolution of semiotic control. In: Workshop on Control Mechanisms for Complex Systems: Issues of Measurement and Semiotic Analysis, Las Cruces, New Mexico, December 1996. Santa Fe Institute Studies in the Sciences of Complexity, Proceedings. Addison-Wesley, Redwood City (1997)

Pattee, H.H.: Causation, control and the evolution of complexity. In: Anderson, P.B., et al. (eds.) Downward Causation. Aarhus University Press, Aarhus (2000)

Piaget, J.: Genetic Epistemology. Columbia University Press, New York (1970)

Piaget, J.: The construction of Reality in the Child. Ballantine Books, New York (1971)

Piaget, J.: L'épistémologie des régulations: introduction. In: Lichnerrowicz, A., Perroux, F., Gadoffre, G. (eds.) L'idée de régulation dans les sciences: 2e vol. des Séminaires interdisciplinaires du Collège de France, A. Paris: Maloine: Doin:I-XIII (1977)

Piaget, J.: Épistémologie des sciences de l'homme. La Pléiade, Paris (1980)

Piaget, J., Garcia, R.: Psychogenesis and the History of Science. Columbia University Press, New York (1989)

Poli, R.: The basic problem of the theory of levels of reality. Axiomathes 12(3-4), 261–283 (2001)

Poli, R.: Ontological methodology. Int. J. Human-Computer Studies 56, 639–664 (2002)

Poli, R.: Three obstructions: forms of causation, chronotopoids and levels of reality. Axiomathes 17(1), 1–18 (2007)

Popper, K.: Natural selection and the emergence of mind. In: Radnitzky, G., Bartley, W.W. (eds.) Evolutionary epistemology, rationality and the sociology of knowledge, Open Court, La Salle (1987)

Rocha, L.M.: Evolution with material symbol systems. Biosystems 60, 95–121 (2001)

Rosen, R.: Life itself: A comprehensive enquiry into the nature, origin and fabrication of life. Columbia University Press, New York (1991)

Turney, P.D.: Increasing evolvability considered as a large-scale trend in evolution. In: Proceedings of the 1999 Genetic and Evolutionary Computation Conference (GECCO 1999) Workshop Program, pp. 43–46 (1999)

von Uexküll, J.: Theoretische Biologie. Frankfurt a. M, Suhrkamp Taschenbuch Wissenschaft (1973)

Varela, F.: Autonomie et Connaissance. Essai sur le vivant. Seuil, Paris (1989)

Würtz, R.P. (ed.): Organic Computing. Understanding Complex Systems. Springer, Heidelberg (2008)

# Chapter 2
# Methods

**Abstract.** The methodology to analyze and manage complex systems is presented here.

The polystochastic models, PSMs, are the considered mathematical tools.

PSMs characterize systems emerging when several stochastic processes occurring at different conditioning levels, are capable to interact with each other, resulting in qualitatively new processes and systems. The modeling hierarchy, which is modeling at several abstraction levels, appears as deep-rooted in the higher categories frames. Models of models, that is, meta-models allowing studying processes of processes, and so on, are presented with case studies from informational systems and statistical methodologies.

Innovative is the introduction of a partial differential model for multiple levels modeling. This imposes making use of unconventional notions of time, space, probabilities and informational entropy.

## 2.1 Polystochastic Models

### *2.1.1 Introduction to Polystochastic Models*

The term polystochastic was created to describe stochastic systems emerging when several stochastic processes are enabled to interplay with each other, resulting in a system that is structurally different from its components. Engineers, who bring together real stochastic processes in their devices, perform a task similar to that of chemists but instead of atoms and molecules they deal with component stochastic processes. Chemists have devices to combine substances and afterwards to separate them. Also the chemists make use of theoretical tools as the periodic table for classification and models as the wave equation from quantum physics or the balance equations describing transfer phenomena. The challenge for engineers is to develop analogous methods, techniques and models for combining and decomposing component processes.

PSMs suggest possible answers to the modern days request to build an engineering science putting the processes and the interactions, on an equal footing as physical or chemical objects.

The first objective of PSMs study was the structure of the stochastic evolution over more conditioning levels. Basic frameworks have been presented in the research monograph from 1987 (Iordache 1987). The initially examined frameworks, restricted to the real field, were later generalized to outline the interaction between real field and other algebraic fields (Iordache 1992) Starting from 1990, the PSMs were applied as theoretical tool for problem solving in the domain of industrial engineering systems. PSMs capabilities have been tested for flowchart generation, for evolutionary computation, for diagnosing and failure analysis, for evolvable designs of experiments (Iordache 2009).

The mathematical tools used initially for PSMs were the so-called random or stochastic systems as for instance "random evolutions" (Hersh 2003) "random dynamical systems", (Arnold 1998) "random systems with complete connections" RSCC, (Iosifescu and Grigorescu 1990), "random iterated function systems" RIFS (Barnsley 1993). More recently PSMs start to be formulated in the unifying framework of category theory (Iordache 2009).

The preliminary PSM presentations kept the structure and the notations as close as possible to that used for general learning model or RSCC but it should be emphasized that the significance of the PSMs elements and the calculus is different. The general learning model or RSCC aims to describe processes in which the subject is repeatedly exposed to an experimental situation that is a condition or an event, for which various responses are possible, and each trial can modify the subject's response tendencies. The set of all possible states is denoted by s, the set of conditions by k (Fig. 2.1). Let $k_n$ be the condition in the trial n when the subject state is $s_n$. Let $p(k_n|s_n)$ be the probability of the condition $k_n$, conditional on the state $s_n$. The state at the n+1 trial will be $s_{n+1}=u(k_n,s_n)$. Denote by $p(k_{n+1}|s_{n+1})$ be the probability of the condition $k_{n+1}$, conditional on the state $s_{n+1}$. The resulting state at the first trial will be $s_{n+2}=u(k_{n+1},s_{n+1})$ and so on.

The operator $u$: $k \times s \to s$ characterizes the system tendency. The elements of the general learning model are quadruples $\{s, k, u, p\}$. This basic model has been comprehensively studied in automata theory, control theory, learning theory, and mathematical theory of RSCC (Iosifescu and Grigorescu 1990).

PSM intend to study stochastic systems that go beyond adaptation and learning, more specifically to cognitive, intelligent and evolvable systems. Such objectives imply higher dimensionality.

Learning systems, finite automata, RSCC, grammars, cognitive nets, have been often developed as low dimensions structures. This means that these studies are limited to two levels. To clarify the dimensionality issue let us consider the learning model or RSCC case. These are systems formed by pairs of stochastic chains evolving in an inter-related manner. There are two chains associable to two levels (Fig. 2.1).

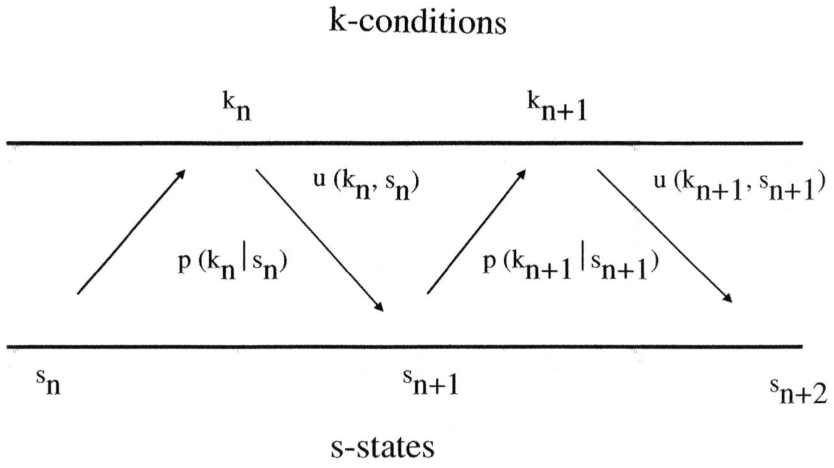


**Fig. 2.1** Learning model

Of the two chains, one is Markov that is with one step memory, but with a complicated alphabet, or transition functions. This is the chain of states. The other chain is of infinite order that is with infinite memory, with a simpler alphabet. This is the chain of conditions. The latter chain is, used to infer properties of the Markov chain.

One time scale sequential models as RSCC may characterize systems that learn and the adaptability but are unable to describe emergence and evolvability, showing a too low dimensionality for this goal.

The high complexity imposes to develop higher dimensional modeling. The impediment to study complexity and emergence in a system is linked to the difficulty to create a higher dimensional theory for that system.

Several aspects illustrating the specificity and originality claims of the PSM approach are emphasized in the followings. The first aspect is the vectorial or more general, tensorial character. This is related to system hierarchical organization in a certain number of conditioning levels and scales. The organization in levels is the strategy to understand and confront complexity in several systems. It may be a configuration in major levels as for instance physical, biological, cognitive followed by sub-levels.

Most real-world complex systems have both temporal and spatial heterogeneous structure and are hierarchical in nature.

The focused elements of PSM are quadruple of vectors {S, K, U, P} called "SKUP". The notations are: S-states, K-conditions, U-operators, and P-possibilities.

The SKUP elements will be vectors. This enables to describe sequential and parallel evolutions.

The states S, the conditions K, the operators U and the possibilities P are vectors denoted as follows: S = ($s^0$, $s^1$,...,$s^m$,...,$s^M$); K = ($k^0$, $k^1$,...,$k^m$,...,$k^M$); U = ($u^0$, $u^1$,...,$u^m$,...,$u^M$); P = ($p^0$, $p^1$,...,$p^m$,...,$p^M$).

Here $s^m$ represents the particular state at the conditioning level m, and $k^m$ represents the particular condition at the conditioning level m≤M. Typically upper indices are reserved to conditioning levels, while lower indices are reserved to time steps. The components of U may be operators such as $u^m$: $k^m$ x $s^{m'} \rightarrow s^{m''}$ or $u^m$: $k^m \rightarrow s^{m'}$, for example. The operator U should be able to accommodate change of conditioning level and the splitting of these levels.

Replacing the elements s, k, u, p of an RSCC by vectors doesn't make this automatically a PSM since the definition of the elements of PSM need to take into account the passage of time at different levels and to verify concurrency and coherence conditions.

Observe that S and K may be defined on different types of algebraic frameworks. Frequently S is defined on the real field R, whereas K is defined on "other than real" algebraic frameworks, for instance over finite fields, local fields or algebras. The difference in frameworks is associated to closure and to evolvability issues.

The role of basic frameworks differentiation for biosystems and cognitive systems, for evolvability and for modeling relations was highlighted by Pattee (Pattee 1995, 2000) and by Rosen (Rosen 1991). These authors refer to two levels only.

Despite algebraic framework differences, S and K should be interconnected. This interconnection is described by operators U and possibilities P. U characterizes the K to S transition and P characterizes the S to K transitions, that is: U: K→S and P: S→K.

Another innovative aspect for PSM concerns the differential model for K process. The elements of K are resulting as solutions of a partial differential equation. It may be functional as a meta-model that is a generic model producing other models or a meta-meta-model that is a generic model producing meta-models. The prefix "meta", is used to mean "information about".

The partial differential model proposed in this book is a wave equation, WE. Its solutions help in describing the space of conditions K, ensuring the highly economic character of transfer of copied information from one level to another for evolvable systems.

Another innovative aspect for PSM concerns the possibilities P. It consists in complementing or replacing probabilities p, by possibilities as for example those defined in fuzzy logic (Dubois and Prade 2001) or by some less conventional measures (Keane 1972, van Rooij 1978). The set theory and corresponding probability theory are inadequate frameworks to capture the full scope of the concept of uncertainty for complexity. Uncertainty in set theory means non-specificity and exactly this specificity is important for complex systems. Probabilities may be of interest when it is not detrimental to flat individual features while they are not adequate to account for strong individual deviations.

Conventional probabilities are inappropriate to illustrate qualitative concepts as plausibility, beliefs, anticipation, partial truth and opportunities, all having significant role in complexity studies. Another reason of this replacement is the need for simple calculations.

## 2.1.2 Frameworks for Polystochastic Models

### 2.1.2.1 Basic Categorical Frameworks

Category theory was developed as a way of studying different mathematical structures in terms of their admissible transformations (MacLane 1971).

Category theory provides a common language and a set of unifying concepts to various branches of sciences and engineering. Using these unifying concepts, the analogous results are represented by a single result which provides deeper understanding of the problem involved, since it may reveal the underlying mechanism.

The conventional learning models outlined a set of states s, a set of conditions k, and transitions relations between them, expressed by operators as u, and probabilities p.

Fig. 2.1 shows the two chains of random variables s, and k and their interconnection by u and p. The categorical approach for PSM appears as a categorification of such transition systems for increasingly higher dimensional problems. Categorification associates category-theoretic concepts to set-theoretic notions (Appendix A4).

Categories are linked to the different levels of reality. The notion of level or reality which was firstly studied from an intuitive point of view may be approached from a more formal point of view based on category theory. The levels and sub-levels of reality are characterized and distinguished by their categories and sub-categories (Poli 2008).

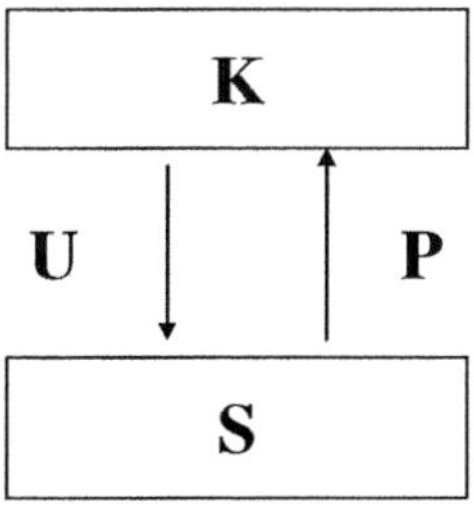


**Fig. 2.2** Two levels framework

Fig. 2.2 shows the basic SKUP framework that contains in the categorical interpretation two categories S and K, and two functors, U and P between these categories.

The SKUP associated to PSM outlines the general architecture, shared by numerous adaptive and evolvable systems (Iordache 2009).

For PSM framework the conditions K represent the category describing the types of component processes. In this case, the processes types are the objects of category. Interactions among types can be modeled as morphisms. Categorical constructions such as colimits characterize fusion. S is the category, associated to the detailed states of the component processes. The arrows that is, the morphisms describe the transition relations between the states of the component processes. Different algebraic frameworks for states-S (dynamical, analogical, and natural) and conditions-K (symbolic, digital, and formal) have to be considered.

Functors as U are accounting for interactions in K, and between K and S.

Functors as the possibilities P, supplements the probabilities to express potentiality, fuzziness, uncertainty, and emergence.

K corresponds with our notation to the conditioning level m=0, and S corresponds to the conditioning level m=1.

The diagram shown in Fig. 2.2 may be viewed as a general method of problem solving also. For this view, K denotes the problems space and S the solving problem space. The problem solving is illustrated in Fig. 2.3. The process in K is that from problem formulation towards problem solutions. Confronted to the problem to solve, the investigator tries different generalization of the elements revealed by that problem. Solving the problem means to select and formulate problem solution.

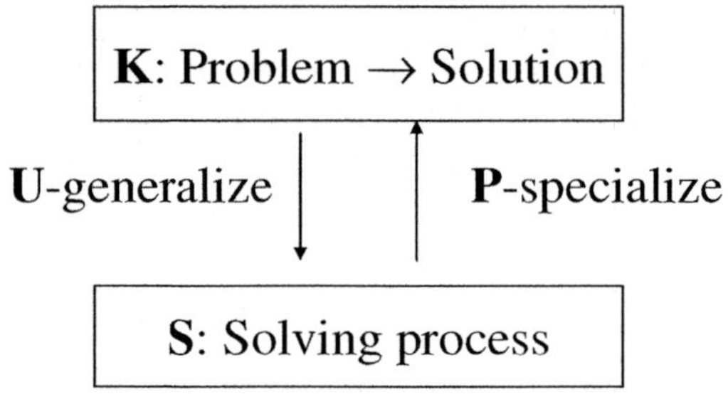


**Fig. 2.3** Two levels framework for problem solving

In this situation the operators U correspond to generalization, whereas possibilities P perform the specialization.

Observe that the SKUP framework from Fig. 2.2 or Fig. 2.3 still involves only two levels or realms, S and K. Advancements in modelling higher complexity, the evolvability request, impose to take into account multiple levels and multiple SKUPs interaction.

Any two levels SKUP may have more complicated relations to other two level SKUPs than can be functional in the multiple levels situation.

In such cases it is necessary to consider 2-categories and more general n-categories (Appendix A2). This means in fact to continue the categorification process (Appendix A4).

Horizontal, vertical, diagonal and parallel composition of elementary two level SKUPs should be taken into account.

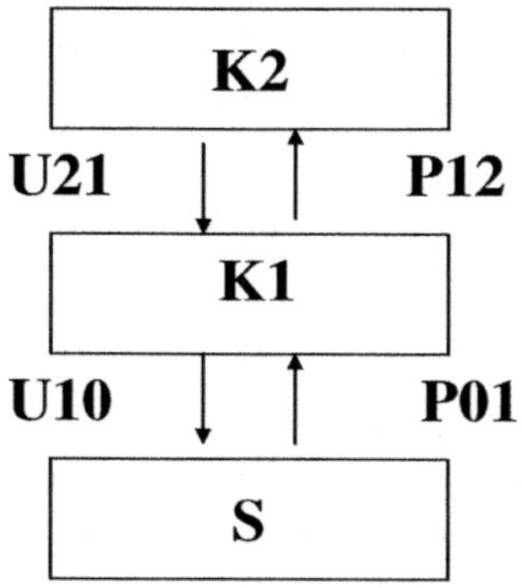


**Fig. 2.4** Three levels hierarchical framework

Fig. 2.4 and Fig. 2.5 show three categories S, K1 and K2 and their interconnections by operators and possibilities.

Fig. 2.4 shows the elements of the SKUPs, the lower cell {S, K1, U10, P01} and the upper cell denoted by {K1, K2, U21, P12}.

The upper cell appears as a second structuring for the category K mirroring in a specific sense the basic SKUP structure.

The highest level, K2 corresponds in our notations to the conditioning level m=0, and the next level K1 to the conditioning level, m=1. S corresponds to m=2.

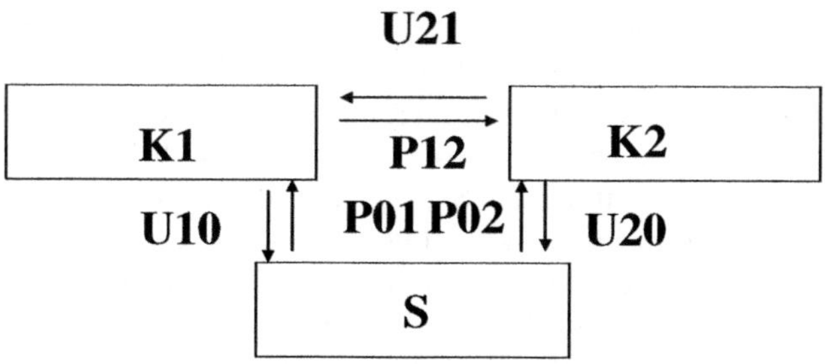


**Fig. 2.5** Three realms network

Fig. 2.5 shows the elements of three SKUPs, the left cell {S, K1, U10, P01}, the right cell {S, K2, U20, P02} and also the SKUP cell denoted by {K1, K2, U21, P12}.

Since we replaced K by two categories K1 and K2, there are two possible operations for the conditions K.

We will refer to operation in K as the tensor product, "$*$".

There are various tensor products we can consider for categories. The Cartesian product is a special case. The categorical product "$\times$" and the coproduct, "$\cup$" are other examples.

Fig. 2.4 and Fig. 2.5 outlines two possible ways in the conditions category K, for instance, the coproduct, "$\cup$", way in K1 the categorical product "$\times$" way in K2.

The switch from product "$\times$" to the coproduct, "$\cup$" and reverse is possible but the two types of categorical product cannot be observed simultaneously.

The interaction between S and K2 as shown in Fig. 2.5 allows an integrative closure including in the loop S, K1, K2 and again S and this may make the whole system evolvable and autonomous.

The Gray tensor product and its generalizations are of interest for 3-categorical constructions (Appendix A4). In such cases, instead of category K we consider three categories K1, K2 and K3.

The tensor product proves to play a significant role for the emergence and evolvability mechanisms.

#### 2.1.2.2 Conditional Stochastic Processes

An example of categorical framework for PSM is related to conditional stochastic chains categorization. PSM study starts by considering systems to be organized hierarchically by conditioning levels as arrays of systems within systems (Iordache 1987). As for other approaches to complexity, it was assumed that the complexity can be managed through hierarchical layering. The conditioning levels are correlated to time and to space scales. For simplicity of presentation we make use here of Markov stochastic processes, with one step memory, as the basic brick for the construction of PSM.

The first conditioning level, m=0, is considered as non-conditioned. Suppose that the process described on K has k states. This is the so-called control process. Let C be the corresponding transition matrix. There are k different choices for the component processes at the next conditioning level, m=1. A "component" process defined on $S_1$, $S_2 \ldots S_k$ corresponds to each state for the control process described by C. The process on $S_i$ is described by $D_i$ as$_i$ x $s_i$ stochastic matrix. The evolution at the conditioning level m=1 depends on the evolution on the conditioning level m=0. This illustrates the term of conditional level and that of conditional process. The process defined on K appears as the process of conditions that switch the evolution from a component process to another. This switching may be induced by external or internal changes. The state in component processes evolves in time and on the other hand their hierarchical scheme evolves too. The evolution at m=0 is on a slower time scale, than the evolution at m=1.

The PSM structure is portrayed by indicating the conditioning levels, 0, 1...m, and the space of states to each level. For instance the states (1,...,k) at the first conditioning level m=0, the states (1,...,$s_1$, 1,...,$s_2$, 1,...,$s_k$) for different component processes at m=1. Suppose $s_1=s_2=\ldots=s_k=\ldots=s$. Let D denotes the common transition matrix of the component processes. The transition matrix C∗D with "∗" a tensor product describes the whole process at the conditioning level m=1. It is a stochastic matrix too. The resulting processes have been studied as conditional stochastic processes.

The tensor product "∗" connects different conditioning levels. The same method of construction, via the tensor product "∗", may be continued for a number of conditioning levels.

Interesting PSMs arise if K and S-processes are tensors and are defined on different types of algebraic frames, that is, for different categories. This implies parallelism, multiple scales, interactions and closure. The operators U, generalized as functors, characterize the conditioned transitions for different component process, $S_1$, $S_2$,. . . ,Sk. They help to detect interactions between K and S. To prove that U is functor it is necessary to show that it preserve identity morphisms, the composition of morphisms and commutativity between categories. The interactions are critical for PSM definition and applications. Mechanisms to characterize emergence and to limit the number of states in K as in S, are mandatory in PSM study. The techniques are based on the interpretation of tensor product "∗" as categorical product "×", categorical coproduct, "⊔", Gray tensor product and so on.

### 2.1.2.3 Lumped Stochastic Processes

Another example of categorical framework for PSM is related to the study of lumpability for stochastic chains associated to compartmental models (Kemeny and Snell 1960, Iordache and Corbu 1987). In such models the category K, may corresponds to the transition matrices for the classes or the lumps, while the category S corresponds to the transition matrices for the elementary states of the component stochastic processes. Observe that the evolution in K is on a slower time scale, than the evolution in S.

To prove that S and K are categories it is necessary to show that the identity, the associativity and commutativity of the morphisms, hold in S and K. These properties are evident for the stochastic matrices, that is, in S. In K, the chosen lumping of states method should ensure similar properties for morphisms.

The challenge is to define functors that will be a natural way to transform the category K of lumped matrices in the category S of stochastic matrices and also functors allowing to transform S back to K.

For the studied lumping models the functors P and U are product of matrices. To give an example, let us denote the stochastic matrix in S by R and the lumped matrix in K by $\hat{R}$. The definition of P is, P: R → $\hat{R}$ with $\hat{R}$=BRC. Here B is a distribute matrix such that its k-th row is a probability vector whose no null components describe the repartition of the initial states in the lumped chain. C is a collect matrix such that the no null components of its k-th columns are equal to "1" and corresponds to the states in the initial chain (Kemeny and Snell 1960). The Markov character of the S chains doesn't ensure the Markov character of the K chain of lumped states, except in very special cases that is if CBRC=$\hat{R}$C.

To prove that P and U are functors it is necessary to show that they preserve identity morphisms, the composition of morphisms as well as commutativity between categories.

## *2.1.3 Higher Dimensional Modeling*

### 2.1.3.1 Hierarchy of Informational Models

Since the complex systems are structured in levels or realms, associated to multiple scales, it is expected that the modeling methods will adopt a similar hierarchical or network architecture.

This means to anticipate a similar granularity multi-level to model complexity itself, rather than just to the solution methods and to described reality.

The way to confront complexity of modeling development is by raising the level of abstraction. Higher complexity imposes to develop higher dimensional modeling. The difficulties arising in the study higher complexity and emergence in a system may be rooted in the conceptual difficulty to develop a higher dimensional theory for that system.

The informational systems start to be studied and managed by different levels of modelling in a hierarchy in which the model of some level is described in terms of a model of the hierarchically upper level and it also describes one or more models in the hierarchically lower level. A standard hierarchy of modelling levels was developed in computer science (ISO 1993, Rossiter and Heather 2003, Del Vecchio 2003).

The international standardization organization, ISO, denotes a family of standards that are widely accepted and successfully used for cooperative work for interoperability and security purposes. OSI-Open System Interconnection Standards is a developed tool for interoperability.

Fig. 2.6 shows a typical modeling hierarchy for complex informational databases, DB, processing. Starting from the data that has to be described, called in this case zero reality level, n=0, there are at the next level, n=1, the so-called schemas, followed at the second reality level, n=2 by the models and finally at the higher reality level n=3 by the models of models or in other words the meta-models, that contains the methodologies that produce other methods.

A model is an abstraction of phenomena in the real world, and a meta-model is a supplementary abstraction highlighting the properties of the model itself.

A meta-model is a model that defines the language for expressing models. It is a precise definition of the constructs and rules needed for creating models.

Names as data, schema models and meta-models associated to different reality levels correspond to the conventional data-bases, DB, studies (Fig. 2.6).

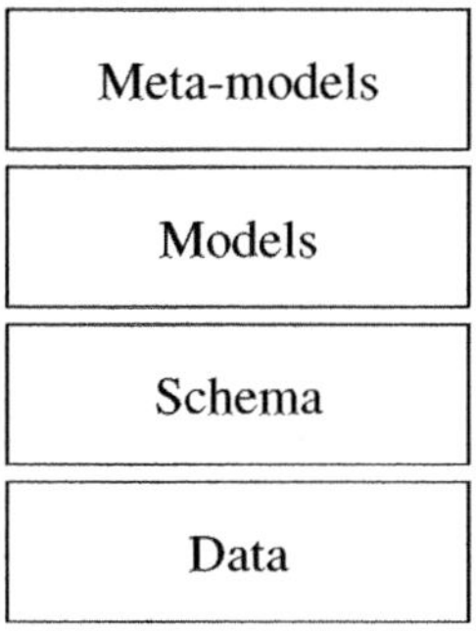


**Fig. 2.6** Modelling levels hierarchy for data-bases

In data processing literature, often the schemas are represented by models, the models are called meta-models and consequently the meta-models are called meta-meta-models (Crawley et al. 1997).

This means that modeling starts at the first level of abstraction.

A model can be treated as an instance of a meta-model and a meta-model as an instance of a meta-meta-model. As a meta-model completely defines a language, it may be not necessary to make a distinction between a meta-model and a meta-meta-model. Because a meta-meta-model is a language itself, it must be defined in terms of a meta-model.

Fig. 2.7 shows the taxonomy proposed by the object management group, OMG, studies. OMG is an organisation for the standardization in the object oriented field of the software development in which many of the leading software production companies in the world participate.

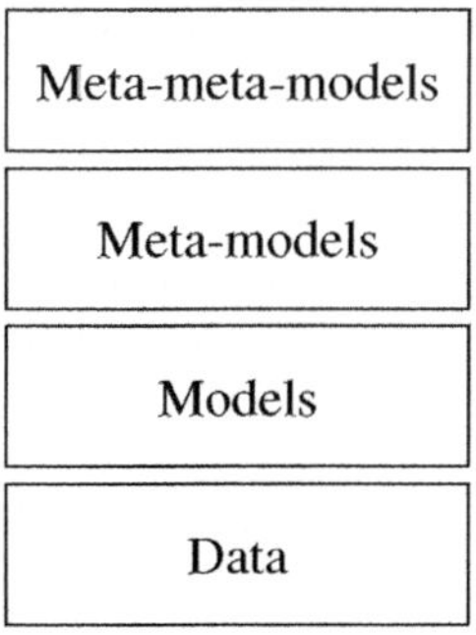


**Fig. 2.7** Modelling levels hierarchy for OMG

The higher we go up in this hierarchy, the more abstract and general becomes the modelling. The top layer of the modeling architecture encompasses general tools and methods and defines possible structural schemes, oriented towards computer science.

The hierarchical principle turns out to be useful in the conventional statistical system modelling. In fact, the hierarchy of levels allows a link between aspects such as the internal architecture of the information system, languages, roles of involved agents, not necessarily human, processing capabilities, and so on.

#### 2.1.3.2 Data Processing Standards and Applications

According to the idea of a hierarchy of models, an international standard was proposed by ISO/ANSI for the design and implementation of a generic Information Resource Dictionary System IRDS (ISO 1993). An IRDS is an information system that describes another information system.

**Fig. 2.8** Hierarchical IRDS system

The information about information is also called meta-information and so on. The IRDS can be considered as a meta-meta-informational system (Nissen and Jarke 1999).

The ISO 1993 is based on a multi-layered structure consisting in four levels in which every level has the purpose of defining the immediately lower level (Fig. 2.8). The first reality level, n=0, that is the application data APP, may be considered as external to the IRDS. The second level, n=1, that is the Information Resource Dictionary, IRD, has the purpose of defining the data and is considered as the content of the IRDS. The third level n=2, that is the Information Resource Dictionary Definition, IRDD, contains the structure or in other words, the model of the IRDS and is in turn defined by the fourth level, n=3, the so-called Information Resource Dictionary Definition Schema, IRDDS, fixed by the standard.

An application of the same principle is the four level structures proposed by the OMG (Fig. 2.9). For OMG the first layer refers to data, the second layer to subject matter models that is models of data, the third layer to statistical methodologies that is to meta-models and the fourth layer to methodologies that define methodologies that is to meta-meta-models. Additionally an n=0 layer representing physical reality may join the OMG architecture.

Following suggestions of the OMG standard (OMG 2000), the four levels will be denoted here by K3, K2, K1, and K0, respectively. K3 is the fourth level, the meta-meta-models level. One model at level K3 is necessary to define all the K2 level models. The OMG standard for the K3 model, also called MOF, Meta Object Facility, is able to define itself (Crawley et al. 1997). MOF is a common framework that is used to define other modeling frameworks within the OMG. Another example of meta-modeling framework is EMF, Eclipse Modeling Framework used by IBM.

K3-model is the language used by MOF to build meta-models, called also K2-models. Examples of the third level, K2-models are the UML Universal-Modelling Language model and the relational models. UML has been accepted as a standard notation for modeling object-oriented software systems. Correspondently, at the second level, K1, there are UML models and relational models relevant to a specific subject. K1 is based on user concepts. First level, K0, contains user runtime data or objects. It may be used to describe the real world.

MOF allows a strict multi-level modelling architecture. Every model element on every layer is a mean to define the structure of a language or of data.

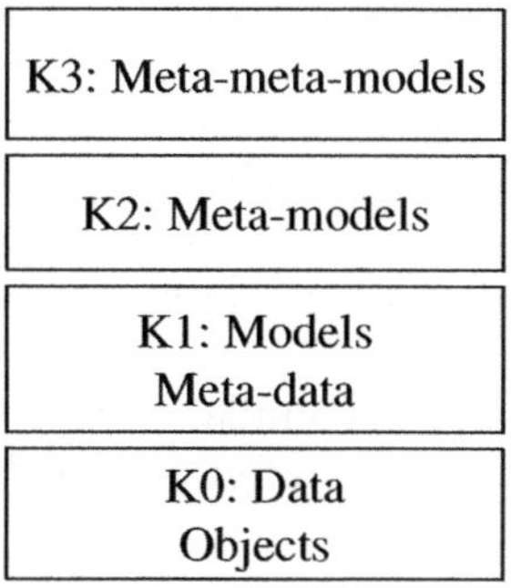


**Fig. 2.9** OMG Meta-meta-model frameworks

MOF is a meta-meta-modeling architecture. It defines a K3-model which conforms to it. MOF allows a strict modeling architecture since every model element on every layer is strictly in correspondence with a model element of the layer above. It only provides a means to define the structure of a language or of data.

Different other meta-meta-model architectures have been considered as for instance that shown in Fig. 2.10. In this case the linear or hierarchical architecture may be developed in cyclic or self-similar architectures (Alvarez et al. 2001). The top and bottom levels in architectures are different. In the hierarchical meta-meta-model architecture every element should be an instance of exactly one element of a model in the immediate next level.

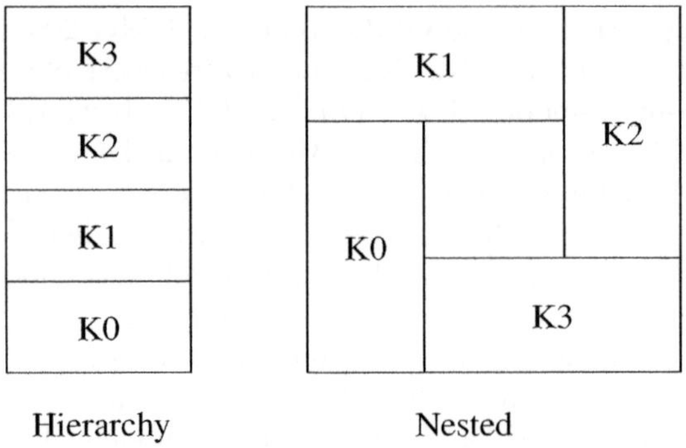


**Fig. 2.10** Different meta-meta-models frameworks

For example, the level K3 could describe elements from the UML meta-model K2 but not elements from the user models K1. More flexibility is allowed by the nested architectures shown in Fig. 2.10. In the nested architecture a model can describe elements from every meta-model below it. This is a significant feature since it means that if a tool implements the K3 meta-meta-model than it can be used to define languages such as UML but user models and objects as well.

The nested Meta-Modelling Language, MML, architecture shown in Fig. 2.10 and Fig. 2.11 outlines the possibility of integrative closure including the critical link between K0 and K3 and allowing evolvability and autonomy.

Theoretically the nested structure architecture is not restrained to four realms.

Fig. 2.11 shows nested and self-similar spiral architectures.

A similar structure is repeated to four sub-realms denoted here by k0, k1, k2 and k3.

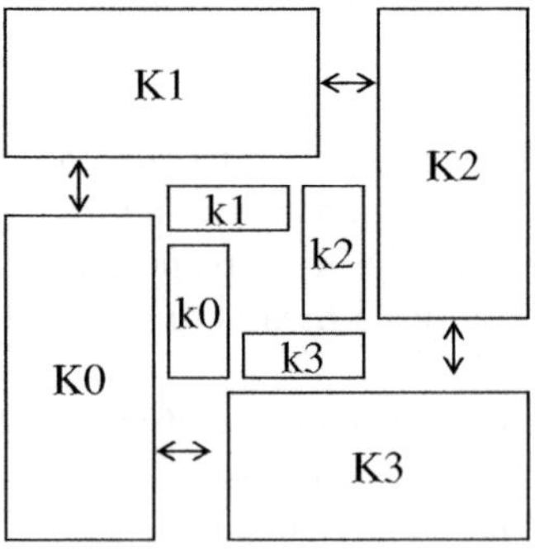


**Fig. 2.11** Nested generic meta-meta-model frameworks

Fig. 2.11 suggests that an integrative closure does not have to be seen as a final stage or a balance due to equilibrium, but rather as a process that can develop self-similar patterns. Nested structures may unify large systems.

The initial frame offers a generic, four-fold, relational model whose elements are configured as a self-similar structure. This means that it can be re-scaled from the smallest to the largest or reversely without compromising its form. This has the advantage that it is a configuration that is shareable across different domains. The self-similarity allows parallel processing with similar software.

#### 2.1.3.3 Modelling Statistical Information

The categorical and multi-layered structure starts to be practically used in statistical data processing implementations (McCullagh 2002). The relational database introduced in the statistical information systems is a significant tool used to store statistical data and to design the data structure of a specific statistical data processing application, as an IRDS.

The OMG standards and four level hierarchies of models were applied to the conceptual modeling of the statistical information systems that support the activity of financial institutions (Del Vecchio 2003, Grossmann 2003).

As in other data processing examples, in the statistics case a fourth level-model has the purpose of defining structures suitable to define third level-models (Fig. 2.12). Such structures are not specific to statistics, but they are more general and functional also in other fields as for example, the operational systems. A fourth level-model contains structures able to define any kind of methodology, possibly shared by all of them. An important feature of a fourth level-model is its self-describing property, that is, the ability of some structures to describe themselves and, consequently, to make the existence of levels higher than four unnecessary.

The specificity of the statistical field is located at the third level. A statistical third level-model is considered as the formal representation of a methodology for statistical description of the reality that is, a descriptive statistic methodology. A third level-model contains structures able to give a concrete and possibly formal shape to statistical methodological rules. The existence of a statistical third level is a consequence of the recognition of the specificity of the statistical methodology with respect to others methodologies.

A model of the second level can be considered the definition of a specific statistical information segment, that is, the definition of data and processes relevant to a specific subject. Second level-models are specific subject-matter models produced using a certain statistical methodology that is, a third level model.

As shown in Fig. 2.12 a model of the first level is the extension of a statistical information segment, that is, an occurrence of a second level model. It is a set of values that correspond to a definition. Different sets of values, therefore, are different first level-models. More than one extension may correspond to the same definition such as in the case that the measurement process generating the data extension is performed more than once. An update in the data content gives origin to a new extension, different from the previous one but with the same definition. The reality to be described is located at the n=0 level.

Therefore K3 are methodologies that define methodologies, K2 are descriptive statistic methodologies, K1 are statistical information definitions, and K0 are data.

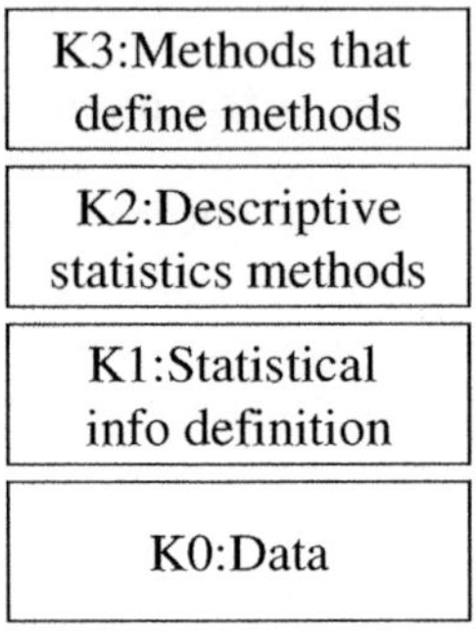


**Fig. 2.12** Levels hierarchy in statistical modeling

### 2.1.3.4 Hierarchy of Models and Interoperability

The hierarchy of models is seen as the conceptual tool to deal with complexity of the statistical information systems. It appears to be useful in the design and the operation of the information systems as well as in the analysis of existing ones and in the effort of matching and standardizing them, independently of how the implementation is done.

The practical application of this idea leads to identify many models on every level and the type instance relationships between models belonging to consecutive levels.

Such a schema of decomposition and description, applicable to a single information system and different information systems of different organisations alike, provides a guideline for interoperability efforts.

The interoperability should takes place within each level. Models at different levels have different purposes and their objects are different because the goal of a certain level is to describe the lower one. On the contrary, it makes sense to compare and match models at the same level when their objects are also partly the same.

Interoperability is simpler if models are defined by means of the same upper level model that is, using the same modelling method. For this reason, the ideal situation would consist in having only a single fourth level-model, as suggested by the ISO and the OMG standards, able to model any other kind of third level-methodology.

For statistical modelling, a recommended situation would consist in having a unique descriptive statistic methodology, able to model every type of subject-matter statistical information segment.

The existence of many competing modelling methodologies cannot be always avoided. In fact, a unique methodology may be not the optimal way to satisfy different needs. Methodologies can evolve, can be inherited from the past experience, the reason and the power to unify them may lack, especially if they are developed and owned by different units or organisations.

The interoperability between different models in levels K1 and K2 is very important. The mapping between different models enables to convert a model into another, to exchange their contents, that is data and definitions, to share parts of the model and to ensure some degree of coherence between them.

Every model in the hierarchy gives rise to a language used for defining and naming its structures, and to the possible operations on them. In defining models, terms can be from the natural language, but they assume a more specific and formal meaning in the model context. The same natural term can be used in different models, assuming different meanings in each one.

Therefore, for the proper comprehension of the meaning of the terms, knowledge of the context in which they are used that is, the natural language, or a more formal model, is necessary. Meaning of a term belongs to the model in which the term is defined.

This principle, applied to the hierarchy of models, leads to four levels of languages each one corresponding to a modelling level:

- A generic modelling language for each fourth level model
- A statistic methodology language for each third level model, in which the terms are derived from the descriptive statistic methodology
- A subject matter definition language for each definition model in which the terms are basically derived from the discipline to which the model refers
- A subject matter extension language for each extension model, in which the terms are the symbols used in the extensional representation.

Models pertaining to two different levels are oriented to describe different aspects. Consequently the meaning of a term possibly used in two models, owing to the different levels, cannot be the same. This is valid also if two models of the same level use the same term with different meanings. When it is not convenient to use different terms, the term use should be accompanied by the indication of the context to which it refers, that is, the model. On the contrary, it is possible and desirable that two different models in the same level may share the same term with the same meaning. The structure of the terminology of an information system appears to be composed by a matched group of terminologies, strictly corresponding to the structure of the models.

Finally the four level hierarchies of models is a model of its own and has its proper language that contains terms with a particular meaning according to the models defined in the hierarchy. An example is the term level that, in this context, means modeling level, but in a third level-model may mean classification level.

#### 2.1.3.5 Modelling Network

A network perspective is expected to replace and enhance the hierarchical schemes capabilities. It represents a natural development of the hierarchical perspective.

An example of network approach is represented by the study of equilibration and regulations for multiple level cognitive systems (Piaget 1977, Piaget and Garcia 1989).

Fig. 2.13 shows a network presentation of a four level architecture encountered in IRDS information systems (Rossiter et al. 2000). The four levels or realms are: APP, IRD, IRDD and IRDDS.

In categorical terms each of the four realms is defined as a category. Between each level there is a functor which ensures that certain consistency conditions are met in the mapping between the source and target categories.

The signification of the functors U and P are as follows (Rossiter et al. 2000):

U10-Data, U21-Organization, U32-Policy, U30-Platform

P01-Name, P12-Meta, P23-Meta-meta, P03-System

Observe that: U10: IRD→APP, U21: IRDD→IRD, U32: IRDDS→IRDD, and

U30: IRDDS→APP. In this case U30=U10oU21oU32 while P03= P01oP12oP23

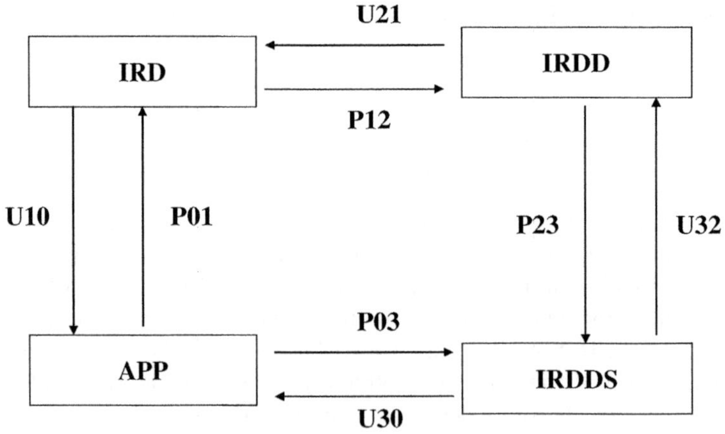


**Fig. 2.13** IRDS system network

The purpose of such structures is on one side to emphasize the self-documentation of the information system. A model cannot exist if there isn't an upper level model that describes it. On the other side, the interoperability of the information system, that is, the existence of a generic model, in the upper level or realm, that describes many others in fact its instances in the lower level.

Critical are the interactions between realms IRDDS and APP.

Between the reality levels n=3 and n=0, that is RDDS and APP, there is an operator U30 called "Platform" (Rossiter et al. 2000). This gives a direct implementation of abstractions to data values.

The possibility P03, called "Sys" from system, directly transfer information from real data to conceptual level.

These interactions described by U30 and P03 allow the integrative closure and system evolvability and autonomy.

A similar structure is proposed for the OMG or for the levels hierarchy in statistical architecture.

The network form is shown in Fig. 2.14.

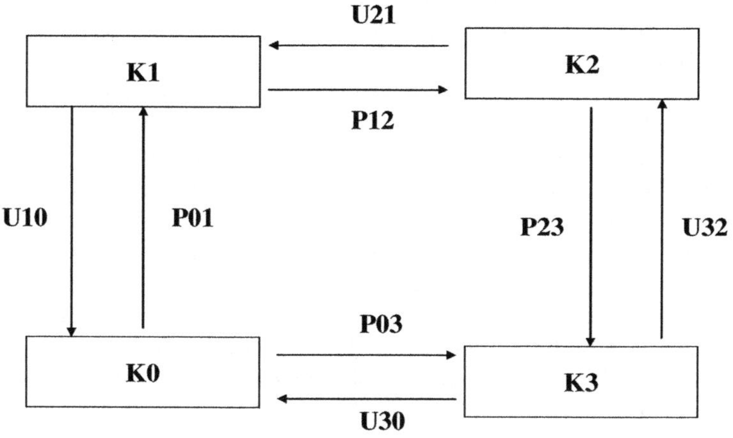


**Fig. 2.14** Statistical modelling networks

The notations are K0 for data, K1 for statistical information definitions and K2 for descriptive statistic methodologies. K3 denotes methodologies that define methodologies.

The $1^{st}$ order evolutionary step characterizes the K1 emergence, $2^{nd}$ order evolutionary step the K2 emergence, and so on (Fig. 2.15).

The network can be used to distinguish different roles in the information system. Basically, the idea is that a role consists in using the model on a certain level in order to produce models in the lower level.

According to the order of reality levels in the diagram shown in Fig. 2.14 it results that the generic modeller or statistical methodologist U32: K3→K2 uses a general purpose model to produce statistical methodologies. This corresponds to the $3^{rd}$ order evolutionary step.

The statistics definer U21: K2→K1 uses a statistical methodology to produce subject matter definitions. The statistics producer and user

U10: K1→ K0 uses a subject matter definition to produce statistics to understand the reality and possibly produce actions on it. The interaction between K0-Data and K3-Methods that define methods allows integrative closure and may make the statistical models system evolvable. This corresponds to the $4^{th}$ order evolutionary step.

The roles defined above are independent of the nature of the agent, the role of executor, that can be human or software artefacts. The same role can be played in principle by people or by machine. In this role-playing, the upper level model supplies specifications to the agent that interprets and applies them in order to produce the lower level model. When this behaviour is enforced in practice, the system is active because the upper level model drives agent behaviour.

The schema shows how software artefacts can be made active: they have to be driven by the respective upper level model. The major data processing software packages in the statistical system should be founded on the idea of active models in hierarchy or network. To process a level, the software is driven by upper level. For example, to produce a level K1-model, a set of statistical data, software is driven by its level K2-model, expressed in a formalized subject matter language and therefore highly independent from the technical aspect of the implementation. To produce a level K2-model, the activity of subject matter experts is supported by software tools, driven by the level K3-model in use. The specificity of the statistical field is located at the level of K2. The K2-model is considered as the formal representation of a descriptive statistic methodology.

It should be observed that any of the realms in statistical methodologies may be deeply evaluated at the sub-realms level of detail. The K2-model contains structures able to give a concrete and possibly formal shape to statistical methodological rules.

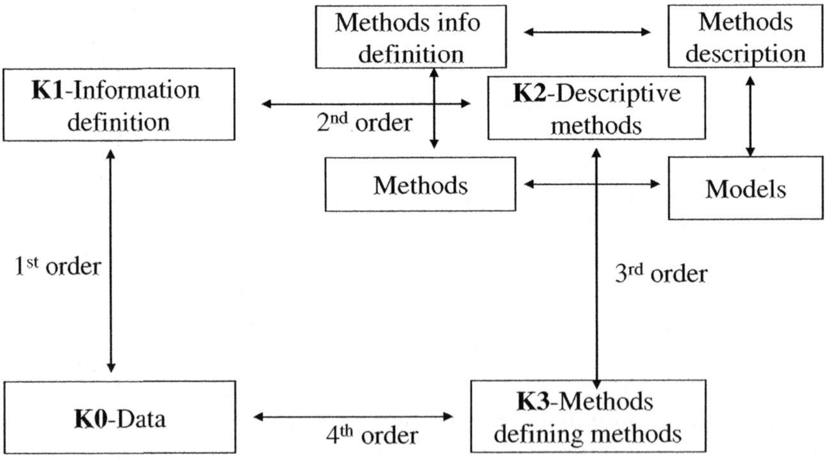


**Fig. 2.15** Four realms and sub-realms for methodology to define methods

Fig. 2.15 clarifies such situations. The basic framework from Fig. 2.14 is recapitulated for the K2 realm only. The sub-realms are methods, methods information definition, methods description and lastly models that may be considered an instantiation of the descriptive methods.

The self-similarity shown in Fig. 2.15 suggests possibility of making use of similar software for the whole system as for K2 instantiation.

Modelling frameworks as shown in Fig. 2.15 incorporate evolvability in order to selectively manipulate the models and to incorporate details and complexity only in those areas of the models which are critical to provide an adequate solution and remove such details and complexity were it is not (Fraga et al. 2006).

Thus we can imagine a modelling and simulation capability within which the following hold:

- A model consists of a hierarchy of layers or realms of increasing detail, complexity and sophistication, spanning the entire set of length and time scales from molecules to business chains
- Each layer or realm contains a model definition and a number of parameters
- Each layer accepts parameters from below and calculates the parameters required by the layer above
- Evolvability capabilities such as ontologies, languages, agents and so on, may be incorporated at any point to define and modify the models, parameters and solutions.

Such architecture has a number of advantages as for instance:

- It is flexible and extensible
- It provides a rational and consistent basis for multiple scale models
- It may incorporate external modules, models, codes and may be integrated with laboratory and plant systems
- Fitness for purpose can be indicated to the user

This architecture of smart models offers a potential way to evolve in a business environment changing at an ever-increasing rate (Farga et al., 2006).

#### 2.1.3.6 Multiple Scale Information and n-Graphs

Often the reality or description levels are associated to specific scales.

Representation of information at different resolution levels can be approached in terms of n-graphs (Appendix A5).

Fig. 2.16 illustrates the n-graphs associated to multiple scale information systems.

The reality level n=0 corresponds to the 0-graphs or sets. In real systems this may be associated to the objects or to areas of interest. They are called also 0-cells, or nodes. The reality level n=1 corresponds to the 1-graphs.

These are directed graphs including the morphisms that is, relations between different objects or areas of interest. The morphisms are 1-cells.

The level n=2 corresponds to the 2-graphs. These are graphs plus the so-called 2-cells between paths of same source and target. The 2-cells describe relations between relations.

The reality level n=3 corresponds to the 3-graphs. These are 2-graphs that include 3-cells that is, the cells between 2-cells. They represent graphs modification and are subjected to conditions of natural transformations. More than this level n=3, may be in theory imagined but as observed from numerous case studies, just a modification of modifications for n=4 seems to not bring new information (Cruz et al. 2006).

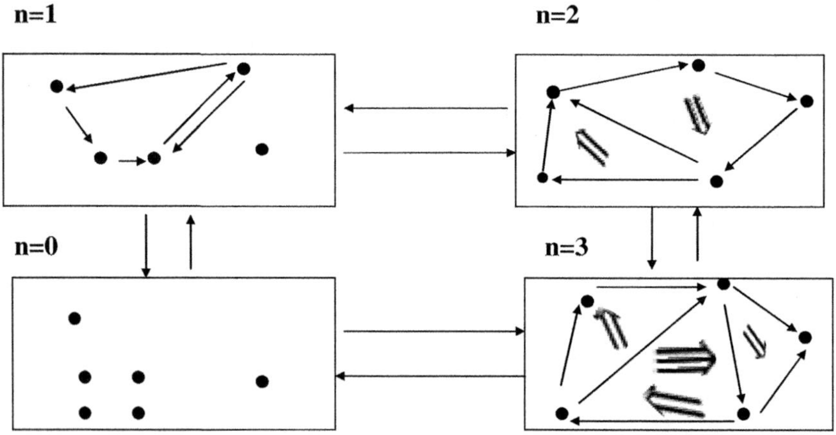


**Fig. 2.16** n-graphs multiple scales networks

The integrative closure between the top level n=3 of 3-graphs and the lower level n=0 of sets is the challenge for evolvable and autonomous systems.

### 2.1.3.7 Standpoint

The approach of the multi-layered hierarchy of models to describe information systems meets many objectives at once. It appears to be a method to give formal structure to more information systems, decomposing them in self-consistent parts, to interoperate standardize and define terms, to distinguish roles, to obtain closure and to make active information systems.

The approach offers a synthetic and high-level vision of statistical information systems, connecting different perspectives in which they are usually seen.

Several open problems are revealed by the study of modeling layers.

A significant concern is the number of levels.

The number of reality levels, observation levels or abstraction levels is finite. Four levels are probably necessary and sufficient for data analysis. In fact, the maximum four level architectures are shared by several operational structures of evolutionary devices, by the functional organization of organisms as informational and cognitive systems, and also by some scientific and engineering methods.

Support for the four level architectures in data processing is given by the neurodynamics (Cowan 2000). According to Cowan the capacity of short-term memory is limited to the number of four items to which attention can be simultaneously directed. There exists a central capacity limit of four chunks in short-term explicit memory presumably corresponding to the focus of attention. This theory assumes that attention is depending on oscillation of cortical potentials. A cortical wave of about 10 Hz is supposed to select items from a large short terms store. Other wavelets at a frequency at about 40 Hz then select one item each. Such considerations don't exclude to direct attention to more than four items but the resulting processes are expected to be transient. The number four is correlated to the study limit in processing capacity as defined by Halford et al. (1998) and to the anatomical studies of cerebellum and of cerebral rhythms (Freeman 2000).

Hummel and Holyoak (2003) correlated the four levels of memory in the neurocognitive framework to the limits in mental storage capacity.

Implicit support for the four level architectures in data processing is given by category theory too (Baez and Dolan 1995, Rossiter and Heather 2003).

In category theory, four levels are required to define morphism as unique up to isomorphism. The four levels are the objects, that is, the elements within a category, the category comparing the objects, the functors comparing categories and the natural transformation comparing the functors.

These four constituents represent the categorification of the four corresponding elements in set theory namely: the elements of sets, the sets, the functions and the equations between morphisms (Appendix A4). This fact suggests to associates the realms or levels in modeling to categorification process from K0 to K1, then K2 and then K3 or in other notations from S to K1, K2 and K3. It should be emphasized that in our previous notation, the highest reality level, for instance K3, corresponds to the conditioning level m=0, the reality level K2 to the conditioning level, m=1, the reality level K1 to the conditioning level m=2 and the reality level K0 to the conditioning level m=3.

Four levels seem to be necessary for interoperability (Rossiter and Heather 2003). Less than four offers only local interoperability. More than four may be formally defined but yields no benefits for interoperability. The practical consequence of a fifth level is equivalent to an alternative of the fourth level.

Limiting the study to four levels means to limit the categorical approach to 3-categories (Power 1995). There exists attempts to define 4-categories but the problem is that the knowledge about 4-categories is still controversial. The 4-categorical diagram techniques are as yet entirely non-existent. The

4-categories, in classical logics where we can duplicate and delete information, are categories with finite products the basic example for these being again the category Set. This observation may be considered as a supplementary reason to restrict the present study to 3-categories, without excluding higher categories in the long run.

For the time being we may focus the investigations to four levels that is, to the successive levels of complexity indexed by 0, 1, 2 or 3, and to 3-categories.

The single level S or K0 describes simple systems. Then the n-tuples (S, K1), (S, K1, K2) and (S, K1, K2, K3) characterize the successive levels of complexity.

In terms of Hartmann hierarchy this sequence may be linked to material, biological, cognitive and intelligence levels. Other, more detailed areas of reality may be studied by formally similar four sublevels splitting.

Several architectures studied in computer science are based on less than four levels and may be considered as still incompletely developed (Rossiter and Heather 2003).

The semantic web appears to lack elements of the top levels but the use of ontologies and agents may compensate, in part, for some of this loss (Berners-Lee et al. 2001).

The conventional Grid computing lacks the top two levels for data addressing (Foster and Kesselman 1999). The development of the Grid outlined the great difficulty of employing data held in formal data bases as opposed to operating systems files.

IRDS has four levels but use only one way mapping and neglects the two-way mapping of the general framework.

MOF outlines four-levels but the top-levels are not linked to data that is to the basic level.

The observation that some conceptual levels or some mapping are missing may guide the research efforts.

Mappings between the top level K3 and the lower level denoted here by K0 or S represents the main challenge for understanding and managing the higher complexity.

## 2.2 Wave Equation

### *2.2.1 Frameworks for "Time" and "Space"*

Classification and the judgment of similarity are fundamental in cognition, serving as the basis for actions. The classification and pattern recognition are the key ingredients in data processing and in problem solving for both natural and artificial evolvable systems.

Living or artificially living systems do not survive in environments that they don't recognize or misclassify. Living supposes identification, classification or categorization.

Preliminary attempts for classification or pattern recognition modeling by differential equations outlined the major role of orthogonal arrays (Iordache 1992, 1996, 2009). A significant result was that the pattern recognition methods parallel screening procedures in experiment design and in problem solving. In particular cases one obtained as solutions of the first order wave equation orthogonal arrays matrices, Walsh matrices, or Latin squares. Models of cognitive processes such as pattern recognition prove to have as solutions logical thinking methods as that applied in designs of experiments. The result emphasized the deep relation between cognition and evolvability as presented in constructivist perspective and the assertion that both cognition and evolution are based on similar sets of techniques (Piaget and Garcia 1989).

To establish the analogous of a dynamical model for classification or pattern recognition, the concept of time and of space in the conditions space K, will be revisited and adapted to the objectives of the present study.

The point of view adopted here is that the significance of the process parameters should agree firstly with the mechanism and the nature of analysis of the studied system.

Less-conventional mathematical frameworks are acceptable if these frames can naturally describe the system evolution and the system analysis can proceed on this basis.

Numerous studies defend and elaborate the idea that there are different kinds of time and different kinds of space to be considered in modeling (Nehaniv 2005, Poli 2007). Specific "chronotopoids" should be associated to different reality levels.

Evolvable systems study needs appropriate concepts for time and space.

For numerous transport processes the basic material balance is represented by first order wave equations as for instance equation 2.1 (Rhee et al. 1989):

$$\frac{\partial y}{\partial t} + v\frac{\partial y}{\partial z} + q(y) = 0 \tag{2.1}$$

Here y(t, z) may denote a concentration variable in t-time and z-space, v denotes the velocity, and q(y) denotes a separation rate.

The model (2.1) known also under the name of Euler's equation describes the incompressible fluid flow and many other phenomena of physical and technological interest.

The basic model (2.1) shows that the variation of concentration in time is due to a convective process with velocity v, and to a kinetic process of interaction, q(y).

For material level studies, the real field plays the dominant role.

This may be unsuitable for some studies of other ontological levels as biological, cognitive or intelligent levels.

Algebraic finite frames such as Galois fields and rings or cyclic groups, with trivial valuation, represent mathematical frameworks that have been used to describe the finite, logical or cyclic type of cognition processes.

Multiple time scales and time cycles should be taken into account. A cyclic framework complementing the usual linear one from classical physics proves to be necessary. Evolvability description requires both the quiescent "timeless" or "cyclical" K-processes and the relatively fast, "dynamical" or "linear" S-processes (Pattee 1995). Algebraic finite fields represent a common choice for K, whereas the real field is the commonplace structure for S. There is a natural hierarchical or cyclic structure associated to finite fields and this explains, in part, why they are considered as the appropriate tool for systems structured in conditional levels.

An equation describing the cognitive and evolvable systems would contain parameter analogues to the space and the time from the dynamical mathematical model (2.1) known from physics.

Consider for example the space Z of vectors describing the properties of an object to be classified and the time T describing the degree of advancement of the pattern recognition, classification, or development, for that object. For the classification process it is possible to associate to different steps in a classification scheme digits such as "0" or "1", with significance as "no" or "yes", "true" or "false" "separated" or "non-separated", "identified" or "non-identified" "developed" or "non-developed" (Iordache et al. 1993a, 1993b, 1993c).

To any object to be classified, a vector Z will be associated in which the properties are specified by digits in the hierarchical order of significance for classification. The same vector Z will give a description of the classification stages in the associated pattern recognition scheme. Z describes pattern recognition or stages in problem solving or development stages for organisms and so forth. Denote $Z=z_0z_1....z_j$.

The component $z_j$ should specify the presence of an attribute in classification step, its absence but also partial or uncertain results.

The mathematical framework for Z can't be limited to that of dyadic that is to Boolean calculus. The need for multi-valued characterization of classification steps and of objects, the uncertainty, imposes mathematical tools completing the better studied dyadic calculations. Detailed description of dynamical systems need vector characterization corresponding to multi-valued logic such as: "0", "1", "2" and so on, meaning for instance, low, average, high and so on. The coordinate "$z_j$" characterizes the properties and also it is naturally associated to a stage of classification schemes that make use of the difference in properties noted by "$z_j$" to perform that kind of classification, pattern recognition, or development.

The degree of advancing in the classification, pattern recognition or in development, denoted by T, was defined as the necessary level of similarity T, between two objects representation, to be classified in the same class (Iordache et al. 1993 a, 1993c). It may be an expansion of the type: $T=t_0t_1...t_j$ with the digits $t_j$=0, 1, 2 and so on. Denote also this vector by $T=(t_j)$. Each value of T corresponds to another potential step in pattern recognition or in development. Single component vectors with modulo-m algebra structure will be presented as a first example. This is one of the weakest algebraic structures for T and Z still providing a mathematically tractable model adequate to classification and pattern recognition operations or to development study. A slightly different framework to be considered is that of Galois finite fields. Recall that finite fields with the same number of elements are isomorphic.

Examples of addition and product tables are presented in Table 2.1 and Table 2.2. Here "$\oplus$" denotes the addition and "$\otimes$" denotes the product. The sum and product refers to component-wise operations for vectors as Z or T in K.C (m) denotes the modulo-m algebraic framework, and GF (m) the Galois field of order m.

C (m) enables to fit the physical intuition concerning the cyclic character of the classification operations in m steps and to justify this first choice for algebraic framework. If m=2 the sum $\oplus$ is defined as follows: for any two elements $T=(t_j)$ and $S=(s_j)$ the dyadic sum is: $(t\oplus s)_j=((t_j+s_j) \bmod 2)$. This means that $1\oplus 1 = 0$, $1\oplus 0 = 1$. The sum is the dyadic addition, $\oplus$ equivalent to the dyadic difference. The rule of addition $\oplus$ signifies that two identical digits have no effects for classification. Only the difference in digits makes a contribution. This addition appears rather as a comparison than as a sum. The product $\otimes$ is introduced in a way related to cyclic operations too. Product definition takes into account that after m steps the classification process restarts. For the "time" T or "space" Z, no change should happen after completion of a cycle of classification operations.

Another algebraic framework to be considered is the finite field, GF (m). If m is not a prime number we are faced with rings instead of fields. This algebraic framework was extensively applied in formal logics. For illustration purposes, the operations in GF (3) and GF (4) are presented in Table 2.2.

Let Y denotes, the range, the output of a system that performs classification based on features or property examination. Y are element of the same algebraic frames as T or Z. Y may be single-dimensional vector and may assume values 0, 1, 2 and so on, corresponding to various outputs. Multi-dimensional values like $Y=y_0y_1y_2\ldots y_j$ should be examined too. Y, as T or Z is represented by finite strings. Y definition needs to ensure the logical consistency of the framework. Appropriate algebraic structures for the range of Y are algebras or fields such as the field of real numbers the modulo-m algebras, or the finite Galois field, GF (m) that provides physically significant and mathematically tractable models. The dyadic differential calculus was initially developed for GF (2) situations (Harmuth 1977). If m is a prime-p, the range Y is the standard framework for multi-valued logic. Single dimensional vectors T, Z, Y are useful if the classification process is based on a single property. For the multiple level situations T, Z and Y will be tensor products, "$*$", of single-levels cyclic groups (Bochmann and Posthoff 1981, Yanushkevich 1998).

**Table 2.1** Sum and product in C (m)

C (2)

| (x+y) mod2 | | | (x.y) mod2 | | |
|---|---|---|---|---|---|
| ⊕ | 0 | 1 | ⊗ | 0 | 1 |
| 0 | 0 | 1 | 0 | 0 | 0 |
| 1 | 1 | 0 | 1 | 0 | 1 |

C (3)

| (x+y) mod3 | | | | (x.y) mod3 | | | |
|---|---|---|---|---|---|---|---|
| ⊕ | 0 | 1 | 2 | ⊗ | 0 | 1 | 2 |
| 0 | 0 | 1 | 2 | 0 | 0 | 0 | 0 |
| 1 | 1 | 2 | 0 | 1 | 0 | 1 | 2 |
| 2 | 2 | 0 | 1 | 2 | 0 | 2 | 1 |

C (4)

| (x+y)mod4 | | | | | (x.y)mod4 | | | | |
|---|---|---|---|---|---|---|---|---|---|
| ⊕ | 0 | 1 | 2 | 3 | ⊗ | 0 | 1 | 2 | 3 |
| 0 | 0 | 1 | 2 | 3 | 0 | 0 | 0 | 0 | 0 |
| 1 | 1 | 2 | 3 | 0 | 1 | 0 | 1 | 2 | 3 |
| 2 | 2 | 3 | 0 | 1 | 2 | 0 | 2 | 0 | 2 |
| 3 | 3 | 0 | 1 | 2 | 3 | 0 | 3 | 2 | 1 |

**Table 2.2** Sum and product in GF (m)

GF (2)

| (x⊕y) | | | (x⊗y) | | |
|---|---|---|---|---|---|
| ⊕ | 0 | 1 | ⊗ | 0 | 1 |
| 0 | 0 | 1 | 0 | 0 | 0 |
| 1 | 1 | 0 | 1 | 0 | 1 |

GF (3)

| (x⊕y) | | | | (x⊗y) | | | |
|---|---|---|---|---|---|---|---|
| ⊕ | 0 | 1 | 2 | ⊗ | 0 | 1 | 2 |
| 0 | 0 | 1 | 2 | 0 | 0 | 0 | 0 |
| 1 | 1 | 2 | 0 | 1 | 0 | 1 | 2 |
| 2 | 2 | 0 | 1 | 2 | 0 | 2 | 1 |

GF (4)

| (x⊕y) | | | | | (x⊗y) | | | | |
|---|---|---|---|---|---|---|---|---|---|
| ⊕ | 0 | 1 | 2 | 3 | ⊗ | 0 | 1 | 2 | 3 |
| 0 | 0 | 1 | 2 | 3 | 0 | 0 | 0 | 0 | 0 |
| 1 | 1 | 0 | 3 | 2 | 1 | 0 | 1 | 2 | 3 |
| 2 | 2 | 3 | 0 | 1 | 2 | 0 | 2 | 3 | 1 |
| 3 | 3 | 2 | 1 | 0 | 3 | 0 | 3 | 1 | 2 |

### 2.2.2 First Order Wave Equation

One of the simplest mechanisms of pattern recognition, classification or development is that in which "small" changes of the degree of pattern recognition, $\partial$ T, are associated to "small" changes of the answer, $\partial$ Y. It should be emphasized that the differential is in fact a difference since T and Y are discrete.

Moreover, the change of answer $\partial$ Y depends on both the existing answer Y and the change $\partial$ T of T that is:

$$\frac{\partial Y}{\partial T} \propto Q(Y) \tag{2.2}$$

It is supposed that $\partial$ T is non-null. Otherwise the differential equations are replaced by difference equations. The rate of pattern recognition or classification is denoted by Q. This mechanism is of "kinetic" type.

Another classification mechanism, takes into account that the variation of the answer Y, along the degree of recognition T is proportional to the answer variation along the features space Z. Classification, pattern recognition and development means in fact travel in time T, along the space of properties, Z. As Z is screened with a velocity V, the degree of pattern recognition varies proportionally.

This means that:

$$\frac{\partial Y}{\partial T} \propto V \otimes \frac{\partial Y}{\partial Z} \tag{2.3}$$

Here the velocity is a vector $V = v_0 v_1 v_2 ... v_{jor}$ $V = (v_j)$. This mechanism is of "convection" or "drift" type.

The general model of the pattern recognition process including both types of recognition processes, corresponding to the evolution according to T and Z, is the first order wave equation, WE:

$$\frac{\partial Y}{\partial T} \oplus V \otimes \frac{\partial Y}{\partial Z} \oplus Q(Y) = 0 \tag{2.4}$$

The initial condition is:

$$Y(Z, 0) = F(Z) \tag{2.5}$$

Obviously V and Q may depend on T and Z.

The fact that the addition is equivalent to the difference suggests that a second order wave equation doesn't gives new solutions in K, as defined.

The mathematical formalism for modeling conditions K apparently follows that of the real states S as a first order wave equation, WE, but with different addition and product operations taking into account the corresponding domain. Symbolic models, in K, and dynamical models, in S, are complementary in the sense that, neither type of model is reducible to the other.

Both are necessary for understanding the whole system including classification schemes and dynamic processes.

The first-order wave equation, WE, is formally similar to the basic model (2.1) applied in different domains by chemical engineers. For this reason the developed methodology may be considered as a kind of "artificial chemical engineering". It may be related to chemical engineering as the "artificial chemistry" is related to chemistry (Dittrich et al. 2001) or "artificial life" to natural life. The physical or biological domains offer inspiration for the artificial domains, for calculus and for artifacts.

### *2.2.3 Kinetic Model*

For V=0 the first order wave equation, WE, reduces to:

$$\frac{\partial Y}{\partial T} \oplus Q(Y) = 0 \tag{2.6}$$

The solution in GF (2) is presented here for illustration purposes. In GF (2), "0", denotes the null element. The real product and the sum were translated towards GF (2) operations.

Suppose that the rate of pattern recognition, Q, is the constant expansion denoted by

$$Q = q_0 q_1 q_2 \ldots q_j \text{ or } Q = (q_j).$$

The solution similar to Euler solution for differential equations will be:

$$Y(T) = Y(0) \oplus Q \otimes T \tag{2.7}$$

Recall that the sum $\oplus$ is equivalent to the dyadic difference.

Suppose that, Y(0) = 1. In this case the solution of the first order wave equation, WE, for different Q is Y (T, Q) as shown in Table 2.3.

The detailed equations for m=0 are:

$$\frac{\partial y_0}{\partial t_0} \oplus q_0 = 0 \tag{2.8}$$

$$y_0(0) = f_0 \tag{2.9}$$

Denote, the resulting "0" by "-1" with the same logical signification, for instance "no". Table 2.4 replaces Table 2.3.

Suppose that Y, T, Q are vectors with two components: $Y=y_0y_1$, $T=t_0t_1$, $Q=q_0q_1$ and $F=f_0f_1$.

This corresponds to two conditioning levels. The first order wave equation, WE, reduces in fact to two similar equations, one for each level. For m=0 the model is given by equation (2.8) with initial condition (2.9).

**Table 2.3** Kinetic model, m=0

| Q\T | 0 | 1 |
|---|---|---|
| 0 | 1 | 1 |
| 1 | 1 | 0 |

**Table 2.4** Kinetic model, modified, m=0

| Q\T | 0 | 1 |
|---|---|---|
| 0 | 1 | 1 |
| 1 | 1 | -1 |

For m=0 and m=1 a new equation and initial condition should be added:

$$\frac{\partial y_1}{\partial t_1} \oplus q_1 = 0 \tag{2.10}$$

$$y_1(0) = f_1 \tag{2.11}$$

The fact that one equation in K is replaced by two differential equations, one for each conditioning level, outlines one of the differences between models in K and in S.

The following procedures are suggested by constructions in the categorical framework associated to SKUP. This refers to tensor product "$*$" as for instance the categorical product "$\times$" and the coproduct "$\cup$" connecting different conditioning levels.

It should be emphasized that these are different from component-wise product $\otimes$ and sum $\oplus$ that refers to elements of the same level. The tensor product "$*$" connects the levels not the conditions at the same level.

Consider the initial condition:

$$Y(Z,0) = F(Z) = f_0 \times f_1 \tag{2.12}$$

The solution of the model will be:

$$Y(T) = y_0 \times y_1 \tag{2.13}$$

The particular case $f_0=f_1$, $q_0=q_1$, implies $y_0=y_1$.

Table 2.5 shows the product solution.

This represents the Walsh-Hadamard, WH, matrices in DOE. With more coordinates in Y, T, Z it is possible to obtain Walsh matrices with 8, 16, 32, and so on elements.

Table 2.6 shows the three conditioning levels solution for $Y=y_0y_1y_2$ and $T=t_0t_1t_2$ a Walsh-Hadamard DOE matrix.

It was considered that $f_0=f_1=f_2$, $q_0=q_1=q_2$, and this imposes $y_0=y_1=y_2$.

**Table 2.5** Kinetic model, product Y(T)=$y_0 \times y_1$

| | | | |
|---|---|---|---|
| 1 | 1 | 1 | 1 |
| 1 | -1 | 1 | -1 |
| 1 | 1 | -1 | -1 |
| 1 | -1 | -1 | 1 |

**Table 2.6** Kinetic model, product Y(T)=$y_0 \times y_1 \times y_2$

| | | | | | | | |
|---|---|---|---|---|---|---|---|
| 1 | 1 | 1 | 1 | 1 | 1 | 1 | 1 |
| 1 | -1 | 1 | -1 | 1 | -1 | 1 | -1 |
| 1 | 1 | -1 | -1 | 1 | 1 | -1 | -1 |
| 1 | -1 | -1 | 1 | 1 | -1 | -1 | 1 |
| 1 | 1 | 1 | 1 | -1 | -1 | -1 | -1 |
| 1 | -1 | 1 | -1 | -1 | 1 | -1 | 1 |
| 1 | 1 | -1 | -1 | -1 | -1 | 1 | 1 |
| 1 | -1 | -1 | 1 | -1 | 1 | 1 | -1 |

Walsh series as solution of differential equations in dyadic field have been obtained by Iordache (Iordache 1992).

For the obtained solutions, the tensor product was interpreted as a categorical product, denoted by "×".

### *2.2.4 Convection Model*

Consider now the convective part of the first order wave equation, WE:

$$\frac{\partial Y}{\partial T} \oplus V \otimes \frac{\partial Y}{\partial Z} = 0 \tag{2.14}$$

The initial condition is:

$$Y(Z,0) = F(Z) \tag{2.15}$$

The operations in K are the sum $\oplus$ and the product $\otimes$ from GF (m).

The general solution of the partial first order wave equation, WE, is:

$$Y(Z,T) = F(Z \oplus (V \otimes T)) \tag{2.16}$$

Consider the initial condition:

$$F(Z) = Z \tag{2.17}$$

This means that at T=0, the output Y of the classification scheme at the distance Z in scheme is exactly Z. The scheme is one in which each classification level activates a new difference in properties allowing classification. This kind of initial condition ensures that the wave of the classification process is initiated and is going on.

It results the characteristic:

$$Y = Z \oplus (V \otimes T) \tag{2.18}$$

The GF (3) solution is presented in detail. For T=0 the solution Y is shown in Table 2.7.

**Table 2.7** Convection model, m=3: Y(0, Z)

| z\v | 0 | 1 | 2 |
|---|---|---|---|
| 0 | 0 | 0 | 0 |
| 1 | 1 | 1 | 1 |
| 2 | 2 | 2 | 2 |

For T=1 the solution is shown in Table 2.8.

**Table 2.8** Convection model, m=3: Y(1, Z)

| z\v | 0 | 1 | 2 |
|---|---|---|---|
| 0 | 0 | 1 | 2 |
| 1 | 1 | 2 | 0 |
| 2 | 2 | 0 | 1 |

For T=2 the solution is shown in Table 2.9:

**Table 2.9** Convection model, m=3: Y(2, Z)

| z\v | 0 | 1 | 2 |
|---|---|---|---|
| 0 | 0 | 2 | 1 |
| 1 | 1 | 0 | 2 |
| 2 | 2 | 1 | 0 |

There is a relation between different solutions of the first order wave equation, WE, and DOE matrices.

For different values of T, T=1, T=2 one obtained different (3x3) Latin-squares.

Latin squares close association to DOE is well-known (Hedayat et al. 1999)

The following procedures to obtain DOE are suggested by universal constructions in categorical framework. There are several DOE to be obtained by combining the solutions obtained for different values of T.

**Table 2.10** Concatenated solutions, m=3

| | | |
|---|---|---|
| 000 | 012 | 021 |
| 111 | 120 | 102 |
| 222 | 201 | 210 |

Superposing by concatenation the elements of the Table 2.7, Table 2.8, and Table 2.9, the Table 2.10 will result.

Pasting down the 3-digit numbers from Table 2.10, column after column, the Table 2.11 is obtained. It is a well known DOE with 9 experiments for 3 factors, F0, F1, and F2.

Columns in Table 2.11 are orthogonal. Each column corresponds to first order wave equation, WE, solutions at different velocities V. Associating one supplementary digit for each column in the Table 2.10, the four digit numbers as in Table 2.12 results. Here the index (0) corresponds to the first column in Table 2.10, (1) to the second column and (2) to the third column.

The resulting 4 digits numbers from Table 2.12 correspond to columns of well-known orthogonal design with 9 experiments and 4 factors (Taguchi 1986, Hedayat et al. 1999).

Concatenation and pasting down operations are, in line with the coproduct "∪" type of operation in categorical framework.

The previously obtained matrices are linked to the tensor product interpretation as coproduct "∪". Obviously making use of tensor products as categorical product "×" will offer other class of solutions, asking for

**Table 2.11** Pasting down columns, m=3

| F0 | F1 | F2 |
|---|---|---|
| 0 | 0 | 0 |
| 1 | 1 | 1 |
| 2 | 2 | 2 |
| 0 | 1 | 2 |
| 1 | 2 | 0 |
| 2 | 0 | 1 |
| 0 | 2 | 1 |
| 1 | 0 | 2 |
| 2 | 1 | 0 |

**Table 2.12** Indexed concatenated solutions, m=3

| | | |
|---|---|---|
| (0)000 | (1)012 | (2)021 |
| (0)111 | (1)120 | (2)102 |
| (0)222 | (1)201 | (2)210 |

significantly more experiments. It is known that the product of two Latin squares gives another Latin square. For instance, the Kronecker product of two (3x3) Latin squares gives a (9x9) Latin square. Moreover the Kronecker product of a difference scheme as that resulting by the described here pasting down procedure, to an orthogonal array gives another orthogonal array (Hedayat et al. 1999). This means that switching from categorical product to categorical coproduct, maintains orthogonality.

## 2.3 Possibilities

### *2.3.1 Similarities and Classification*

Complex systems are multiple component and multiple level systems that are not predictable in a conventional sense. For this reason they need appropriate notions to complement or replace conventional probabilities.

Classification methods parallel and substantially simplify partition in classes for stochastic processes, exchanging transition probabilities by similarities.

The classification procedure may be applied to the rows or columns vectors of the matrices obtained as solutions of the wave equation, WE.

These solutions have been used as matrices for design of experiment, DOE (Iordache 2009).

Each line in the matrix of DOE corresponds to an experiment that is to specific settings, each column to a factor. Statistical DOE makes use of such matrices and the corresponding responds to establish the effects of different factors, focusing on columns. In DOE method a supplementary objective is the classification of experiments or samples, focusing on rows.

The procedure is as follows: let $i=[i_1,...,i_k,...]$ and $j=[j_1,....,j_k,...]$ be two row vectors, that is two runs. To any pair of vectors the weighted similarity $r_{ij}$ is associated.

$$r_{ij} = \Sigma_k t_k (a_k)^k \tag{2.19}$$

With $0<a_k<1$, a constant, $t_k=1$ if $i_k=j_k$, $t_k=0$ if $i_k \neq j_k$. One use $a_k=0.5$ here. This means that the first factor has the weight 0.5, the next 0.25, and the next 0.125 and so on.

Similarities may replace and remarkably simplify the real-field defined probability calculus.

To any matrix of design corresponds a similarity matrix $R=[r_{ij}]$. This is the analogous of transition probabilities matrix associated to Markov stochastic processes.

The stabilization procedure allowing partition in disjoint classes is mandatory. To obtain the stable matrix, denoted by $R(n)=[r_{ij}(n)]$, the composition rule R(n)=R(n-1)oR is applied starting from R(2)=RoR towards the stable matrix R(n)=R(n+1)=R(n+2)=.... For two similarity matrices $R=[r_{ij}]$ and $Q=[q_{ij}]$, the composition rule "o" is defined by:

$$(RoQ)_{ij} = max_k[min(r_{ik}, q_{kj})] \tag{2.20}$$

This gives the ij-th element of the composed matrix. The classification algorithm is as follows: two experiments i and j are assigned to the same class at the classification level T, $0 \leq T \leq 1$, if their similarity in the stable matrix is larger than T that is if: $r_{ij}(n) > T$.

Notice that T denotes one of the valuations of the previously defined time T.

If $T = t_0 t_1 ... t_j$ with $t_j = 0$ or 1 then:

$$T = \Sigma_k t_k (0.5)^k \tag{2.21}$$

In a more general way, it is possible to start by introducing a real valued distance $d_{ij}$ between any two vectors, i and j.

The Minkowski distance is:

$$d_{ij} = (\Sigma_k (i_k - j_k)^k)^{1/k} \tag{2.22}$$

If k=2 this gives the Euclidean distances. The Manhattan or city-block distance is also of interest. This is defined as:

$$d_{ij} = \Sigma_k |i_k - j_k| \tag{2.23}$$

It is equivalent to the number of settings, which are different in the experiments i and j.

The Hamming distance is defined by:

$$d_{ij} = \Sigma_k XOR(i_k, j_k) \tag{2.24}$$

XOR is the exclusive or logical function. For Boolean vectors the Manhattan and the Hamming distances are equivalent.

The elements of the basic categorical framework SKUP may be easily revealed in classification procedures. The category S corresponds to the similarity matrices for experiments while the category K corresponds to the similarity matrices for the classes or lumps. S and K are categories since the identity, the associativity and commutativity of the morphisms hold in S and in K. The problem is to define functors U, that will be a natural way to transform the category K of lumped matrices in the category S of similarity matrices, and functors P, that will allow to transform S back to K. The study of classification schemes outlined the advantage of fuzzy measures replacing probabilities in defining the functors U and P (Iordache et al. 1993c).

Similarities defined by category theory constructions known as colimits, may be useful to predict biosystems similarity.

Bisimilarity represents a categorification of the notion of similarity (Blute et al. 1997).

To clarify the meaning of bisimilarity, consider a set of conditioned or labeled Markov chains. The labels process pertains to the category K. There is also a category S of states. The system is in a state at a given time and moves between states. Which state it moves to is governed by which interaction with the set of conditions and this is indicated by labels. The system evolves according to probabilistic laws. If the system interacts with the conditions by synchronizing on a label, it makes a transition to a new state governed by a transition probability distribution. This is the concept of bisimilarity model developed by Larsen and Skou (Larsen and Skou 1991). It specifies the transition by giving for each condition a probability for going from one state to another. Bisimulation then amounts to matching the moves with matching probabilities as well. Two states are bisimilar if they are related to by a bisimulation relation.

In the case of Markov chains bisimulation corresponds to standard lumpability of chains condition (Kemeny and Snell 1960).

## 2.3.2 Ultrametric Measures

The case of dyadic frames is presented here as an attempt to extend real valued probabilities. The constructions are intended to be valid for m-adic and p-adic situations.

The tentative definitions make use of some elements of the ultrametric integral theory developed by Monna and Springer (Van Rooij 1978). It takes into account the relation between the probabilities and the measure. The construction exploits also Harris (1955) results on chains of infinite order or equivalently Keane (1972) results on g-measures.

Let X denotes the space of all elementary events that is atoms in X. Denote by $X$ the Borel ring of all compact subsets of X. An event U is a subset of $X$. Consider as events the dyadic expansions of the type U=($u_1$, $u_2$,) where $u_j \in\{0, 1\}$ are the digits. To clarify the construction let us consider the hierarchical tree associated to U. The representation of U begins at an arbitrary point. When $u_i$ takes the value "0", a left-down branch is assigned. When $u_i$ takes the value "1" a right-down branch is assigned. This way the nodes are (0) and (1) on the first indexed conditioning level under the arbitrary point. Then, at the next level the nodes are (00), (01), (11) and (11), then (000), (001), (010), (011), (100), (101), (110), and (111) and so on. In the associated hierarchical tree it is easy to visualize operations as reunion, intersection arising in probability calculations.

There are several ways to define a possibility P: $X \rightarrow$K, with K dyadic ring. For instance it is possible to define a possibility for any digit in expansion as another digit that is: P($u_i$)=$v_i \in\{0, 1\}$.

Then the possibility P: $X \rightarrow$K is a map having the following property: if U=($u_1$, $u_2$,. . . ) then P(U)=(P($u_1$), P($u_2$),. . . )=($v_1 v_2$...). The function P may maintain the digit $u_i$ value as "0" or "1" or may change this to its contrary.

The result is a dyadic number too. Since the possibility of a digit is another digit, P(U) may be conceived as resulting by a component-wise operation of comparison (for instance addition) between U and a reference event as for instance (000,...,0) or (111,...,1). The operation in K may be the GF (2) addition $\oplus$, defined by $0\oplus1=1\oplus0=1$, $0\oplus0=1\oplus1=0$, or the GF (2) multiplication $0\otimes1=1\otimes0=0\otimes0=0$, $1\otimes1=1$. In this way the possibility appears as a similarity between two events.

The contact of the possibility as defined by P (U) with real field data is allowed by valuations and norms.

Denote $V=(v_1v_2...)$ with $v_i \in \{0, 1\}$, $V\in K$. It is possible to associate to the sequence V the sub-unitary number: $||V|| = \Sigma_n v_n 2^{-n}$. This is a real valuation that makes probabilistically significant the apparently arbitrary definition of possibility on K. Observe that $0\leq||V|| \leq 1$.

An ultrametric valuation in K is given by: $|V|=2^{-m}$; $m=\min\{j;\ v_j \neq 0\}$. The ultrametric valuation associated to the possibility is: $|P\ (U)| = |V|$.

To illustrate additionally the probabilistic content of the definition let us consider the hierarchical tree associated to dyadic numbers. Let $P\ (U)=p_k$, $p_k\in V$ where k is the level allowing the transition from an initial state in X, say $U_o=(000\ldots)$ to the state U in X in the hierarchical tree naturally associated to X. Denote $p_1=(100...)$, $p_2=(010...)$, $p_3=(001...)$ and so on. Observe that $p_k \in V$ are in fact analogous to transition probabilities and to possibilities. Moreover with GF (2) sum and product:

$$|p_k \oplus p_j| \leq max\{|p_k|, |p_j|\}; |p_k \otimes p_j| = |p_k||p_j| \tag{2.25}$$

The relation with the theory of integral is mediated by the function N introduced by Monna and Springer (van Rooij and Schikhof 1969) as follows:

$$[W] = max\{|P(U)|, U \in X, U \subset W\} \tag{2.26}$$

$$N_P(A) = min\{[W], A \in W\} \tag{2.27}$$

$N_P$ (.) is a real number for any A, and could be interpreted as a normalized distance.

Therefore, 1-$N_P$ is a measure of possibility and is related to coincidences in vector descriptions of two events.

$N_P$ allows a partition of K in disjoint classes. The events $A\in W$ are assigned to the same class D included in $X$ if $N_P$ (A)$\geq$T for some $0\leq T\leq 1$. Taking for instance $T=|p_2|$ it results D={001, 010, 011}. Observe that:

i) $$0 \leq N_P(A) \leq 1 \tag{2.28}$$

ii) If $T_1<T_2$ then $D_2\subseteq D_1$ where $D_j$ is the class corresponding to $T_j$, j=1, 2.

Definitions of possibility for finite frames may be based also on g-measures (Keane 1972, Stenflo 2003). This g-measure is a notion similar to that of chain with complete connections (Iosifescu and Grigorescu 1990). Let the

K={1, 2,..., N}, $K^N$ = {1, 2,..., N}$^N$and introduce a topology on $K^N$ by the ultrametric as follows:

Let u=$u_1$,$u_2$,..., and v==$v_1$,$v_2$,... are two words or paths, u, v∈$K^N$.

$\rho$(u, v)=$2^{-n}$ if u and v differs for the first time in the n-th digit and $\rho$(u, v)=0 if u=v.

The space ($K^N$, $\rho$) is an ultrametric space. Let s=$s_1$, $s_2$,..., in $K^N$ and v∈{1, 2,..., N} and let s'=vs=v,$s_1$,$s_2$,...,

Consider the function g: $K^N$ →(0, ∞) normalized in the sense that:

$\Sigma_K$ g(vs)=1 for any path s in $K^N$. Such a function is called a g-function. Invariant probability measures associated to the RIFS (Barnsley 1993) associated in turn to a g-function are called g-measures. The g measures are analogous to probabilities of transitions from the path s to the path vs. The g-measures are in fact the conditional probabilities, for instance P(s, v) and can be thought as the possibility of transition to the state v ∈ K conditional on the previous path s in $K^N$. To P(s, v) it is possible to associate the number s' = (s,v), or s' = (v,s) in $K^N$. This is equivalent to the pasting new digit in the dyadic expansions after or before the existing digits. Uniqueness in g-measures has been studied by Keane (1972) and by Stenflo (2003).

## 2.4 Entropy

### *2.4.1 Informational Entropy*

The entropy associated to elements in K will be studied in the following.

The elements in K appear as solutions of WE. They are matrices and the problem of comparing such matrices appeared in the study of evolvable DOE, EDOE (Iordache 2009). The aim was to select the set of significant experiments associated to a DOE matrix. The solution takes into account informational criteria. To any matrix of design a similarity matrix R is associated, and to this informational entropy:

$$H(R) = -\Sigma r_{ij} ln r_{ij} - \Sigma(1 - r_{ij}) ln(1 - r_{ij}) \quad (2.29)$$

This expresses the quantity of information associated to the matrix of design.

The defined entropy is a measure of the imprecision in classifying the experiments. This entropy approach the maximum value for equal similarities, 0.5 (Iordache et al. 1990). The higher entropy is an objective since it offers significance to experiments to be performed.

To have a choice between more possible matrices of DOE, M, a matrix can be utilized, say Q, as reference and the distance between M and Q as an optimality criterion.

Suppose that to the design matrices M and Q correspond respectively the similarity matrices R=[$r_{ij}$] and S=[$s_{ij}$]. The distance is:

$$DD(M, Q) = -\Sigma r_{ij} ln(r_{ij}/s_{ij}) - \Sigma(1 - r_{ij}) ln((1 - r_{ij})/(1 - s_{ij})) \quad (2.30)$$

This informational distance helps in selecting the design matrices.

## *2.4.2 Informational Results*

### 2.4.2.1 Entropy Variation with the Number of Levels

Denote by Wn,m,s the Walsh DOE matrix with n rows, that is runs or experiments, m columns that is factors, and s maximum settings that is levels for parameter values. Frequently, n varies from 3 to 16, m varies from 2 to 15 while the number of settings s is 2. DOE with more than 2 settings of each factor have been considered too.

The similarity with $a_k$=0.5 and the entropy as defined by (2.29) is calculated in what follows. These choices limit the generality of the numerical simulation results.

Consider the number of experiments, n=$2^k$, fixed. As m increases from m=1 to m=$2^k$, the entropy of Walsh-Paley, WP, matrices decreases and evolves towards minimal informational entropy.

$$\frac{\Delta H}{\Delta m} \leq 0 \tag{2.31}$$

Some examples are presented in Table 2.13, Table 2.14 and Table 2.15.

**Table 2.13** WP4 entropy variation for n constant

| m | 1 | 2 | 3 | 4 |
|---|---|---|---|---|
| H (WP4) | 8.318 | 7.795 | 7.542 | 7.542 |

**Table 2.14** WP8 entropy variation for n constant

| m | 1 | 2 | 3 | 4 | 5 | 6 | 7 | 8 |
|---|---|---|---|---|---|---|---|---|
| H (WP8) | 38.816 | 35.677 | 33.182 | 32.038 | 30.867 | 30.537 | 30.371 | 30.371 |

**Table 2.15** WP16 entropy variation for n constant

| m | 1 | 2 | 3 | 4 | 5 | 6 | 7 | 8 |
|---|---|---|---|---|---|---|---|---|
| H (WP16) | 166.365 | 151.704 | 138.755 | 131.892 | 125.694 | 123.437 | 122.216 | 121.894 |

| m | 9 | 10 | 11 | 12 | 13 | 14 | 15 | 16 |
|---|---|---|---|---|---|---|---|---|
| H (WP16) | 121.381 | 121.209 | 121.122 | 121.101 | 121.079 | 121.073 | 121.710 | 121.710 |

Observe that the production of entropy tends to zero as m tends to n-1.

Since the defined entropy is a measure of the imprecision in classifying the experiments it results that for fixed number of experiments, n, it would be beneficial to have fewer factors, m, to be tested. The number of factors or experiments included in a DOE may be optionally selected to ensure an imposed minimal variation of informational entropy for new factors.

#### 2.4.2.2 Entropy Variation with the Number of Trials

Consider the number of factors m=$2^k$, constant. As n increases the informational entropy increases too.

$$\frac{\Delta H}{\Delta n} \geq 0 \tag{2.32}$$

This means that for an imposed number of factors to be tested, the WP design offers more imprecision in classifying and more significance to the test if the number of trials increases.

Table 2.16 and Table 2.17 show the entropy behavior.

**Table 2.16** WP4 entropy variation for m constant

| n | 1 | 2 | 3 | 4 |
|---|---|---|---|---|
| H (WP4) | 0 | 1.125 | 3.771 | 7.542 |

**Table 2.17** WP8 entropy variation for m constant

| n | 1 | 2 | 3 | 4 | 5 | 6 | 7 | 8 |
|---|---|---|---|---|---|---|---|---|
| H (WP8) | 0 | 0.468 | 2.486 | 4.973 | 10.079 | 15.653 | 22.778 | 30.371 |

In experimental activity there is a natural trend to obtain maximum of significance that is maximum entropy with limited imposed resources. The strategy depends on similarity definition and on the set of matrices used in building EDOE.

An experimental strategy is to fill up DOE matrices with new experiments that is, new rows, since the goal is to increase entropy. If the system is not in the state of maximum entropy it would be necessary to perform new experiments at the same level, to produce more information. If the production of informational entropy by specific new experiments becomes lower than an imposed threshold, new levels that is, new factors, should be involved and this would allow increasing of the total system entropy. Informational results may be of help in initial choice of matrices for DOE.

## References

Alvarez, J., Evans, A., Sammut, P.: MML and the meta-model architecture. In: Workshop on Transformations in UML (WTUML 2001), Genoa (2001)

Arnold, L.: Random Dynamical Systems. Spinger, Berlin (1998)

Baez, J., Dolan, J.: Higher-dimensional algebra and topological quantum field theory. Jour. Math. Phys. 36, 6073–6105 (1995)

Barnsley, M.F.: Fractals Everywhere. Academic Press, New York (1993)

Berners-Lee, T., Hendler, J., Lassila, O.: The semantic web. Sci. Am. 284(5), 35–43 (2001)

Blute, R., Desharnais, J., Edalat, A., Panangaden, P.: Bisimulation for Labelled Markov Processes. In: Proceedings of 12th Annual IEEE Symposium on Logic in Computer Science, pp. 149–158 (1997)

Bochmann, D., Posthoff, C.: Binare Dynamische Systeme. Akademieverlag, Berlin (1981)

Cowan, N.: The magical number 4 in short-term memory: A reconsideration of mental storage capacity. Behavioral and Brain Sciences 24, 87–185 (2000)

Crawley, S., Davis, S., Indulska, J., McBride, S., Raymond, K.: Meta-meta is better-better. In: International Working Conference on Distributed Applications and Interoperable Systems, DAIS 1997, Cottbus, Germany (1997)

Cruz, G., Lewandowicz, E., Oziewicz, Z.: Multiscale geographic information with multigraph of multigraphs. In: The 12th International Symposium on Data Handling, Univ. of Vienna, Austria (July 2006)

Del Vecchio, V.: Modelling levels in the statistical information system of the bank of Italy. In: Papageorgiou, H. (ed.) Proceedings of the Final MetaNet Conference organised by the University of Athens. University of Athens, Greece (June 2003)

Dittrich, P., Ziegler, J., Banzhaf, W.: Artificial chemistries-a review. Artificial Life 7(3), 225–275 (2001)

Dubois, D., Prade, H.: Possibility theory, probability theory and multiple-valued logics: a clarification. Annals of Mathematics and Artificial Intelligence 32, 35–66 (2001)

Foster, I., Kesselman, C.: The Grid: Blueprint for a New Computing Infrastructure. Morgan Kaufmann, San Francisco (1999)

Fraga, E.S., Wills, G., Fairweather, M., Perris, T.: Smart Models - a framework for adaptive multi-scale modelling. In: 16th European Symposium on Computer Aided Process Engineering and 9th International Symposium on Process Systems Engineering, Garmisch-Partenkirchen, Germany (2006)

Freeman, W.J.: Neurodynamics An exploration of mesoscopic brain dynamics. Springer, London (2000)

Grossmann, W.: Metadata Usage in Statistical Computing. In: Braverman, A., Hesterberg, T., Minotte, M., Symanizik, J. (eds.) Proceedings of the 35th Symposium on the Interface, pp. 648–663. Interface Foundation of North America (2003)

Halford, G.S., Wilson, W.H., Phillips, S.: Processing capacity defined by relational complexity. Implications for comparative, developmental and cognitive psychology. Behavioural and Brain Sciences 21(6), 803–831 (1998)

Harmuth, H.F.: Sequency Theory, Foundations and Applications. Academic Press, New York (1977)

Harris, T.E.: On chains of infinite order. Pacific J. Math. 5, 702–724 (1955)

Hedayat, A.S., Sloane, N.J.A., Stufken, J.: Orthogonal Arrays. Theory and Applications. Springer, New York (1999)
Hersh, R.: The Birth of Random Evolutions. The Mathematical Intelligencer 25(1), 53–60 (2003)
Hummel, J.E., Holyoak, K.J.: A symbolic-connectionist theory of relational inference and generalization. Psychological Review 110(2), 220–264 (2003)
IBM, An architectural blueprint for automatic computing (2005)
Iordache, O.: Polystochastic Models in Chemical Engineering. VNU Science Press, Utrecht (1987)
Iordache, O.: Dyadic frames for intermittency. Perturbed models. In: Bayod, J.M., De Grande De Kimpe, N., Schikhof, W.H. (eds.) Proceedings of the Conference on p-adic Functional Analysis, Laredo (Spain), May 1990. Lecture Notes in Pure and Applied Mathematics, vol. 137, pp. 89–99. Marcel Dekker, New York (1992)
Iordache, O.: Theoretical frames for smart structures. Material Science and Engineering C 4, 143–148 (1996)
Iordache, O.: Evolvable Designs of Experiments Applications for Circuits. J. Wiley VCH, Weinheim (2009)
Iordache, O., Corbu, S.: A stochastic model of lumping. Chem. Engng. Sci. 42, 125–132 (1987)
Iordache, O., Bucurescu, I., Pascu, A.: Lumpability in compartmental models. Journ. Math. Anal. Appl. 146(2), 306–317 (1990)
Iordache, O., Corriou, J.P., Garrido-Sanchez, L., Fonteix, C., Tondeur, D.: Neural network frames applied for biochemical kinetic diagnosis. Comp. Chem. Engng. 17, 1101–1113 (1993a)
Iordache, O., Valentin, G., Corriou, J.P., Pons, M.N., Pethö, A.: Intermittent Interfacial Transfer. A Dyadic Model. Acta Chemica Hungarica. Models in Chemistry 1(130), 1–18 (1993b)
Iordache, O., Corriou, J.P., Tondeur, D.: Separation Sequencing. Use of Information Distance. Canad. Journ. of Chem. Engng. 71, 955–966 (1993c)
Iosifescu, M., Grigorescu, S.: Dependence with complete connections and applications. Cambridge Univ. Press, Cambridge (1990)
ISO, ISO/IEC 10728:1993 Information technology – Information Resource Dictionary System, IRDS (1993)
Keane, M.: Strongly mixing g-measures. Invent. Math. 16, 309–324 (1972)
Kemeny, J.G., Snell, J.L.: Finite Markov Chains. Van Nostrand, New York (1960)
Larsen, K.G., Skou, A.: Bisimulation through probabilistic testing. Information and Computation 94, 1–28 (1991)
MacLane, S.: Categories for the Working Mathematician. Springer, New York (1971)
McCullagh, P.: What is a statistical model? Ann. Statist. 30, 1225–1310 (2002)
Nehaniv, C.L.: Self-Replication, Evolvability, and Asynchronicity in Stochastic Worlds. In: Lupanov, O.B., Kasim-Zade, O.M., Chaskin, A.V., Steinhöfel, K. (eds.) SAGA 2005. LNCS, vol. 3777, pp. 126–169. Springer, Heidelberg (2005)
Nissen, H.W., Jarke, M.: Repository Support for Multi-Perspective Requirements Engineering. Inf. Syst. 24(2), 131–158 (1999)
OMG (2000) Meta Object Facility (MOF) Specification. Version 1.3 (March 2000)
Pattee, H.H.: Evolving self-reference: matter, symbols, and semantic closure. Communication and Cognition –Artificial Intelligence 12(1-2), 9–25 (1995)
Pattee, H.H.: Causation, control and the evolution of complexity. In: Anderson, P.B., et al. (eds.) Downward Causation, pp. 63–67. Aarhus University Press, Aarhus (2000)

Piaget, J.: L'épistémologie des régulations: introduction. In: Lichnerrowicz, A., Perroux, F., Gadoffre, G. (eds.) L'idée de régulation dans les sciences: In: 2e vol. des Séminaires interdisciplinaires du Collège de France: A. Paris: Maloine: Doin: I-XIII (1977)

Piaget, J., Garcia, R.: Psychogenesis and the History of Science. Columbia University Press, New York (1989)

Poli, R.: Three obstructions: forms of causation, chronotopoids, and levels of reality. Axiomathes 17(1), 1–18 (2007)

Poli, R.: Ontology: The categorial stance. In: Poli, R., Seibt, J. (eds.) TAO-Theory and Applications of Ontology, vol. I. Springer, Dordrecht (2008)

Power, A.J.: Why Tricategories? Information and Computation 120(22), 251–262 (1995)

Rhee, H.K., Aris, R., Amundson, N.R.: First Order Partial Differential Equations II. In: Theory and applications of hyperbolic systems of quasilinear equations. Prentice-Hall, Englewood Cliffs (1989)

van Rooij, A.C.M.: Non-Archimedean Functional Analysis. Marcel Dekker, New York (1978)

van Rooij, A.C.M., Schikhof, W.H.: Non-Archimedean integration theory. Indag. Math. 31, 190–199 (1969)

Rosen, R.: Life itself: A Comprehensive Inquiry into the Nature. Origin and Fabrication of Life. Columbia University Press, New York (1991)

Rossiter, N., Heather, M., Nelson, D.A.: Categorical Formalism for Interoperability based on the Information Resource Dictionary Standard (IRDS). Computing Science Technical Report no.717, University of Newcastle upon Tyne (2000)

Rossiter, N., Heather, M.: Four-level Architecture for Closure. In: Interoperability, EFIS 2003, 5th International Workshop on Engineering Federated Information Systems, Coventry, UK, July 17-18, pp. 83–88 (2003)

Stenflo, O.: Uniqueness in g-measures. Nonlinearity 16, 404–410 (2003)

Taguchi, G.: Introduction to Quality Engineering: Design Quality into Products and Processes. Asian Productivity Organization, Tokyo (1986)

Yanushkevich, S.: Logic Differential Calculus in Multi-Valued Logic Design. Tech Univ. Szczecin, Szczecin, Poland (1998)

**Table 3.2** Boolean Input Information-Separations of Xylene Isomers

| Data | Name | Volatility | Dipole, D | Freezing point, K |
|---|---|---|---|---|
| 1 | m-xylene | 1 | 1 | 0 |
| 2 | o-xylene | 0 | 1 | 1 |
| 3 | p-xylene | 1 | 0 | 1 |
| 4 | Ethyl Benzene | 1 | 1 | 0 |

The "similarity index" that may characterize the relative positions or properties of any two species in a separation scheme is introduced. There are many ways of defining such an index. The one used here is based on the physical and chemical properties of the species that allow their separation (such as volatility, partition coefficient, density, dipole moment, diameter of particles, etc). Some of the properties are related to thermodynamics while other ones come from operating conditions. This similarity index is pertinent when the flow-sheet is to be generated.

#### 3.1.2.2 Similarity Based on Boolean Vectors

First, define a similarity index $r_{ij}$, between two different species i and j as:

$$r_{ij} = \Sigma_k t_k (a_k)^k ; k = 1, 2, .... \tag{3.1}$$

Here: $0 \leq a_k \leq 1$ and $t_k=1$ if $i_k = j_k$, $t_k=0$ if $i_k \neq j_k$ for all k. The entire system is then defined by the matrix $R=[r_{ij}]$. The similarity index should possess the natural properties of reflexivity ($r_{ij}=1$) and of symmetry ($r_{ij}=r_{ji}$). This definition assigns a coefficient of weight $a_k$ to any property involved in the description of the species i and j provided that the Boolean values $i_k$ and $j_k$ are the same for these two species. Instead of similarity, the distance $d_{ij}=1-r_{ij}$ could be used. It verifies the usual triangular inequality:

$$d_{ij} \leq d_{ik} + d_{kj} \tag{3.2}$$

It should be noted the parallel between the similarities as defined by (3.2), with $a_k=0.5$, and the probabilities introduced in Sect. 2.3.1.

#### 3.1.2.3 Similarity Based on Actual Values

The concept of similarity of molecules has important ramifications for chemical engineering theory of separation systems and for chemical physics. In addition there have appeared in literature a number of ad-hoc definitions of molecular similarity. Difficulties related with attempting to obtain precise numerical indices for qualitative molecular and structural concepts have already been extensively discussed in the chemical-physics literature (Varmuza 1980). Separation processes are based on differences of properties for species. The use of quantitative structure-property relationships in separation methods is based on the assumption that similar molecules have similar properties and will be separated together in the same class of compounds. One can associate to any substances a description by their properties significant for separation scheme generation. Consider for instance for components x and y:

$\underline{x}=<x_1,x_2,...,x_k,...>$; $\underline{y}=<y_1,y_2,...,y_k,...>$ where $x_k$, $y_k$ are the real values of the k-th property in describing the component x and y respectively. Thus the component could be represented as points in a space having the dimensionality exactly the number of dimensions of the vector component in equation (3.1). It is straightforward to calculate various measures of the distance $d_{xy}$ between x and y.

Frequently used is the Minkowski distance:

$$d_M = (\Sigma_k(x_k - y_k)^k)^{(1/k)} \tag{3.3}$$

If k=2 this gives the Euclidean distances. The "city block" distance is also of interest:

$$d_{cb} = \Sigma_k|x_k - y_k| \tag{3.4}$$

If all features for x and y are binary encoded (only the values "0" and "1" are allowed) the city block distance is called the Hamming distance. It is equivalent to the number of features which are different in x and y. Binary codification is accomplished for instance by associating the value "0" or "1" to the properties under and over the mean respectively. Vectors as x and y will contain the normalized value of different properties. Then on account of the distance definition from the first example one defines a normalized distance between the component x and y as:

$$d_{xy} = \Sigma_k|x_k - y_k|(a_k)^k \tag{3.5}$$

Here: $0 \leq a_k \leq 1$. The normalized values $x_{norm}$ for a given property x are obtained replacing this value by:

$$x_{norm} = (x - x_{min})/(x - x_{max}) \tag{3.6}$$

Here $x_{min}$ and $x_{max}$ are respectively the minimum and the maximum value of x.

The definition presented in equation (3.5) assigns a weight $w_k=(a_k)^k$ to the property k involved in the description of the component x and y. If $x_k$ and $y_k$ are binary encoded and if $a_k$=0.5 then the definition (3.5) gives the so-called "addition modulo-2" or "dyadic" distance of the vectors x and y. Distances are measures of dissimilarity.

It is easy to draw conclusions about similarity from them. The similarity index $r_{xy}$, could be obtained using for instance the definition:

$$r_{xy} = 1 - \frac{d_{xy}}{max\ d_{xy}} \tag{3.7}$$

#### 3.1.2.4 Classification

The fact that the relation described by $r_{ij}$ is reflexive and symmetric allows a partition of the set of components in classes that are not necessarily disjoint. A class consists of a number of similar chemical species gathered together. Limit here the study to split separations corresponding to a partition into disjoint classes. Hence the defined similarity must be transitive that is: $\min_k (r_{ik}, r_{kj}) \leq r_{ij}$. The procedure to ensure transitivity is that the classification algorithm starts from the "stable" matrix of similarity. To obtain such a matrix, the sequence R, R(2),..., R(k),... with R(2)=RoR and R(k)=R(k-1)oR is calculated. The composition rule "o" is given by:

$$(RoW)_{ij} = max_k[min(r_{ik}, w_{kj})] \tag{3.8}$$

Here $R=[r_{ij}]$, $W=[w_{ij}]$ are two arbitrary matrices of the same type. Equation (3.8) calculates the $(i,j)$-th element of the matrix RoW. It consists in taking the smallest of the two elements $r_{ik}$ and $w_{kj}$, for a given row i of R and a column $j$ of W, then repeat the procedure for all $k$, and select the largest of all such resulting elements. There exists an integer n such that from n on, the matrix is stable to the composition rule "o" so that R(n)=R(n+1) and so on. The elements of the stable similarity matrix R(n) verify symmetry, reflexivity and transitivity. Denote by $r_{ij}(n)$ the elements of the stable matrix R(n). Though the following calculations can be performed with original matrices without stabilization, they have been performed with the stable matrices, since the present aim is to obtain partitions in disjoint classes. The partition is established on the base of the "degree of separation" $T$ with $0 \leq T \leq 1$.

The classification rule is the following: two species i and j, are assigned to the same class if $r_{ij}(n) \geq T$. Applying the rule, the set of classes at the degree of separation $T$ is obtained. For $T$=0, a unique class results including all species, whereas for $T$=1 each class includes only one species. When $T$ varies from 0 to 1, different sets of classes arise. Actually a new set of classes arises every time $T$ crosses the value of one similarity index $r_{ij}$ of the matrix R. In this way a general tree of classes is built, which is nothing but the expected flow-sheet. The class of i, noted $\hat{\imath}$, is the set of species $j$ which satisfies the rule: $r_{ij} \geq T$. The similarity matrix of classes $\hat{R}$ is constructed as follows:

$$\hat{R}_{\hat{\imath}\hat{\jmath}} = max(r_{wt}); w \in \hat{\imath}, t \in \hat{\jmath} \tag{3.9}$$

Here w designates any index of species belonging to the class of i and similarly t any index referring to the class of $j$. The definition of matrix of classes suggests finding the maximum similarity between elements of two different classes. Denote also by $\hat{R}_T$ the matrix of classes at degree of separation $T$.

#### 3.1.2.5 Adaptability Procedures

Different procedures of adaptability are implemented in this algorithm as follows.

Suppose a given flow-sheet as optimal or as the best from practical considerations. It is easy to associate a matrix of similarity denoted in the following by $W=[w_{ij}]$ to this best flow-sheet. For instance, following the optimal separation scheme, fictitious vectors of properties can be introduced so that this optimal flow-sheet is conveniently described. Thus a vector representation of the species is obtained, and then equation (3.1) was used to calculate the associated similarity $w_{ij}$. This is obtained using equal weights $a_1=a_2=...=a$. For instance, in order to perform a separation from (1,2,3,4) to (1,2), (3,4), define the vectors $\underline{1}=<1,...>$, $\underline{2}=<1,...>$ while $\underline{3}=<0,...>$ and $\underline{4}=<0,...>$; thus the class (1,2) is separated from the class (3,4) with respect to the first property. In this way, the desired first step separation is obtained and the procedure can be extended to obtain any prescribed separation scheme.

The number of properties for R and W may differ. Consider the same set of species as in the optimal flow-sheet. Some actual properties that are of interest for the purpose of separation are taken into account. The vectors of properties are associated to each species. The similarity degree $r_{ij}$ is then calculated with equation (3.1), giving the matrix R. In general, with identical weights for each property the flow-sheet deduced from R will be different from the optimal one described by the matrix W.

a. Changing the weights
The adaptability procedure consists in trying to find flow-sheets resulting from the similarity matrix R as close as possible to the optimal flow-sheet. To obtain this result, a modification of the weights of the properties occurring in the vectors of properties has been used to compute R. Due to the fact that the values of the weights $a_k$ are conventionally selected as equal, the first weight $a_1$are considered as constant and only the following weights $a_2$, $a_3$,... are subject to random variations. Suppose that after the first random step, the set of weights $a_1$, $a^*{}_2$, $a^*{}_3$,... results. The new similarity matrix is $R = [r_{ij}]$. It is calculated using equation (3.1) and the new weights. Now, it is necessary to define a distance between the flow-sheets characterized by matrices R and W. On account of methods developed in the theory of information (Kullback 1980) the distance DD is defined:

$$DD = -\Sigma_{i,j} r_{ij} ln(r_{ij}/w_{ij}) - \Sigma_{i,j}(1-r_{ij}) ln((1-r_{ij})/(1-wij)); 0 < r_{ij}, w_{ij} < 1 \quad (3.10)$$

Observe that this distance is null if R=W. Every step yielding a decreasing distance DD is taken into account in the evolution of weights. By random search in a given domain around an initial set of weights and by observing a large number of sets, a better set of weights is retained. Iterating this procedure and decreasing the search domain, a nearly-optimal set of weights is retained.

b. Changing the hierarchical order of properties
It may happen that some properties assumed as secondary are more important in classification than it was initially expected. A drastic procedure of learning based on the change of the order of properties should be applied in this case. Suppose that there exist m properties indexed from 1 to m. Every permutation of the indices 1,..., m represents a possible hierarchy. The problem is to find a permutation for which the distance DD to a good training example is minimal. Trying all permutations, undesired combinatorial optimization problems result. A simple method based on the above presented procedure-a, of changing the weights could be used. If the effective weight $(a_k)^k$ is inferior to the effective weight $(a_{k+1})^{k+1}$ this signifies that the properties k and k+1 must be interchanged.

c. Changing the number of digits in the property vectors
Another adaptability feature of the algorithm could be realized by assigning more than one digit to a given property in an associated vector. This means in fact that a more detailed characterization of the property is necessary according to its significance in the classification. For instance process variables that may be low, normal, high and very high were represented by pairs of data as (00), (01), (10), and (11) respectively. Procedure-c and procedure-a have similar effects.

## *3.1.3 Illustrative Examples*

### 3.1.3.1 Flow-Sheets for Olefin-Paraffin Separation

The components of the mixture are: 1: ethane, 2: propylene, 3: propane, 4: i-butene, 5: n-butan 6: n-pentane. The relative volatilities are: $\alpha_{12}$=3.5; $\alpha_{23}$=1.2; $\alpha_{34}$=2.7; $\alpha_{45}$=1.21; $\alpha_{56}$=3. The flows (kg.mol/hr) are respectively: 9.1; 6.8; 9.1; 6.8; 6.8; 6.8.

This is a well-known example of separation in pure products system. Fig. 3.1 shows a possible separation scheme. The property based similarities defined by equation (3.1) will be used. Suppose that the properties of hydrocarbons can be ranged according to their contribution to the separation process in the following order of decreasing importance: relative volatility (heuristics H1), flow (heuristics H2), number of carbon atoms. The first two properties are used in most heuristic analysis. The third property is useful when separation products are imposed by the demand for uniform product. It is also closely related to other physical or chemical properties. Other set of properties could be useful in practice for different reasons, but the aim here is to illustrate the method. At this stage, volatility is divided into two domains, the higher values (for ethane, propylene, propane) being indexed by "1" and the lower values (for i-butene, n-butane, n-pentane) by "0". The second property, the flow rate has only two values 9.1 indexed by "1" and 6.8 indexed by "0". Taking into account the third property, the index "1" is assigned to a species if a species with the same number of atoms of carbon exists and "0" otherwise. This simplification of the data consisting in a binary encoding should be based on numerical values of thresholds obtained by adaptability procedures based in turn, on statistical analysis of large sets of data. Here, the binary encoding is based on the hypothesis that the data are uniformly distributed. The associated vectors to the six species are respectively:

1=<110>; 2=<101>; 3=<111>; 4=<001>; 5=<001>; 6=<000>.

The initial matrix R corresponding to equal weights $a_1=a_2=a_3=0.5$ is:

$$R = \begin{matrix} 1 & .5 & .75 & 0 & 0 & .125 \\ .5 & 1 & .625 & .375 & .375 & .25 \\ .75 & .625 & 1 & .125 & .125 & 0 \\ 0 & .375 & .125 & 1 & .875 & .75 \\ 0 & .375 & .125 & .875 & 1 & .75 \\ .125 & .25 & 0 & .75 & .75 & 1 \end{matrix} \qquad (3.11)$$

Notice that the elements of the matrix are among the possible values of T in its dyadic expansion.

Looking at the vectors: 2=<101>, 3=<111>, it may be checked according to equation 3.1 that for instance: $r_{23} = t_1\ a_1 + t_2\ a_{22} + t_3\ a_{33}=0.625$; with $t_1 = 1$; $t_2 = 0$; $t_3 = 1$ and; $a_m = 0.5$ for all m. The stabilized matrix obtained after 3 iterations using the composition rule given by equation (3.8) is in this case:

$$R(3) = \begin{matrix} 1 & .625 & .75 & .375 & .375 & .375 \\ .625 & 1 & .625 & .375 & .375 & .375 \\ .75 & .625 & 1 & .375 & .375 & .375 \\ .375 & .375 & .375 & 1 & .875 & .75 \\ .375 & .375 & .375 & .875 & 1 & .75 \\ .375 & .25 & .375 & .75 & .75 & 1 \end{matrix} \qquad (3.12)$$

Using the classification procedure it results:

- For $0 < T \leq 0.375$ the unique class (1, 2, 3, 4, 5, 6)
- For $0.375 < T \leq 0.625$ two classes (1, 2, 3), (4, 5, 6)
- For $0.625 < T \leq 0.75$ three classes (1, 3), (2), (4, 5, 6)
- For $0.75 < T \leq 0.875$ five classes (1), (2), (3), (4, 5), (6)
- For $0.875 < T \leq 1$ six classes (1), (2), (3), (4), (5), (6)

Details about the calculations for a step of the degree of separation $\Delta T=0.25$ will be presented in the following. Taking: $a_1=a_2=a_3=0.5$ the V variant of the separation scheme shown in Fig. 3.1 is obtained:

V: (1,2,3,4,5,6) → (1,2,3),(4,5,6) → (1,3),(2), (4,5,6) → (1), (2),(3), (4),(5), (6)

Indeed at the degree of separation T=0.25, the classification rule gives a unique $\hat{1}$ = (1, 2, 3, 4, 5, 6) since all $r_{ij} \geq T$. At the degree of separation

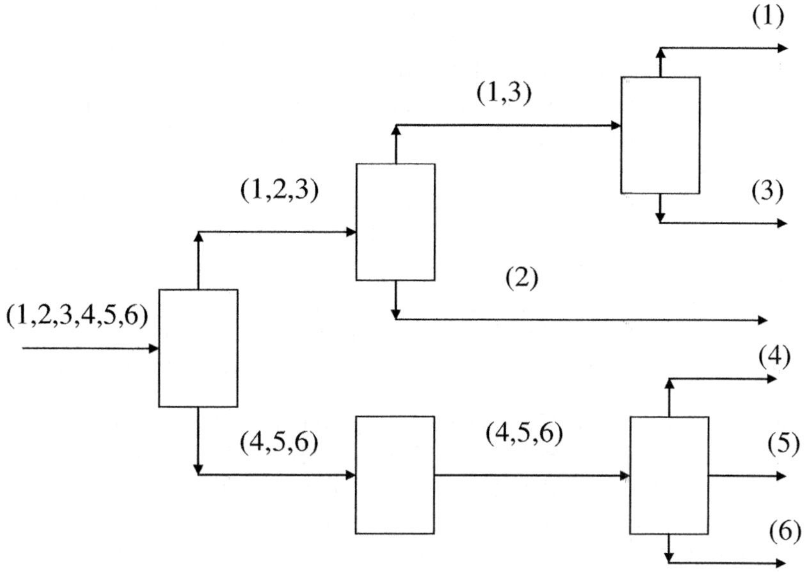


**Fig. 3.1** Flow-sheet for olefin-paraffin separation. Example.

T=0.50, the classification rule gives two classes $\hat{1} = (1, 2, 3)$ and $\hat{4} = (4, 5, 6)$. In this case: 0.375<T≤0.625.

Notice also that: $\hat{1} = \hat{2} = \hat{3}$ and similarly $\hat{4} = \hat{5} = \hat{6}$. The corresponding matrix of classes is: $\hat{R}_{0.5}$

$$\hat{R}_{0.5} = \begin{matrix} 1 & .375 \\ .375 & 1 \end{matrix} \tag{3.13}$$

Observe that the propane is less volatile then propylene, so that the split (1,2,3) → (1,3), (2) cannot be done by distillation but by another separation method (as extraction or adsorption). The calculations have been performed at the degree of separation: 0, 0.25, 0.50, 0.75, and 1. This step $\Delta$T allows to obtain a separation scheme in three stages, as in the optimal example presented by Seader and Westerberg (1977). Due to this step the separation in five classes: (1),(2),(3),(4,5),(6) was by-passed by the algorithm. A smaller step would result into a separation sequence in more than three stages.

### 3.1.3.2 Adaptability Procedures

The optimal scheme proposed by Seader and Westerberg (1977) was:

$V_b$: (1,2,3,4,5,6) → (1,2,3),(4,5,6) → (1),(2,3),(4,5),(6) → (1),(2,(3),(4),(5),(6)

The adaptability procedure -a. with scheme Vb as the optimal goal will be used now. The matrix W giving the scheme Vb is obtained starting from the conventional vectors of properties as follows: 1=<110>, 2=<101>, 3=<100>, 4=<011>, 5=<010>, 6=<000> using equation (3.1) with equal weights a=0.5. Using the adaptability procedure -a, we modify $a_2$ and $a_3$, and a nearly optimal separation scheme results for $a_1$=0.5, $a_2$=0.34, $a_3$=0.63. In this case, the procedure-a, gives the minimum distance DD=0.08. At this distance, despite differences between $w_{ij}$ and $r_{ij}$, the optimal scheme $V_b$ proposed by Seader and Westerberg is obtained. Thus, for fixed weight of the second property, $a_3$=0.35 (close to the optimum) changing by small steps the weights of the third property $a_3$, three different schemes have been obtained:

$V_a$: (1,2,3,4,5,6) → (1,2,3),(4,5,6) → (1),(2),(3),(4,5),(6) → (1),(2),(3),(4),(5),(6)
$V_b$: (1,2,3,4,5,6) → (1,2,3),(4,5,6) → (1),(2,3),(4,5),(6) → (1),(2,(3),(4),(5),(6)
$V_c$: (1,2,3,4,5,6) → (1),(2,3),(4,5),(6) → (1),(2),(3),(4,5),(6) → (1),(2),(3),(4),(5),(6)

A transition from $V_a$ to $V_b$ appears for $a_3$=0.630 and a transition from $V_b$ to $V_c$ appears at $a_3$=0.728. Variant $V_b$ is just the optimal separation scheme proposed by Seader and Westerberg (1977). The distance DD between the actual scheme and the optimal scheme given by equation (3.8) is continuous and extreme about the optimal scheme $V_b$. Near the extreme, this distance varies very little and schemes $V_a$ and $V_b$ are almost equivalent. The adaptability procedure was made in direction of $V_b$ which was assumed to be optimal. Recall that for this study of the influence of the weight $a_3$, the other weight $a_2$ was fixed different from the "optimal" weight. The result of the procedure-a, is a set of weights $a_1$=0.5, $a_2$=0.34, $a_3$=0.63 which ensures an optimal separation from the point of view proposed here if the hierarchy of properties used to generate the flow-sheet is: volatility, flow, number of carbons. However one observes that the procedure-a, gives $a_2{}^2$ inferior to $a_3{}^3$. This signifies that the third property, that is the number of carbon atoms could be more significant than the second property that is the flow rate. The more drastic procedure-b, of adaptability based on the change of the hierarchy of properties should be applied. A new set of vectors is used, in which the second and the third properties are interchanged, namely: 1=<101>; 2=<110>; 3=<111>; 4=<010>; 5=<010>; 6=<000>. Applying again the optimization procedure-a, one observes that the weight of the second property, number of carbon atoms, tends to remain nearly 0.5 while the weight of the third property, flow rate, tends to zero and the distance DD between R and W tends to zero, too. Therefore only two properties: volatility and number of carbon atoms are sufficient to obtain this separation scheme. The small difference between flows could explain this result. In the studied case the retained property vectors are: 1=<10>; 2=<11>; 3=<11>;4=<01>;5=<01>;6=<00>.

The weights are equal: $a_1$=$a_2$=0.5. This new hierarchical order of properties as well as the set of weights for optimal separations represents the final product of the adaptability method.

To obtain a global characterization of the generated flow-sheet the informational entropy H(R) could be associated to the flow-sheet described by the similarity matrix: R=[rij].

$$H = -\Sigma_i \Sigma_j r_{ij} ln r_{ij} - \Sigma_i \Sigma_j (1 - r_{ij}) ln(1 - r_{ij}) \qquad (3.14)$$

In system synthesis to each variant of a flow-sheet, a dependence of entropy versus the grouping level T and thus an H-T diagram is associated. To the time T=0 corresponds H=0 (a unique class), to T=1 corresponds the maximum entropy $H_{max}$ (every class contains a compound).

The dramatic change in entropy when a new separation scheme appears suggests that the entropy H may be a good indicator of novelty.

Few other illustrative examples of flow-sheets for hydrocarbon separations will be briefly presented. Components are presented in the order of volatility, 1 is the most volatile.

#### 3.1.3.3 Hydrocarbons Separation Based Only on Volatility

To illustrate the fact that more detailed description of a given property (procedure-c), could be used a Henley and Seader (1981) separation problem is examined. The mixture is: 1: propane, 2: isobutane, 3: n-butane, 4: isopentane, 5: n-pentane. A set of approximate relative volatilities is: $\alpha_{12}$=2; $\alpha_{23}$=1.33; $\alpha_{34}$=2.40; $\alpha_{45}$=1.25. The procedure to classifying could be limited according to the relative volatility only. These volatilities can be separated for example in three classes: high, medium and low. Two components (digits) will be associated in the vectors corresponding to every hydrocarbon so that to a high volatility correspond the vector <10>, to a medium volatility the vector <01> and to a low volatility the vector <00>. Notice that this binarisation of data is objective being based on actual values of the considered uniformly distributed property. Thus the vectors of properties of the components are:

1=<10>, 2=<01>, 3=<01>, 4=<00>, 5=<00>

The resulting separation scheme is:

(1,2,3,4,5) → (1), (2,3,4,5) → (1), (2,3), (4,5) → (1), (2), (3), (4), (5).

This scheme was assumed optimal by Henley and Seader (1981).

#### 3.1.3.4 Ternary Hydrocarbon Separation

The data are the following: species: 1 n-hexane, 2 n-octane, 3 n-decane: component flows (kg.mol/hr): 40, 30, 30 (Gomez and Seader 1985). The vectors of properties of the components are: 1=<1101>, 2=<1010>, 3=<0011>. The first element of the vector corresponds to volatility. The volatility of n-hexane and of n-octane is considered as large with respect to the volatility of n-decane. The second element of the vector corresponds to flows. Only the flow of the first component is indexed as large. The last two components correspond to the number of atoms of carbon. Taking "01" (meaning "low") for hexane, "10" ("average") for octane and "11" ("large") for decane to account for different number of carbon atoms the resulting separation scheme is: (1,2,3) → (1), (2,3) → (1), (2), (3).

#### 3.1.3.5 Five-Component Hydrocarbons Separation

Consider the following data: species:1: n-pentane, 2: n-hexane, 3: n-heptane, 4: n-octane, 5: n-decane; component flows (kg.mol/hr): 15, 20, 30, 20, 15 (Gomez and Seader 1985). The vectors of properties of the components are: 1=<1000>, 2=<1001>, 3=<0110>, 4=<0011>, 5=<0011>. The signification of the vector elements is the same as above. The resulting separation scheme is:

(1,2,3,4,5) → (1,2), (3,4,5) → (1,2), (3),(4,5) → (1), (2), (3), (4), (5)

The schemes shown are considered as optimal by Gomez and Seader (1985) in their thermodynamic analysis. In fact the result is not surprising since the properties used to generate the flow-sheet are thermodynamically significant.

#### 3.1.3.6 Xylene Isomers Purification

The corresponding vectors of properties have been shown in Table 3.1 and Table 3.2. The feed mixture is a group of four closely-boiling compounds. Some plausible sequences to separate this mixture have been suggested by Garg et al. (1991). The separation methods can be distillation, adsorption, crystallization, molecular sieving. With a less of 1-A size difference between compounds, the separation by molecular sieving is impossible. By applying the classification algorithm with weights $a_k$=0.5 for weight coefficients, it results the separation scheme: (1,2,3,4) → (1,3,4),(2) → (1,4),(2)(3) → (1),(2),(3),(4).

This corresponds to the scheme with the first split by simple distillation, 2 resulting as a product, followed by an adsorption, 3 being the product, followed by a crystallization allowing separation of 1 and 4. Therefore the order of used properties is: volatility, polarizability, freezing point. This hierarchical order results by applying the adaptability procedure-b.

#### 3.1.3.7 Lime Production

This is a solid-liquid separation problem in which three un-dissolved solids are mixed in one liquid. Some of the separation methods of interest are drying, screening, flotation. The property vectors are shown in Table 3.3. They are: relative surface wettability in water, average particle size (microns), and specific gravity (Garg et al. 1991).

Table 3.4 contains corresponding digits for data shown in Table 3.3. As above "0" for low values and "1" for large values relative to the mean will

**Table 3.3** Input Information-Separations in Lime Production

| Data | Name | Relative Wettability | Size, mm | Specific gravity |
|---|---|---|---|---|
| 1 | CaCO3 | 2.0 | 200 | 2.5 |
| 2 | Stones | 1.0 | 1000 | 2.0 |
| 3 | Powder | 1.0 | 50 | 1.9 |
| 4 | Water | 100 | 0 | 1 |

**Table 3.4** Boolean Input Information-Separations in Lime Production

| Data | Name | Relative Wettability | Size, mm | Specific Gravity |
|---|---|---|---|---|
| 1 | CaCO3 | 0 | 0 | 1 |
| 2 | Stones | 0 | 1 | 1 |
| 3 | Powder | 0 | 0 | 1 |
| 4 | Water | 1 | 0 | 0 |

be used. The classification algorithm gives for equal $a_k$=0.5 the separation scheme: (1,2,3,4) → (1,2,3)(4) → (1,3)(2)(4) → (1)(2)(3)(4).

This is similar to that proposed by Garg et al. (1991) in which a first split is achieved by drying (separating 4), the next split results by screening (separating 2), the next split results by flotation (separating 1). The hierarchical order of properties used for separation are: wettability, particle size, specific gravity.

### *3.1.4 Differential Model*

Consider separation as a process in which the usual time is the degree of separation T.

Notice that it is possible to make equivalent use of the valuation $T=\Sigma_j t_j 2^{-j}$ or the valuation $T=\Sigma_j t_j 2^j$.

Denote by Y(T) the output or answer function for the separation process giving the "yes" or "not" response to the separation process for a given set of species. Physically this could be interpreted as separation in the "light" phase and respectively the "heavy" phase. This is a two-valued function taking only Boolean values denoted, and considered in calculus as {-1,1}, as {0,1} or as any other pair of values. The two values ensure the Boolean, logical significance of the output. A batch type process is that which stored in Table 3.4 the given set of vectors characterizing the species subjected to separation. At different values of T different vectors are grouped together according to a rule imposing to put in the same class compounds having the distance smaller than T. Observe that to apply this classification rule we considered $T=\Sigma_j t_j 2^{-j}$ and $0<T<1$.

Considering now a continuous type of separation processes, the new recorded data entering the separation process are compared with data stored in a Table as 3.2 or 3.4. Physically this corresponds to new compounds entering in the batch of a separation scheme. The answer for this separation process depends on the data storing table structure that is on the flow-sheet structure. One of the simplest separation processes is that in which small changes of distance T determines small changes of the answer proportional to the existing answer Y(T) and to the change of T, ∂T. This is the kinetic dependence. The model governing the evolution of the answer Y(T) with the distance T is written by analogy with real linear processes:

$$\frac{\partial Y}{\partial T} \oplus Q \otimes Y = 0 \tag{3.15}$$

Q denotes the separation rate.

The solution in GF (2) is presented here for illustration purposes. In GF (2), 0, denotes the null element. The real product and the sum were translated towards GF (2) operations.

The solution similar to Euler solution for differential equations will be:

$$Y(T) = Y(0) \oplus (Q \otimes T \otimes Y(0)) \tag{3.16}$$

Suppose that Y(0) = 1. The solutions of 3.16 are Walsh functions. With more coordinates in Y, T, Z it is possible to obtain Walsh functions with 8, 16, 32, and so on elements (Iordache 2009).

Table 2.6 from Sect. 2.2.3 shows the three levels solution for $Y=Y=y_0y_1y_2$, $T=t_0t_1t_2$ a Walsh matrix.

In general:

$$Y(T) = W_M(Q,T) \tag{3.17}$$

Here:

$$M = 2^m, Q = \Sigma_j q_j 2^j, T = \Sigma_j t_j 2^j, 0 \le j \le m-1 \tag{3.18}$$

The ordering is that corresponding to Walsh-Paley functions if we make use of the categorical product "×" (Sect. 2.1.2). Resulting matrices are well known in conventional DOE as Walsh-Hadamard designs.

Instead of "0", the "-1" with the same signification appears in such conventional designs.

The Walsh functions are the rows of the Walsh matrix. Other definitions of the categorical product give different orderings of the Walsh-type functions.

Based on solutions as (3.17) the expansion in series of Walsh functions for any vector Y has to be used:

$$Y(T) = \Sigma_Q q_Q W_M(Q,T) \tag{3.19}$$

Here $M=2^m$; $0\leq Q \leq M-1$. The Walsh coefficient $q_Q$ depends on the contribution of the rate of separation Q. The coefficients $q_Q$ could be obtained from (3.19) as usually in Walsh-Fourier analysis:

$$q_Q \Sigma_T W_M^2(Q,T) = \Sigma_T W_M(Q,T)Y(T) \tag{3.20}$$

Here Y(T) with $0\leq Q\leq M-1$, denotes row in a table allowing the separation calculus.

A general approach in pattern recognition is to carry out a discrete Walsh-Fourier transform of the obtained pattern and of its pattern model and to cross-correlate the two transformed sets of values to determine the degree of recognition rather than to attempt cross-correlation of the original vector. Advantages of the procedure are due to the quick computability and simple implementation of Walsh-Fourier transforms.

The procedure was used for data shown in Table 3.5 and Table 3.6. This represents a slight modification of the classification Table 3.2 and Table 3.4. In fact the initial condition Y(0)=1 was pasted as a first column.

The physical signification of the initial condition is the presence of all compounds in the mixture subjected to separation. Then, one maintains the digit "1" and replace in Table 3.2 and Table 3.4 the digit "0" by "-1" to be more close to the Walsh functions standard definition on {-1,1}.

**Table 3.5** Problem 1-Vectors of Properties

| | | | | |
|---|---|---|---|---|
| 1 | 1 | 1 | 1 | -1 |
| 2 | 1 | -1 | 1 | 1 |
| 3 | 1 | 1 | -1 | 1 |
| 4 | 1 | 1 | 1 | -1 |

**Table 3.6** Problem 2-Vectors of Properties

| | | | | |
|---|---|---|---|---|
| 1 | 1 | -1 | -1 | 1 |
| 2 | 1 | -1 | 1 | -1 |
| 3 | 1 | -1 | -1 | -1 |
| 4 | 1 | 1 | -1 | -1 |

Firstly, Walsh-Fourier transform (spectrum) have been obtained. Then a matrix of real coefficients of correlation between spectra of different vectors of properties: $R=[r_{sy}]$ was calculated. On the basis of the described stabilization procedure for similarity matrices, the same algorithm of partition in classes was implemented. The separation rule was the following: two vectors s and y pertain to the same class at the separation index T if the coefficient of correlation of their Walsh-Fourier spectra is greater than T, $0<T<1$. Notice that for this calculus the valuation $T=\Sigma_j t_j 2^{-j}$ replaced the valuation $T=\Sigma_j t_j 2^j$ that was used to write the solutions of the differential equation (3.15).

The result of separation Problem 1 for a step $\Delta$T=0.5 using the data shown in Table 3.5 is: (1234) → (14), (2), (3) → (1), (2), (3), (4).

The result of separation T=0.5 using only the data shown in Table 3.6 for Problem 2, is: (1234) → (123),(4) → (1),(2),(3),(4).

The obtained scheme includes steps of the solution based on the use of other distances. The Walsh-Fourier procedure offers in the studied cases a non-detailed separation schemes.

### 3.1.5 Transfer Function

Any new feed compound, could be represented by a vector of properties and heuristics, X(T), containing only "-1" and "1" ("-1" replaces "0").

This feed should be separated by existing flow-sheets associated to classification tables as Table 3.5 and Table 3.6. The exit vector Y(T) gives an indication about the trajectory followed in a separation scheme by this new compound. In this exit vector "1" corresponds to the separation of the studied compound as the "light" phase while "-1" corresponds to their separation as a "heavy" phase. The trajectory along the flow-sheet of any compound will be characterized using the transfer function.

Consider a linear model as given by equation 3.21.

$$\frac{\partial Y}{\partial T} \oplus Q \otimes Y = X(T) \tag{3.21}$$

X(T) denotes the feed vector in the system.

Denote the Walsh transform of the exit Y(T) and of the feed X(T) by:

$$Y(Q) = \Sigma_T Y(T) W_M(Q,T) \tag{3.22}$$

$$X(Q) = \Sigma_T X(T) W_M(Q,T) \tag{3.23}$$

Note that:

$$\sum_T \frac{dY(T)WM(Q,T)}{dT} = QY(Q) \tag{3.24}$$

A similar property is well-known in the theory of real Fourier transform. It results by applying the Walsh transform to both terms in (3.21), that:

$$QY(Q) + Y(Q) = X(Q) \tag{3.25}$$

Therefore the transfer function of the dyadic model is the diagonal matrix:

$$H(Q) = (1/(Q+1)) \tag{3.26}$$

Let the system feed be the vector X=($x_j$). A Walsh transform of the entrance vector is denoted by:

$$X(Q) = WX(T) \tag{3.27}$$

In this particular case the matrix W is the Walsh-Paley matrix:

$$W = \begin{matrix} 1 & 1 & 1 & 1 \\ 1 & 1 & - & - \\ 1 & - & 1 & - \\ 1 & - & - & 1 \end{matrix} \tag{3.28}$$

According to the definition of the transfer function:

$$Y(Q) = H(Q)X(Q) \tag{3.29}$$

The Walsh transform Y(Q) is obtained and then by reverse transformation the exit Y(T):

$$Y(T) = \left(\frac{1}{M}\right) WY(Q) \tag{3.30}$$

The exit results as a vector. As an example let X=(1 1 -1 -1)'. Here " ' ", denotes the transposed vector. Then X(Q)= WX(T)=(0 4 0 0)'. The transfer function H(Q) is a 4x4 diagonal matrix having as diagonal elements as 1/(Q+1) namely (1 1/2 1/3 1/4). It results that: Y(Q)=(0 2 0 0)' and Y(T)=(1/2)(1 1 -1 -1).

Table (3.7) contains entrance vectors X(T) and the corresponding exit vectors Y(T)) calculated with a transfer function H(Q) associated to linear dyadic models.

Observe that Y(T) is the same as X(T) after renormalization.

**Table 3.7** Input-output Vectors. Dyadic Linear Model.

| Data | X(T) | Y(T) |
|---|---|---|
| 1 | (1 1 1 1) | (1 1 1 1) |
| 2 | (1 1 - -) | (1 1- -) |
| 3 | (1 - 1 -) | (1 - 1 -) |
| 4 | (1- - 1) | (1- - 1) |

A "perfect mixing" separation scheme gives for vectors as 1, 2, 3, 4 in Table 3.7 representing the basis of the solution, the same exit vectors, exempt a constant factor. It is possible to make use of transfer function in the case of "skewed" flow-sheets. Such situations corresponds for instance to splits inside a separation column for species whose vectors have identical digits (for example a first split in Table 3.7 could be (1),(234) not (12),(34) as expected. In such situations the dyadic transfer function is not a diagonal matrix. Suppose for instance that the entrance vector is X(T)=(1 1 -1 -1) (corresponding to species 2) while the exit one is Y(T)=(1 -1 -1 -1) corresponding to the flow-sheet:

(1234) → (1), (234) → (1), (3), (24) → (1), (2), (3), (4)

This signifies that the compound 2 is separated as heavy product in all separation steps. In this case the transfer function results from (3.29) that is from:(-2 2 2 2)'=H(Q)(0 4 0 0)'

$$H(Q) = 1/2 \begin{matrix} 1 & -1 & 1 \\ -1 & 1 & 1 \\ 1 & 1 & 1 & - \\ 1 & 1 & -1 \end{matrix} \quad (3.31)$$

Observe that the transfer matrix verifies the condition of dyadic invariance.

A theory paralleling that of real transfer functions for real linear models may be developed for dyadic transfer functions. In this case the real devices in a complex flow-sheet are replaced by flow-sheet portions. An open problem is to use dyadic transfer functions in describing complex columns for example with side-stream, with more than one feed stream, by-passes, recycling or complex column configurations.

## *3.1.6 Perspectives*

### 3.1.6.1 Enumeration of Flow-Sheets

There exists a close relationship between the coherence studies in category theory and the possible sequences of separation for a mixture (Appendix A4).

Consider that the axiom to be imposed to any possible separation sequence is the associativity. This means that, within a sequence of elements containing two or more of the same sequencing operations in a row, the order that the operations are performed does not matter as long as the sequence to be operated is not changed. Rearranging the parentheses in a sequence will not change sequencing general task.

The resulting associahedrons are studied in category theory as coherence conditions.

Suppose that there are four components and that the order of sequencing is imposed from start, by a heuristic as for instance - sequence the splits in the order of adsorbability. Denote the four components according to that order as 1, 2, 3 and 4. The associahedron K4 known also as MacLane pentagon condition is shown in Fig. 3.2 and in Fig. 3.3.

Recall that a monoid is an algebraic structure with a single associative binary operation and an identity element.

In fact, for tensor product of four objects there are five ways to parenthesize it. The association allows us to build two isomorphisms from the sequence ((12)3)4 to 1(2(34)).

The isomorphism is interpreted in the sense that the direct sequence 1(2(34)) is made equivalent to the reverse sequence ((12)3)4, if one retains the heavy phase instead of the light phase during the separation process. Fig. 3.3 expresses the relations from Fig. 3.2 in a tree-like form.

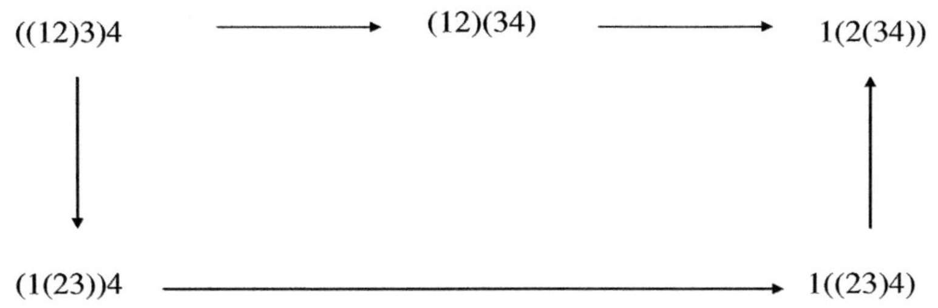


**Fig. 3.2** Monoidal flow-sheets

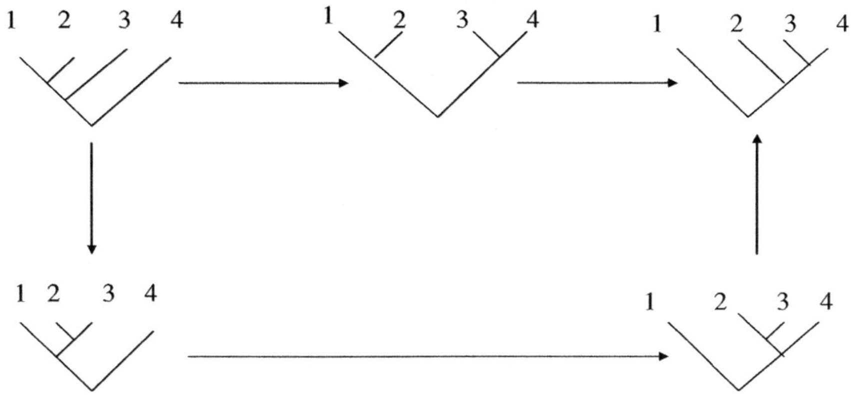


**Fig. 3.3** Monoidal flow-sheets. Tree like form.

For separation processes there are often a process that lets us switch two systems by moving them around each other.

This switching corresponds to braiding, that is to relations as 12=21 (Appendix A4). Consider the case of 3 components denoted by 1, 2 and 3. The hexagon conditions shown in Fig. 3.4 ensures the coherence for braiding conditions.

The first hexagon equation says the switching the component 1 past 23, all at once, is the same as switching it past 2 and then past 3.

The second hexagon is similar. It says that switching 12 past 3 all at once, is the same as doing it in two steps.

This kind of switching called also braiding may be interpreted as a situation in which there is a switch in the separation order. For instance, in a scheme working adsorbability it may be a pressure-swing adsorption, by varying the pressure or a thermal swing adsorption by varying the temperature (Ruthven 1984, Yang 1987). The hexagons outline all the possible separation schemes and their interconnection.

Fig. 3.5 shows flow-sheets that results by parity cube relations (Appendix A4).

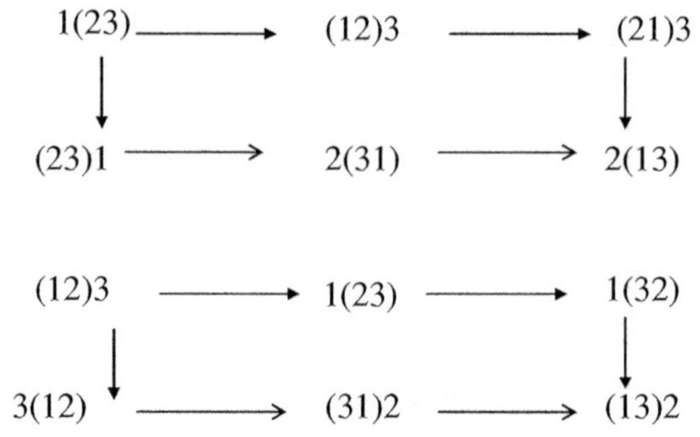


**Fig. 3.4** Braided flow-sheets

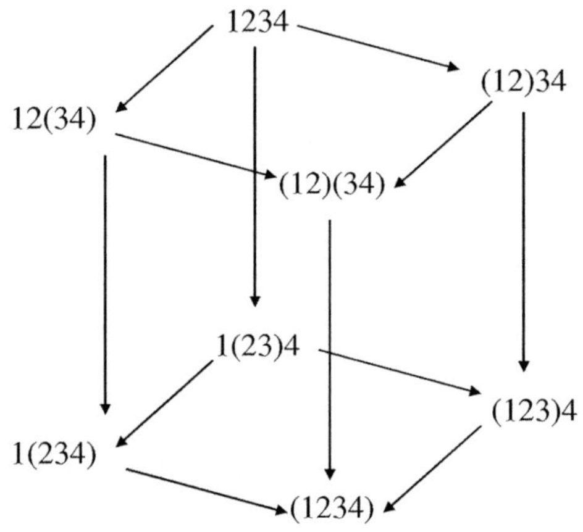


**Fig. 3.5** Parity cube flow-sheets

For coherence the edges of the MacLane pentagon from Fig. 3.2 become five sides of a cube shown in Fig. 3.5. This is the so-called parity 3-cube. This cube shows what happens when an axiom as associativity is replaced by significantly more restrictive ones as required by Gray tensor product definition and tricategorical frame (Appendix A4).

Representation for a tricategory is fundamentally different from the one and two dimensional case of categories and bicategories.

#### 3.1.6.2 Flow-Sheets Evolution

Typical task in chemical process design is to determine the configuration of separation sequences. The problem is that the amount of alternatives leads easily to combinatorial explosion.

Heuristic rules may be of help but synthesis problem is risking to be handled only by heuristic rules.

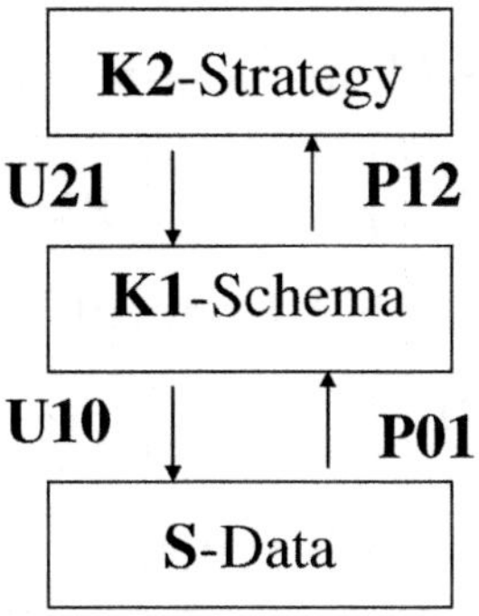


**Fig. 3.6** Three levels scheme for flow-sheet evolution

A method for finding feasible separation process is based on dividing process synthesis in a hierarchy of levels.

The three realms or 2-categorical structure of the process synthesis as shown in Fig. 3.6 may be taken into account for flow-sheet evolution study (Appendix A2).

Observe that there are three levels of the architecture. This is the result of splitting in two-levels the scheme establishing step.

S-may correspond to real data, that is to environment conditions.

K1 and K2 are the two cognitive levels. K1-represents the choice of basic flow-sheets schema and corresponds to the supplementary designed solutions.

K2 allows the revision of the flow-sheets as previously designed in K1. The flow-sheeting schema strategies are defined at the K2-level.

This corresponds to the check and revision of flow-sheeting solutions.

U10: K1→S describes the actions towards the real level while P01: S→K1 summarizes the observations based on S evaluations.

The information changes between the basic K1-level and the K2-level of process synthesis are characterized by the operator U21 and the possibilities P12.

U21 corresponds to the operated solutions while P12 corresponds to retained solutions.

It is possible to run on different time scales for K1 and K2-levels. Several cycles may be performed on a fast scale before the coupling in the larger loops. This allows anticipative control of the process.

Dividing the main levels in sub-levels is beneficial. Sub-levels are for example concept level (such as solvent recovery process), process systems level (such as a distillation column with its auxiliaries) and equipment level (such as selection of a pump).

Re-usable and usable ontology for computer-aided chemical process engineering, CAPE, was extensible studied (Morbach et al. 2008). OntoCAPE is a formal ontology for the domain of CAPE. Fig. 3.7 illustrates the OntoCAPE architecture. The basic layer S, corresponds to the specific application. It

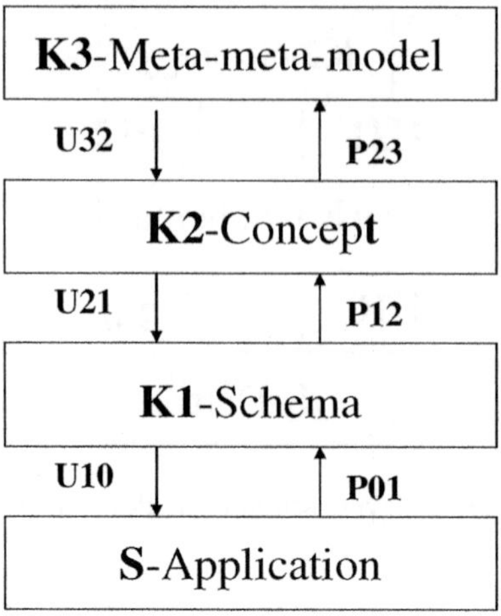


**Fig. 3.7** Detail of CAPE

allows generating an application scheme model K1. The conceptual layer that is the meta-model K2 may be structured in sub-layers. The layer K3 holds a meta-meta-model that introduces fundamental modeling concepts and states the design guidelines for the construction of the actual ontology. Conceptual layer covers unit operations, equipment and machinery, materials and their thermo physical properties, chemical process behaviour and on top on this an upper-layer defining the principles of general systems theory according to which the ontology is organized.

The non-linear coupled differential equations describing transport phenomena will have to be seen as simple stepping stones in the complex hierarchies of layered systems.

Fig. 3.7 outlines that OntoCAPE reduces in fact to an OMG-MOF framework.

OntoCAPE proves to be an ontology validated for use and reuse.

Other ontological based approaches to process synthesis are due to Venkatasubramanian et al. (2006).

What is missing to OntoCAPE system to become part of an evolvable or autonomous system is the interaction and embodiment of the real application domain, that is, the integrative closure.

This development towards organically embedded evolvable CAPE systems may represents the next generation for chemical engineering systems coming after the present CAPE based generation of systems.

#### 3.1.6.3 Evolvability for Flow-Sheets

Development processes are highly creative and therefore can be planned only to a limited extent. The tasks to be performed depend on the product to

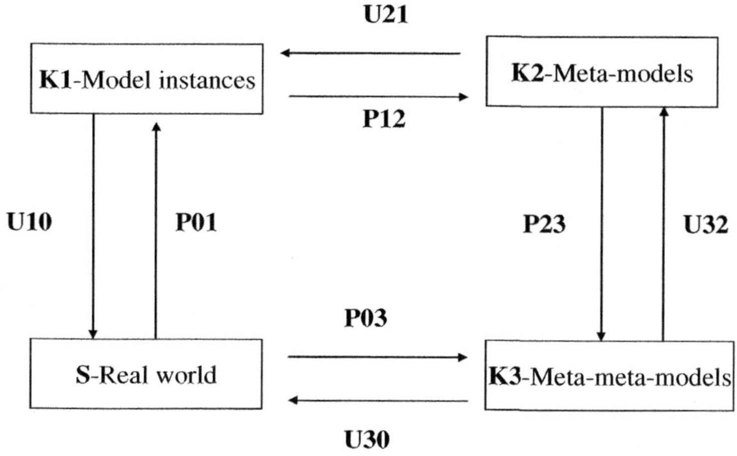


**Fig. 3.8** Network for evolvable development processes

be developed, which is not known in advance. Alternative designs variants should be explored to arrive at a viable solution.

Building effective tools for managing development processes, means to challenge the process evolution. While this has been recognized widely, current industrial systems can cope with process evolution only to a limited extent. In particular, this applies to flow-sheets which were designed for repetitive processes.

For such systems a high number of flow-sheets are executed according to a common definition, ensuring that work is performed following a pre-defined procedure. This approach cannot be transferred to development processes because it does not take process evolution into account:

In response to such problems, a variety of mechanisms have been developed to increase the flexibility of flow-sheets. Flow-sheet definitions are made more flexible, for instance, by relaxing ordering constraints for activities, handling exceptions, or allowing for small deviations. Moreover, mechanisms are provided for changing flow-sheet definitions and instances.

A four-level approach useful for flow-sheet generation was implemented for development processes (Heller et al. 2003). This work offered suggestions for an evolvable framework which distinguishes four levels of modeling (Fig. 3.8). The meaning of these levels as well as their interaction differs from one approach to another. Each level deals with process entities such as products, activities, and resources.

S corresponds to real world processes, K1 to process model instances, K2 to process models definition activities that is to meta-models and K3 to meta-meta-models activities.

P01, P12, P23 and P03 are synthesis steps. Notice that: P03=P01oP12oP23.

U10, U21, U32 and U30 correspond to implementation operations. Observe that: U10: Model instances → Real, U21: Meta-models → Model instances, U32: Meta-meta-models → Meta-models, and U30: Meta-meta-models → Real. In this case U30=U10oU21oU32.

Process evolution may occur on every level. Furthermore, adjacent levels are connected by propagation and analysis relationships. Propagation is performed top-down and constrains the operations that may be performed on the next lower level. Conversely, analysis works bottom-up and aims at providing feedback to the next upper level. The meta-meta-model, K3, introduces the language (or meta-meta-schema) in terms of which process models may be defined at the next lower level. The activity meta-meta-model is based on dynamic task nets. Tasks are organized hierarchically.

Feedback flows are oriented oppositely to control flows; they are used to represent feedback in the development process. Finally, tasks have inputs and outputs which are connected by data flows. Meta-model definitions, K2 are created as instances of process meta-meta-models. Process models are defined in UML, Unified Modeling Language. Tasks and task types are modeled as objects and classes, respectively. At the type level, processes are represented by class diagrams. Process instances, K1 are instantiated from process model definitions. A process model definition represents reusable process knowledge at an abstract level. Multiple process instances may share the same definition. In contrast, there is a one to one correspondence between a process model instance and the respective real-world process, that is, each process model instance represents exactly one real world process. A process model instance is composed of task instances which are created from the task classes provided by the process model definition. Finally, the real-world process, denoted by S or K0, consists of the steps that are actually performed by humans or tools. The process model instance is an abstraction which represents the real world process in the process management system. The process model is used to guide and control process participants. Conversely, process participants provide feedback which is used to update the process model instance. Evolution may occur on each model level. Furthermore, the consequences of evolution have to be propagated vertically between the levels.

Real-world evolution drives the evolution of the upper levels.

The real-world process is represented by a corresponding process model instance. When the real-world process is changed, its model has to be updated accordingly a certain design task. For architecture shown in Fig. 3.8 the closure that is the direct connection between the higher and the basic level represents the challenging request.

This direct connection makes a start by implementing the autonomic computing paradigm. This paradigm refers to increasing environment-awareness and automatic responsiveness.

Autonomic computing methods promise to facilitate case based reasoning tasks and information capture (Montani and Anglano 2006). Flow-sheet generation methods based on evolvable case based reasoning conceptual frames have been evaluated (Surma and Braunschweig 1996, Pajula et al. 2001). The separation of an organic compound, water and some light and heavy impurities was presented in detail (Pajula et al. 2001).

To the components to be separated vectors of properties could be associated. This refers to the type of substances to be separated, for instance hydrocarbons, to the chemical properties, for instance relative volatility, to product purity requirements. Heuristic rules as for instance: if relative volatility is larger than 2 the proposed separation is distillation or if a component is corrosive should be separated in the first step, may be included among features. The similarity between features may be based on digital value as "0" or "1" or on real values (Sect. 2.3.1). Similarity measure based on analogical reasoning, conceptual graphs and object-oriented representations have been studied (Surma and Braunschweig 1996).

Every separation process consists of several pieces of process equipment which have to be selected. The user can make different queries. For instance, the basic query may be to separate the organic compound and water, but also to separate mixtures with nearly similar boiling point or similar chemical type. These are encoded in the hierarchy of features. For different queries specific sequences of distillation and extraction are resulting.

## 3.2 Cyclic Operations of Separation

### 3.2.1 Cyclic Separations

Cycling operation methods are of great consequence in oil chemistry, in pharmaceutical and food industry, isotopes separation, hydrogen purification, desalinization, and so forth. Cyclic separation technologies such as pressure swing adsorption (PSA), temperature swing adsorption (TSA), vacuum swing adsorption (VSA), cyclic zone adsorption (CZA), simulated moving beds (SMB) chromatography, pressure swing reactor (PSR) and reverse flow reactor (RFR), parameter pumping (PP) and so forth, are unsteady non-linear processes difficult to put into practice and to control. Numerous cycling separation schemes, based mainly on intuition have been reported in literature. Well-known examples are the schemes involving 2 or 4-beds, and 2 or 4-step cycles (Ruthven 1984, Yang 1987). An example of SMB scheme is shown in Fig. 3.9.

The SMB consists of four columns or beds, #0, #1, #2 and #3, that are connected in a circular arrangement. The positions of feed, extract, desorbent and raffinate are changed cyclically in four steps corresponding to the four columns. Inlet and outlet points are shifted periodically in the same direction of the liquid flow, bed by bed as shown in Fig. 3.9. An intermittent resin flow in the direction opposite to the liquid flow is created.

For multi-component, and multiple beds systems it is difficult to arrange the process scheduling, to elaborate and to run mathematical models, to

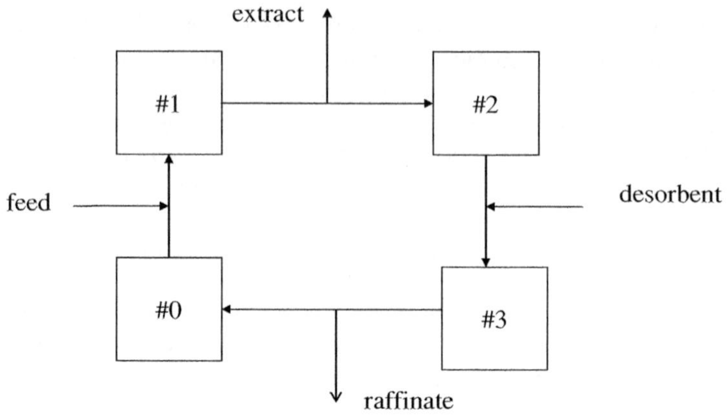


**Fig. 3.9** Basic configurations for SMB

adjust them by experiment. The non-linear interaction of components and the interconnection of beds complicate the cycling operation schedule. As the separation complexity increases it becomes very difficult to formulate a feasible schedule much less an optimal one. The cyclic operations complexity is rooted in the unexpected non-linear interactions of several components, in the random condition of functioning, in the unavoidable uniqueness of each separation device, the incomplete knowledge of inputs and outputs, the variability in time of parameters, in modeling problems, and so forth Some conventional cyclic separation systems are operated in a regime where the complexity may be neglected. However to avoid complexity is neither always possible nor advantageous.

The approach allowing to operate cyclic separations in high complexity conditions is that of evolvable cycling separation systems. These are systems that can change autonomously both the scheme as the dynamic behavior and are capable to control and to take advantage of the unexpected events of their environment in increasingly complex ways. Evolvable devices are separation systems with emergent, not entirely pre-programmed, behavior.

The originality of this approach consists in the elaboration of a differential model for symbolic aspects, the wave equation model, WE, generating separation schemes and schedules. This model complements the dynamic mass transfer model. The wave model is formally similar to the first order wave micro-model used to describe separations (Rhee et al. 1989) but the physical significance of parameters, the factors and the calculus rules are different (Sect. 2.2.1, Sect. 2.2.2).

The WE model allows simple description of flow sheets, characteristics and schedules. Existing separation schemas have to be reconsidered in the new framework and new separation schemes are resulting. The connection with the theory of Latin squares and the designs of experiments is established and illustrated in the study of configurations with variable number of separation units or stages and in the study of operations coupling. Some classes of potentially evolvable separation technologies are discussed in what follows. Cyclic separation device with evolvability based on scheme modification on self-configuring schemes and multi-scale schemes organized by self-similar replication at different conditioning levels are presented.

## *3.2.2 Cyclic Separations Scheduling*

### 3.2.2.1 Conventional Schemes

The solution of the wave equation is

$$Y = Z \oplus V \otimes T \tag{3.32}$$

The velocity V=1 in equation (3.32) will be considered in what follows.

This means that the characteristic solution is:

$$Y = Z \oplus T \tag{3.33}$$

The solutions of the model are in fact the addition operation from the Table 2.1 for different C(m) algebraic frames.

- **Example 1: Three beds PSA scheme**

Smith and Westerberg (1990) considered the following sequence of seven operations:

O1-adsorption and production
O2-adsorption, production and produce purge gas for O5
O3-pressure equalization with low-pressure operation O6
O4-countercurrent depressurization
O5-countercurrent purge with gas from O2
O6-presure equalization with high-pressure operation O3
O7-re-pressurize with feed gas

The Table 3.8 shows the scheme as presented by Smith and Westerberg:

The three beds indexed by #0, #1 and #2 are the effectual and minimal ones.

The reactivation bed #3 was introduced since the adsorbents deactivates slowly.

**Table 3.8** PSA configuration

| | | | | | | | | | |
|---|---|---|---|---|---|---|---|---|---|
| #0 | O1 | | | O2 | O3 | O4 | O5 | O6 | O7 |
| #1 | O5 | O6 | O7 | O1 | | | O2 | O3 | O4 |
| #2 | O2 | O3 | O4 | O5 | O6 | O7 | O1 | | |
| #3 | Reactivation | | | | | | | | |

To run separation in three beds the operation should be lumped in three classes, from the point of view of time spent. It is possible to lump the operations shown in Table 3.8 as follows: O1=0, (O2, O3, O4) = 1, (O5, O6, O7) = 2. This is possible by ensuring that O1 has the same duration as the sequence (O2, O3, O4) or the sequence (O5, O6, O7). This is possible since due to process restrictions O2 and O5, O3 and O6 and implicitly O4 and O7 should have similar duration. For three beds #0, #1 and #2, three groups of operations denoted by 0, 1 and 2 will be considered.

The Table 3.8 may be rewritten as Table 3.9. The PSA scheme from Table 3.9 is identical to that shown in Table 3.8 only the labels of beds are different. This corresponds to a shift in the initial condition.

**Table 3.9** Three beds for PSA

| Z\T | 0 | 1 | 2 |
|---|---|---|---|
| #0 | 0 | 1 | 2 |
| #2 | 1 | 2 | 0 |
| #1 | 2 | 0 | 1 |

- **Example 2: Four beds schedules**

The four-bed scheme is frequently encountered in cyclic operations.

Chiang (1988) presented the cyclic separation scheme for PSA, shown in Table 3.10.

Arrows shows the flow direction. V-vacuum, E-equalization, R-repressurization.

A-adsorption, CD-counter current depressurization, F-feed compression.

Considering the following lumping in blocks: (V, E)=0, (R, F)=1, A=2, (CD, E)=3 the Chiang scheme reduces to the basic C(4) solution shown in Table 3.11.

Notice that in the step 1 the flow direction may change after half period.

The same result may be obtained for the four beds Oxy-Rich process (Smith and Westerberg 1990, Fig. 2).

**Table 3.10** Four beds schedule for PSA

| | | | | | | | | |
|---|---|---|---|---|---|---|---|---|
| #0 | V ↓ | E ↓ | R ↓ | F ↑ | A ↑ | | CD ↑ | E ↑ |
| #1 | CD ↑ | E ↑ | V ↓ | E ↓ | R ↓ | F ↑ | A ↑ | |
| #2 | A ↑ | | CD ↑ | E ↑ | V ↓ | E ↓ | R ↓ | F ↑ |
| #3 | R ↓ | F ↑ | A ↑ | | CD ↑ | E ↑ | V ↓ | E ↓ |

**Table 3.11** Four beds configuration

| Z\T | 0 | 1 | 2 | 3 |
|---|---|---|---|---|
| #0 | 0 | 1 | 2 | 3 |
| #3 | 1 | 2 | 3 | 0 |
| #2 | 2 | 3 | 0 | 1 |
| #1 | 3 | 0 | 1 | 2 |

The notations to be considered in this case are: O1=0, O2=1, O3=2, O4=3.

Similar results are obtained if we consider the cyclic schedule with four beds and seven operations (Smith and Westerberg 1990, Fig. 3).

The lumped operations to be considered are: O1=0, (O2, O3)=1, (O4, O5)=2, (O6, O7)=3. The same result is obtained also for the four beds configuration Wagner-Union Carbide (Tondeur et al. 1985, Fig. 17, operational scheme). The lumping to operate is as follows (1, 2)=0, (3, 4)=1, (5, 6)=2, (7, 8)=3.

Lumping imposes the time restrictions.

The same scheme is resulting for simulated moving bed SMB chromatography (Ruthven and Ching 1989). Suppose that the SMB contains four zones of beds indexed by #0, #1, #2 and #3. The involved elements of the chromatography operation are: F-feed, E-extract (slow mover), D-desorbent (eluant), and R-raffinate (fast mover) (Fig. 3.9). They are denoted by "0", "1", "2", and "3" respectively. F and D are inlets, while R and E are outlets.

The functioning of beds at successive time steps is represented in the Table 3.12.

Observe that the positions of the F, E, D, R are changed in the direction of circular flow at a regularly point in time. At any given time for any zone, only one of the valves corresponding to F, E, D, or R is open.

**Table 3.12** SMB configuration

| Z\T | 0 | 1 | 2 | 3 |
|---|---|---|---|---|
| #0 | F | E | D | R |
| #1 | E | D | R | F |
| #2 | D | R | F | E |
| #3 | R | F | E | D |

If each zone has 2 beds an 8-bed SMB system will results.

• **Example 3: Five beds schedule for PSA with progressive lower-pressure beds**

Chiang (1988) presented the following cyclic separation scheme with 11 steps:

O1-pressure equalization with a high-pressure bed
O2-pressure equalization with other high-pressure bed
O3-re-pressurize with low-quality product from other bed
O4-re-pressurize with feed
O5-adsorption and production
O3'-re-pressurize of another bed with part of all its product
O2'-equalization of its pressure with another bed
O6-partial de-pressurize using the off gas to purge another bed
O1'-equalization rest pressure with freshly purged bed
O7-ventilation
O6'-purge with off gas of another bed

Chiang proposed the scheme shown in Table 3.13.

**Table 3.13** Five beds schedule for PSA with progressive lower-pressure beds

| | | | | | | | | | | | | | | | |
|---|---|---|---|---|---|---|---|---|---|---|---|---|---|---|---|
| #0 | 6 | 1' | 7 | 6' | 1 | 2 | 3 | | 4 | 5 | | | 3' | | 2' |
| #1 | 3' | | 2' | 6 | 1' | 7 | 6' | 1 | 2 | 3 | | 4 | 5 | | |
| #2 | 5 | | | 3' | | 2' | 6 | 1' | 7 | 6' | 1 | 2 | 3 | | 4 |
| #3 | 3 | | 4 | 5 | | | 3' | | 2' | 6 | 1' | 7 | 6' | 1 | 2 |
| #4 | 6' | 1 | 2 | 3 | | 4 | 5 | | | 3' | | 2' | 6 | 1' | 7 |

Consider the lumping in five zones: (6,1', 7)=0, (6', 1, 2)=1, (3, 4)=2, (5)=3, (3', 2')=4.

The scheme shown Table 3.13 is equivalent to that shown in Table 3.14. Time restrictions should be imposed according to the recommended lumping in zones.

**Table 3.14** Five bed configuration

| Z\T | 0 | 1 | 2 | 3 | 4 |
|---|---|---|---|---|---|
| #0 | 0 | 1 | 2 | 3 | 4 |
| #4 | 1 | 2 | 3 | 4 | 0 |
| #3 | 2 | 3 | 4 | 0 | 1 |
| #2 | 3 | 4 | 0 | 1 | 2 |
| #1 | 4 | 0 | 1 | 2 | 3 |

The cyclic separation scheme reduces to the C(5) addition table.

- **Example 4: Polybed PSA scheme**

Smith and Westerberg (1990), considered the following sequence of 13 operations:

O1-adsorption and production
O2-adsorption, equalization with low pressure O17
O3-pressure equalization with low-pressure operation O16
O5- pressure equalization with low-pressure operation O15

O7-produce purge gas for O13
O8-produce purge gas for O12
O9-blowdown counter-current
O12-countercurrent purge with gas from O8
O13-countercurrent purge with gas from O7
O15- pressure equalization with high-pressure operation O5
O16- pressure equalization with high-pressure operation O3
O17- pressure equalization with high-pressure operation O2
O18- re-pressurize with feed gas

The Table 3.15 shows the scheme presented by Smith and Westerberg. Any row in Table 3.15 corresponds to a bed. It is possible to lump all operations in classes containing two successive operations with the exception of O1. O1 is represented as having the duration 4 times higher that of the other operations.

**Table 3.15** Polybed PSA configuration

| | | | | | | | | | | | | | | | | |
|---|---|---|---|---|---|---|---|---|---|---|---|---|---|---|---|---|
| #0 | O1 | | | | O2 | O3 | O5 | O7 | O8 | O9 | O12 | O13 | O15 | O16 | O17 | O18 |
| #1 | O17 | O18 | O1 | | | | O2 | O3 | O5 | O7 | O8 | O9 | O12 | O13 | O15 | O16 |
| #2 | O15 | O16 | O17 | O18 | O1 | | | | O2 | O3 | O5 | O7 | O8 | O9 | O12 | O13 |
| #3 | O12 | O13 | O15 | O16 | O17 | O18 | O1 | | | | O2 | O3 | O5 | O7 | O8 | O9 |
| #4 | O8 | O9 | O12 | O13 | O15 | O16 | O17 | O18 | O1 | | | | O2 | O3 | O5 | O7 |
| #5 | O5 | O7 | O8 | O9 | O12 | O13 | O15 | O16 | O17 | O18 | O1 | | | | O2 | O3 |
| #6 | O2 | O3 | O5 | O7 | O8 | O9 | O12 | O13 | O15 | O16 | O17 | O18 | O1 | | | |
| #7 | O1 | | O2 | O3 | O5 | O7 | O8 | O9 | O12 | O13 | O15 | O16 | O17 | O18 | O1 | |

Moreover all other operations may be coupled in pairs. This suggests associating to O1 two equal steps denoted here by 0 and 1. All other operations need the same one time step for any of two successive operations. In this way all the re-defined eight operations need the same time step.

Denote: O1=0 and 1, (O2, O3)=2, (O5, O7)=3, (O8, O9)=4, (O12, O13)=5, (O15, O16)=6 (O17, O18)=7.

The Polybed scheme is close to the C(8) solution shown in Table 3.16.

This accommodates the 13 steps in which there exist one with duration four times longer than the other.

**Table 3.16** Eight beds configuration Polybed

| Z\T | 0 | 1 | 2 | 3 | 4 | 5 | 6 | 7 |
|---|---|---|---|---|---|---|---|---|
| #0 | 0 | 1 | 2 | 3 | 4 | 5 | 6 | 7 |
| #7 | 1 | 2 | 3 | 4 | 5 | 6 | 7 | 0 |
| #6 | 2 | 3 | 4 | 5 | 6 | 7 | 0 | 1 |
| #5 | 3 | 4 | 5 | 6 | 7 | 0 | 1 | 2 |
| #4 | 4 | 5 | 6 | 7 | 0 | 1 | 2 | 3 |
| #3 | 5 | 6 | 7 | 0 | 1 | 2 | 3 | 4 |
| #2 | 6 | 7 | 0 | 1 | 2 | 2 | 4 | 5 |
| #1 | 7 | 0 | 1 | 2 | 2 | 4 | 5 | 6 |

#### 3.2.2.2 Variable Number of Beds

It was observed that some solutions of the WE are Latin-squares. Consequently well-known results of Latin square theory or of finite group's theory may be applied in the study of separation schemes. The decomposition of a Latin square in Latin rectangles is applied in the followings (Denes and Keedwell 1974, Hedayat et al. 1999). This corresponds to the decomposition of the cyclic group in subgroups and allows the study of possible decreasing of the number of beds for the same number of stages (Tondeur et al. 1985).

- **Example 1: Two beds, four stages**

The main stages in the classical Skarstrom configuration are as follows: "0"-compression, "1"-production at high pressure, "2"-decompression, "3"-purge.

To this configuration corresponds the scheme shown in Table 3.11 if V=1.

Only two beds, denoted here by Z="#0" and Z="#2" may be considered in a simplified separation scheme. This corresponds to the subgroup {0, 2}, in C(4) addition group, supplementing the trivial group and the entire group.

Taking only these values for Z, the C(4) Table 3.11 reduces to Table 3.17:

**Table 3.17** Two beds, four stages

| Z\T | 0 | 1 | 2 | 3 |
|---|---|---|---|---|
| #0 | <u>0</u> | <u>1</u> | 2 | 3 |
| #2 | <u>2</u> | <u>3</u> | 0 | 1 |

In first step, the bed #0 is in compression when bed "#2" is in decompression.

In second step, bed #0 is in production mode when bed "#2" is in purge.

The first two steps, the underlined ones, represent the basic separation process. Then the cycle is completed by bed permutation. This means that the bed #0 pass to operation "2", and the bed #2 pass to "0". Then bed #0 passes to the operation "3" and the bed #2 to operation "1". The underlined elements from Table 3.17 represent one of the most utilized schemes in cycling separations.

- **Example 2: Two beds, six stages**

The subgroups {0, 3} and {0, 2, 4} supplements the trivial group and the entire group in the C(6) addition group. Consequently instead of six bed schemes, some with two beds and with three beds may be considered for six stages separations.

The main steps are: "0"- start of compression by pressure equalization, "1"-finish of compression, "2"-production at high pressure, "3"-decompression, "4"-finish of decompression, "5"-purge. It is the case of Skarstrom scheme with split for compression and decompression.

Only the beds "#0" and "#3" may be considered in a simplified scheme. The C(6) operation reduces to the scheme shown in Table 3.18.

**Table 3.18** Two beds, six stages

| Z\T | 0 | 1 | 2 | 3 | 4 | 5 |
|---|---|---|---|---|---|---|
| #0 | 0 | 1 | 2 | 3 | 4 | 5 |
| #3 | 3 | 4 | 5 | 0 | 1 | 2 |

The first three steps represent the basic separation process. Then the cycle is completed by bed permutation. The bed #0 follows the succession of operations "3", "4", "5" while the bed #3 follows the succession "0", "1", and "2".

- **Example 3: Three beds 6 stages**

The three beds denoted by #0, #2 and #4 would be considered now. The C(6) operation table reduces to the scheme shown in Table 3.19.

**Table 3.19** Three beds, six stages

| Z\T | 0 | 1 | 2 | 3 | 4 | 5 |
|---|---|---|---|---|---|---|
| #0 | 0 | 1 | 2 | 3 | 4 | 5 |
| #2 | 2 | 3 | 4 | 5 | 0 | 1 |
| #4 | 4 | 5 | 0 | 1 | 2 | 3 |

Union Carbide-Batta proposed such configurations (Tondeur et al. 1985). The first two steps represent the basic separation process. Then the cycle is completed by two successive bed permutations.

The subgroups {0, 4} and {0, 2, 4, 6} supplements the trivial group and the entire group in the C(8) addition group. Consequently schemes with two beds and with four beds may be considered for eight stages.

- **Example 4: Four beds, eight stages**

Beds #0, #2, #4, and #6 correspond to the 4-bed scheme.

For only four beds in the C(8) operation from Table 3.16 reduces to Table 3.20.

The first two steps describe the basic process. Three successive bed permutations complete the cycles.

**Table 3.20** Four beds, eight stages

| Z\T | 0 | 1 | 2 | 3 | 4 | 5 | 6 | 7 |
|---|---|---|---|---|---|---|---|---|
| #0 | 0 | 1 | 2 | 3 | 4 | 5 | 6 | 7 |
| #2 | 2 | 3 | 4 | 5 | 6 | 7 | 0 | 1 |
| #4 | 4 | 5 | 6 | 7 | 0 | 1 | 2 | 3 |
| #6 | 6 | 7 | 0 | 1 | 2 | 3 | 4 | 5 |

- **Example 5: Two beds, eight stages**

  The two beds are denoted by #0 and #4.

  The C(8) Table 3.16 reduces in this case to the Table 3.21:

**Table 3.21** Two beds, eight stages

| Z\T | 0 | 1 | 2 | 3 | 4 | 5 | 6 | 7 |
|---|---|---|---|---|---|---|---|---|
| #0 | $\underline{0}$ | $\underline{1}$ | $\underline{2}$ | $\underline{3}$ | 4 | 5 | 6 | 7 |
| #4 | $\underline{4}$ | $\underline{5}$ | $\underline{6}$ | $\underline{7}$ | 0 | 1 | 2 | 3 |

The first four steps describe the basic process.

The subgroups {0, 4} and {0, 2, 4, 6, 8} supplements the trivial group and the entire group in C(10) addition group. Consequently schemes with two beds and five beds may be considered for ten stages.

- **Example 6: Five beds, ten stages**

The beds are denoted by #0, #2, #4, #6 and #8. The scheme is shown in Table 3.22.

**Table 3.22** Five beds, ten stages

| Z\T | 0 | 1 | 2 | 3 | 4 | 5 | 6 | 7 | 8 | 9 |
|---|---|---|---|---|---|---|---|---|---|---|
| #0 | $\underline{0}$ | $\underline{1}$ | $\underline{2}$ | $\underline{3}$ | $\underline{4}$ | 5 | 6 | 7 | 8 | 9 |
| #2 | $\underline{2}$ | $\underline{3}$ | $\underline{4}$ | $\underline{5}$ | $\underline{6}$ | 7 | 8 | 9 | 0 | 1 |
| #4 | $\underline{4}$ | $\underline{5}$ | $\underline{6}$ | $\underline{7}$ | $\underline{8}$ | 9 | 0 | 1 | 2 | 3 |
| #6 | $\underline{6}$ | $\underline{7}$ | $\underline{9}$ | $\underline{0}$ | $\underline{1}$ | 2 | 3 | 4 | 5 | 6 |
| #8 | $\underline{8}$ | $\underline{9}$ | $\underline{0}$ | $\underline{1}$ | $\underline{2}$ | 3 | 4 | 5 | 6 | 7 |

- **Example 7: Two beds, ten stages**

The beds are denoted by #0 and #5. The scheme is shown in Table 3.23.

**Table 3.23** Two beds, ten stages

| Z\T | 0 | 1 | 2 | 3 | 4 | 5 | 6 | 7 | 8 | 9 |
|---|---|---|---|---|---|---|---|---|---|---|
| #0 | $\underline{0}$ | $\underline{1}$ | $\underline{2}$ | $\underline{3}$ | $\underline{4}$ | 5 | 6 | 7 | 8 | 9 |
| #5 | $\underline{5}$ | $\underline{6}$ | $\underline{7}$ | $\underline{8}$ | $\underline{9}$ | 0 | 1 | 2 | 3 | 4 |

Esso-Marsh and Grace proposed such type of configurations (Tondeur et al. 1985).

#### 3.2.2.3 Different Separation Velocities

It is possible to travel along the existing separation schemes with sequencing velocities different from V=1. For the same number of beds as in the initial scheme, this doesn't limit the total number of required operations but may limit the number of operations in a bed. The velocity V refers to the number of beds traveled during a step. The case V≠1 suggests assigning some beds to specific operations only, and proposing innovative separation schemes including operations of regenerations.

• **Example 1: Four beds, two stages in each bed**
The main stages are as follows: "0"-compression, "1"-production at high pressure, "2"-decompression, "3"-purge.

To this corresponds the scheme shown in Table 3.11 if V=1.

Since V=2 only two steps may be performed in any bed.

Taking V=2 in the equation (3.32), it results the following scheme from Table 3.24:

**Table 3.24** Four beds, two stages

| Z\T | 0 | 1 | 2 | 3 |
|---|---|---|---|---|
| #0 | <u>0</u> | <u>2</u> | 0 | 2 |
| #1 | <u>1</u> | <u>3</u> | 1 | 3 |
| #2 | <u>2</u> | <u>0</u> | 2 | 0 |
| #3 | <u>3</u> | <u>1</u> | 3 | 1 |

The steps at T=2 and T=3 repeats the steps at T=0 and T=1. The mixture follows trajectories as for instance #0, #3, #2, and #1 to complete the separation cycle in the natural stage order, 0, 1, 2 and 3. Observe that a bed such as #3 performs only two stages, 3 and 1.

• **Example 2: Six beds two stages**
To this, corresponds the scheme shown in C(6) if V=1. Since V=3 only two steps may be involved in separation. Taking V=3 in the equation (3.32) it results the scheme from Table 3.25:

The steps at T=2, T=3 and T=4, T=5 repeats the steps at T=0 and T=1.

• **Example 3: Three beds**
With V=2, the 3 beds scheme will be traveled in the contrary sense.

This corresponds to a separation scheme in which the raffinate and the extract change their roles.

This is valid for any number of beds.

With V=2 the solution from Table 3.9 becomes that shown in Table 3.26.

**Table 3.25** Six beds, two stages in each bed

| Z\T | 0 | 1 | 2 | 3 | 4 | 5 |
|---|---|---|---|---|---|---|
| #0 | 0 | 3 | 0 | 3 | 0 | 3 |
| #1 | 1 | 4 | 1 | 4 | 1 | 4 |
| #2 | 2 | 5 | 2 | 5 | 2 | 5 |
| #3 | 3 | 0 | 3 | 0 | 3 | 0 |
| #4 | 4 | 1 | 4 | 1 | 4 | 1 |
| #5 | 5 | 2 | 5 | 2 | 5 | 2 |

**Table 3.26** Wave equation solution Y(T, Z), V=2

| Z\T | 0 | 1 | 2 |
|---|---|---|---|
| #0 | 0 | 2 | 1 |
| #1 | 1 | 0 | 2 |
| #2 | 2 | 1 | 0 |

#### 3.2.2.4 Coupling Cycling Operations

Schemes coupling cyclic operation devices, in series or in parallel, start to receive attention in chemical engineering (Chin and Wang 2004, Hur and Wankat 2005, Charpentier 2005). Coupling of different technologies as for instance: SMB with PSA, SMB with crystallization, PSA with separation by sedimentation, fluidization or membrane filtration is also of interest. The multiple and superposed periodic separation fields, encountered in chromatothermography, forced traveling wave cycling adsorption zone, pressure swing adsorption with multiple adsorbents, hierarchical arrays of simulated moving beds represent interesting developments and adaptations in the field of cyclic separations.

The study of combined cycling operation technologies requires the same general conceptual framework resulting from modulo-m algebras and the same wave model.

Coupling of two cycling operations is described in what follows.

It will appear as an application of the categorical product "×" interpretation for the tensorial product "∗". The categorical product "×" involves local enlargement of the scheme while maintaining the old scheme as the coordinator architecture.

- **Example 1: Two cycling operations**

Suppose Y, T, and Z are vectors with two components as for instance $Y=y_0 \times y_1$, $T=t_0 \times t_1$ and $Z=z_0 \times z_1$, and $F=f_0 \times f_1$ For V=1, the wave model equation (2.4) reduces in fact to two equations, one for each scale:

$$\frac{\partial y_0}{\partial t_0} \oplus \frac{\partial y_0}{\partial z_0} = 0 \tag{3.34}$$

$$y_0(0) = f_0 \tag{3.35}$$

$$\frac{\partial y_1}{\partial t_1} \oplus \frac{\partial y_1}{\partial z_1} = 0 \tag{3.36}$$

$$y_1(0) = f_1 \tag{3.37}$$

Suppose that the initial condition is:

$$Y(0, Z) = f_0 \times f_1 \tag{3.38}$$

With this initial condition the solution of the model will be:

$$Y(T, Z) = y_0 \times y_1 \tag{3.39}$$

Consider that the first component corresponds to a two-step cycling operation defined on C(2) while the second component corresponds to a three-step cycling operation defined on C(3). The solutions of equations (3.34-3.39) are shown in Table 3.27 and Table 3.28.

**Table 3.27** Solution $y_0$

| | |
|---|---|
| 0 | 1 |
| 1 | 0 |

**Table 3.28** Solution $y_1$

| | | |
|---|---|---|
| 0 | 1 | 2 |
| 1 | 2 | 0 |
| 2 | 0 | 1 |

Table 3.29 shows the product of the two solutions.

**Table 3.29** Categorical product $Y(T) = y_0 \times y_1$

| | | | | | |
|---|---|---|---|---|---|
| 00 | 01 | 02 | 10 | 11 | 12 |
| 01 | 02 | 00 | 11 | 12 | 10 |
| 02 | 00 | 01 | 12 | 10 | 11 |
| 10 | 11 | 12 | 00 | 01 | 02 |
| 11 | 12 | 10 | 01 | 02 | 00 |
| 12 | 10 | 11 | 02 | 00 | 01 |

The $y_0 \times y_1$ solution shows that for each time step on the first scale operation will be 3 time steps for the second scale operation. For instance the step 0 for the first scale is coupled with the steps 0, 1, and 2 for the second scale. Denote: 00=0; 01=1; 02=2; 10=3; 11=4; 12=5.

Table 3.30 Equivalent form for $y_0 \times y_1$

| Z\T | 0 | 1 | 2 | 3 | 4 | 5 |
|---|---|---|---|---|---|---|
| #0 | 0 | 1 | 2 | 3 | 4 | 5 |
| #1 | 1 | 2 | 0 | 4 | 5 | 3 |
| #2 | 2 | 0 | 1 | 5 | 3 | 4 |
| #3 | 3 | 4 | 5 | 0 | 1 | 2 |
| #4 | 4 | 5 | 3 | 1 | 2 | 0 |
| #5 | 5 | 3 | 4 | 2 | 0 | 1 |

The solution becomes that shown in Table 3.30. The beds and the time steps for the second coordinate have been indicated. As established, the product "×" of two Latin squares is a Latin square too.

## *3.2.3 Evolvability*

### 3.2.3.1 Scheme Modification

Non-linear interactions of many components in multiple beds schemes induce complexity for separation systems. The strategy to face complexity is to elaborate separation schemes having potentiality to evolve, self-adaptability to unavoidable randomness of bed construction, to random change of the feed or to environmental unexpected and un-programmed conditions. Some strategies for evolvable separations will follow.

Let us restrict as a first example, to the C(4) solution shown in Table 3.31. This corresponds to four-bed PSA schemes or to SMB schemes.

Observe that the above examined cycling operations schedules are in fact designs of experiments, DOE. Table 3.31 contains a Latin square. Running DOE based scheme allows fast identification of significant data for separation regime improvement.

The scheme from Table 3.31 is resulting by C(4) calculations and it is in the same time a 4x4 Latin-square. The factors are time steps, beds and operations. The time is multiple of the same time-step denoted by $\Delta$.

Standard DOE table may be developed by indicating the conditions associated to any element of the 4x4 Latin square (Table 3.32).

Table 3.31 Four beds configuration

| Z\T | 0 | 1 | 2 | 3 |
|---|---|---|---|---|
| #0 | O0 | O1 | O2 | O3 |
| #3 | O1 | O2 | O3 | O0 |
| #2 | O2 | O3 | O0 | O1 |
| #1 | O3 | O0 | O1 | O2 |

**Table 3.32** DOE associated to four-beds configuration

| Exp | Bed | Time Step | Operation |
|---|---|---|---|
| 1 | #0 | 0 | O0 |
| 2 | #0 | 1 | O1 |
| 3 | #0 | 2 | O2 |
| 4 | #0 | 3 | O3 |
| 5 | #3 | 0 | O1 |
| 6 | #3 | 1 | O2 |
| 7 | #3 | 2 | O3 |
| 8 | #3 | 3 | O0 |
| 9 | #2 | 0 | O2 |
| 10 | #2 | 1 | O3 |
| 11 | #2 | 2 | O0 |
| 12 | **#2** | **3** | **O1** |
| 13 | #1 | 0 | O3 |
| 14 | #1 | 1 | O0 |
| 15 | #1 | 2 | O1 |
| 16 | #1 | 3 | O2 |

Experimental results of DOE application may be the interesting component exit concentration, the efficiency, the resolution, and so forth. Typical control tasks in separations are to keep a minimum purity of the product, to obtain the highest throughput bed, highest efficiency or to reduce solvent consumption.

The DOE selects the significant results and also the significant factors by standard ANOVA calculations. This is in fact Fourier analysis over the real field, for the device functioning parameters. The evaluation of performances may be based on real data or on real-field dynamical model of the process.

Next step will be to reorganize the scheme or to reproduce the experiment in the direction of beneficial results. The new experiment means a new DOE based on modulo-m algebra calculation and the wave model. Physically this means to generate a new, modified separation scheme. This scheme may be one with a different number of beds or a device with the same number of beds but with modified parameters.

Suppose for instance that the experiment bolded and underlined in Table 3.32 gives the worst result (bed #2, at the third step 3, lumped operation "O1"). Suppose that O1 is the operation offering the product and that bed factor is the only significant factor. In that case the bed may be changed with a modified one, possibly from the same array of beds. The separation scheme is supposed to be redundant. If all factors are significant, an improvement strategy may consists in the modification of the bed (#2 by #2'), of the time step, 3 by 3', of the operation O1 by O1' (for instance by modifying the O1

**Table 3.33** Modified configuration

| Z\T | 0 | 1 | 2 | 3' | 4 |
|---|---|---|---|---|---|
| #0 | O0 | O1' | O2 | O3 | O4 |
| #4 | O1' | O2 | O3 | O4 | O0 |
| #3 | O2 | O3 | O4 | O0 | O1' |
| #2' | O3 | O4 | O0 | O1' | O2 |
| #1 | O4 | O0 | O1' | O2 | O3 |

flow rate, and temperature) followed by the introduction of a new operation O4 in a new bed #4 (Table 3.33).

It is a method similar to stochastic approximation procedures in which as a difference the hierarchy of factors is taken into account. This is less-time consuming that the usual trials and errors procedures applied in cycling procedure start-up and optimization.

Schemes as presented in Table 3.31 or Table 3.33 are solutions of the WE. The separation scheme evolution appears in fact as a continuous oscillation between the finite-field generated DOE and the real field evaluations of the resulting data. The complementarity or disjoint-ness between the finite-field scheme and, the real field data represents the key mechanism for evolvable separation.

Elements of evolvability have been implemented in some of existing cycling operation systems. Conventional systems keep the liquid velocities constant during a step and then switch the inlet/outlet streams at the same time. Challenging conventional schemes for SMB, the Power-Feed systems, allows the fluid velocities to become time-variant. In Modicon SMB variant the feed concentration is varied in time during one switching period (Zhang et al. 2004). Separation schemes in which the duration of some steps is higher than that of other steps are frequently encountered. To the separation step receiving more time than others, a total time is assigned to equal the total time assigned to some faster steps that may be considered as lumped together. It is the case of 3 beds and 7 steps scheme of Smith and Westerberg (1990), of 4 or 5 bed schemes for PSA as presented by Chiang (1988). Asynchronous, periodic cyclic operations have been presented also by Ludemann-Hombourger et al. (2000). Asynchronous SMB process may start as conventional synchronous ones and then may modify the periods of stages by the above described DOE and stochastic approximation procedures.

The separation scheme generation is a discrete digital process governed by the wave model. It is a lap of time necessary to the real separation system to evolve according to the last established separation scheme before the scheme may change again. In synchronous designs the timing is done by a global clock and global time step whereas for asynchronous designs more local co-ordination is required. Synchronous designs indicate insensitivity to variations and inflexibility. Evolvable schemes function best when they combine regularity and randomness in appropriate measure.

#### 3.2.3.2 Self-configuring Arrays

This type of arrays represents an application of the categorical coproduct, "∪".

The categorical coproduct "∪" involves replication of existing schemes and coordinating them as a larger structure.

Making use of the same methods to solve WE but replacing the categorical product by the categorical coproduct it result identical cells or modules: $y_0 = y_1 = y_2 \ldots$ and the solution $Y = y_0 \cup y_1 \cup y_2 \ldots$

The resulting scheme as presented in Table 3.34 corresponds to an array of similar SMB cells. The four steps of any cell are F-feed, E-extract, D-desorbent, and R-raffinate. FEDR rectangles or loops define single cells.

The beds are similar and may be able to run for any of the four steps, but the coupling of cells may be variable. After each step, the feed, F for the next SMB cell may be the extract E or the raffinate R.

In quest of evolvability, transition from R↔F connection, to E↔F connection of two cells may be triggered by the presence of the component of interest in R or E. The output of any SMB cell can be configured to be driven by its output or by signal arriving from a central valve at R or E ports. The overall behavioral effects coming from scheme physical construction are implicitly taken into consideration. Non-uniformity aspects in bed construction or functioning will not be ignored in such schemes. Removing accidental constraints, the scheme evolution is relied on to find a system with expected overall behavior. Nearest–neighbor interconnections between cells were enabled in the scheme from Table 3.28. However multiple scales of SMB cells may be considered in 3-D space. The 3-D scheme organized as Latin cubes based on orthogonal Latins offer interesting suggestions for high compactness separation devices (Denes and Kedwell 1974).

**Table 3.34** Self–configuring array

| | | | | | | | | |
|---|---|---|---|---|---|---|---|---|
| | E | — | D | | E | — | D | |
| | \| | | \| | | \| | | \| | |
| | F | — | R | ↔ | F | — | R | |
| | ↕ | | | | | | ↕ | |
| | E | — | D | | R | — | F | |
| | \| | | \| | | \| | | \| | |
| ↔ | F | — | R | | D | — | E | ↔ |

For 2-D or 3-D schemes, it is possible that there appear recurrent connection paths through the scheme by which a SMB cell output can indirectly affect its own input.

The successive separation operations should take place with a timing that ensures the cyclic functioning at the next scale. Along the spatial structure that is the modularity, the temporal structure that is the synchronization should be evaluated.

Evolution is allowed to exploit the capability of the scheme freely. The strongly interactive multi-component mixtures may be forced to explore their space of possible scheme and may create new unexpected patterns of interconnections. A fitness score may be assigned to the degree to which the output approximates the desired separation degree behavior. The fitness would show by discontinuities when new configurations of coupled SMB cells appear.

#### 3.2.3.3 Multi-scale Evolvable Cyclic Separation Arrays

A general scheme for a three scale evolvable separation scheme is presented in Table 3.35. The beds run at different time scales. The first scale considered as conditioning level m=0 includes the operations O0, O1, O2 and O3. They are represented here as a square or a cycle. Any of these first scale operations may split in another sub-set of cyclic operations, for example O00, O01, O02, and O03 at the second scale indexed by m=1. Then at the third scale, m=2, operations as O00 may split in O000, O001, O002, and O003. In this way the system may pass from 4 stages at the first scale to 16 at the second scale and to 64 stages and beds at the third scale.

Such multi-scale arrays may be obtained as solutions of the WE by the method indicated for cyclic operations coupling. Making use of the same method as in equation 3.34 to equation 3.39, it results identical cells or modules: $y_0 = y_1 = y_2$ and the solution $Y = y_0 \times y_1 \times y_2$.

The m=0, m=1 and m=3 the model solutions, are vectors that shown in Table 3.11 for fixed time, say T=0. Considering that $y_0$ is represented by the square O0O1O2O3, that $y_0 = y_1 = y_2$ and making use of the categorical product $Y(T) = y_0 \times y_1 \times y_2$, the multi-scale array as shown in Table 3.35 results. It shows the status of separation at a specified moment T.

Different scales in array are interconnected. To illustrate the array from Table 3.35, consider the example of SMB in which the first scale SMB operations are O0=F, O1=E, O2=D, and O3=R. In this case O10, O11, O12, and O13 represent the SMB steps in a secondary SMB receiving the extract E=O1 from the first scale SMB as a feed F in the operation O10. At the next scale the extract from the second scale O11 may be the feed at the third scale in the operation denoted by O110.

A more realistic evolvable SMB scheme is shown in Table 3.36. There are three scale illustrated by letters of different types. Scale m=0 corresponds to bolded and underlined letters, <u>**F**</u>, <u>**E**</u>, <u>**D**</u>, and <u>**R**</u>. Scale m=1 correspond to

**Table 3.35** Multi-scale evolvable configuration

| | | | | | | | | | | | | | | |
|---|---|---|---|---|---|---|---|---|---|---|---|---|---|---|
| O111 | | O112 | | O121 | | O122 | | O211 | | O212 | | O221 | | O222 |
| | O11 | | | | O12 | | | | O21 | | | | O22 | |
| O110 | | O113 | | O120 | | O123 | | O210 | | O212 | | O220 | | O223 |
| | | | O1 | | | | | | | | O2 | | | |
| O101 | | O102 | | O131 | | O132 | | O201 | | O202 | | O231 | | O232 |
| | O10 | | | | O13 | | | | O20 | | | | O23 | |
| O100 | | O103 | | O130 | | O133 | | O200 | | O203 | | O230 | | O233 |
| | | | | | | | | | | | | | | |
| O011 | | O012 | | O021 | | O022 | | O311 | | O312 | | O321 | | O322 |
| | O01 | | | | O02 | | | | O31 | | | | O32 | |
| O010 | | O013 | | O020 | | O023 | | O310 | | O313 | | O320 | | O323 |
| | | | O0 | | | | | | | | O3 | | | |
| O001 | | O002 | | O031 | | O032 | | O301 | | O302 | | O330 | | O331 |
| | O00 | | | | O03 | | | | O30 | | | | O33 | |
| O000 | | O003 | | O030 | | O033 | | O300 | | O303 | | O332 | | O333 |

capital letters as F, E, D, R while scale m=2 to small letters, f, e, d, r. The transition from one scale to another is possible by modifying the inputs and outputs in columns.

A small column f, e, d, r, with 4 compartments may become an F column if the input is the same feed F in all 4 compartments. The F, E, D, R may be lumped as a single column E̲. The lumping and the splitting are guided by sensors.

The multi-scale scheme is characterized by high parallelism and redundancy. It contains 64 beds instead of typically 4. Redundancy appears as one of the necessary conditions for evolvability. In multi-scale arrays the evolvability may be ensured by replication of the separation 4-bed scheme triggered by the fitness or control criteria.

The transition from a scale to another may be triggered by the presence of components of interest in the expected product, by specific shapes of the recorded signals, and so forth. The higher-scale separation operations should take place with a timing that ensures and support the cyclic functioning at the previous scale. The separation scheme should have an adjustable or unsettled construction since it contains interacting modules subjected to continuous reorganization after confronting the reality. It is not only a spatial scheme but a temporal one as well.

Due to the size of search space the multi-scale scheme is confronted to the apparent improbability of chance to produce any successful solution of the separation problem. But in fact the separation trajectory in this scheme is not a blind search. The multi-scale scheme allows modifying the search domain and the search velocity by adding more scales to the search process.

**Table 3.36** Evolvable SMB configurations

| | | | | | | | | | | |
|---|---|---|---|---|---|---|---|---|---|---|
| | | D | | | | | | D | | |
| | | | | | | | | | | |
| E | | E | | R | ↔ | E | | D | | R |
| | | | | | | | | | | |
| | | F | | | | | | F | | |
| | | ↕ | | | | | | ↕ | | |
| | | D | | | | | | D | | |
| | | | | | | | | | | |
| E | | F | | R | ↔ | E | | R | | R |
| | | | | | | | | | | |
| | | e d f r | | | | | | F | | |

Any new separation scale appears as adding more sensors to the system. Moreover, interaction with real data accelerates discovering new separation trajectories.

### 3.2.3.4 Schema for Multiple Scales

The elements of the SKUP for multi-scale scheme will be presented in what follows. This is of help for separation schemes design and processes visualization.

A section of the general Table 3.35, illustrating PSM structure at two scales only, m=0 and m=1 will be considered. The SKUP elements are:

$S = (s^0, s^1); K = (k^0, k^1); U = (u^0, u^1); P = (p^0, p^1)$

The scheme includes at the first scale m=0 the operations O0, O1, O2, O3 and at the second scale, m=1, the operations O30, O31, O32 and O33. In this particular case, the second scale is resulting by a separate re-cycle processing after operation O3.

Table 3.37 includes the conditions K and the real valued states S. The conditions at the scale m=0 are O0, O1, O2 and O3 are. Let O0=$k_0^0$, O1=$k_1^0$, O2=$k_2^0$ and O3=$k_3^0$. The upper index refers to scale while the lower index refers to the time step. Time steps at different scales are different. The states and the conditions at the level m=0 are represented by high thickness border cells.

The system initial state is $s_0^0$. With possibility $p^0(k_0^0|s_0^0)$ the condition $k_0^0$ is selected. This is a digit symbolizing a specific operation O0. Based on this, the operator $u^0(k_0^0, s_0^0) = s_1^0$ allows the transition to the new state $s_1^0$. Then with possibility $p^0(k_1^0| s_1^0)$ the new condition, $k_1^0$ arises. This condition symbolized by a digit corresponds to the selection of O1. In the new condition the operator $u^0(k_1^0, s_1^0) = s_2^0$ allows the system reach the state $s_2^0$. With possibility $p^0(k_2^0|s_2^0)$ the operation, $k_2^0$ that is O2, is selected and finally the product $u^0(k_2^0, s_2^0) = s_3^0$ results. It will be operated at the scale m=0 in the condition O3 denoted by $k_3^0$.

**Table 3.37** Example of schema with two-scales

| | | | | | | | | |
|---|---|---|---|---|---|---|---|---|
| | | | | $s_2^0$ | | | | |
| | | | | | | | | |
| | | O1 | | | | O2 | | |
| | | | | | | | | |
| $s_1^0$ | | | | | | $s_2^1$ | | $s_3^0$ |
| | | | | | O31 | | O32 | |
| | | O0 | | $s_1^1$ | | O3 | | $s_3^1$ |
| | | | | | O30 | | O33 | |
| | | | | $s_0^0$ | | $s_0^1$ | | |

Then the state is $s_4^0$. The states at the scale m=0 are represented by the square: $s_0^0$, $s_1^0$, $s_2^0$, $s_3^0$. The conditions at the scale m=0 are represented by the square O0, O1, O2, O3 that is: $k_0^0$, $k_1^0$, $k_2^0$, $k_3^0$.

The states at the scale m=0 are: $s_0^0$, $s_1^0$, $s_2^0$, $s_3^0$. The interpretation of the high thickness trajectory is the process description as follows: from the state $s_0^0$ through condition $k_0^0$ towards the state $s_1^0$, then through condition $k_1^0$ towards the state $s_2^0$, and so on.

Piaget (Piaget 1971) uses to represents this kind of circular schema by cycles as: $(s_0^0,k_0^0)\rightarrow(s_1^0, k_1^0)\rightarrow(s_2^0, k_2^0)\rightarrow(s_3^0, k_3^0)\rightarrow(s_0^0, k_0^0)$.

Correspondence with the categorical frames is of interest too.

If experiments shows that O3 is the critical operation, the separation may be continued at the scale m=1 for the operations O30=$k_0^1$, O31=$k_1^1$, O32=$k_2^1$ and O33=$k_3^1$.

The states and the conditions at the level m=1 are represented by medium thickness border cells.

The system initial state at the scale m=1 is $s^1_0$. With possibility $p^1(k^1_0|s^1_0)$ the condition $k^1_0$ arises. This is a digit symbolizing a specific operation. Based on this the operator $u^1(k^1_0, s^1_0) = s^1_1$ describes the transition to the new state $s^1_1$ and so on. Each condition supposes the selection of other condition for operations.

The states at the scale m=1 are represented by the square: $s^1_0$, $s^1_1$, $s^1_2$, $s^1_3$. The conditions at the scale m=0 are represented by the square O30, O31, O32, O33 that is: $k^1_0$, $k^1_1$, $k^1_2$, $k^1_3$.

The potentialities are defined by vectors as $P = (p^0, p^1)$. The component p $(k^m)$ is an evaluation of the condition $k^m$. An example of evaluation is to take p $(k^m)$ equal to 0 or 1. The value zero corresponds to situation in which that condition is ineffective while the value 1, corresponds to active conditions.

Transfer between scales may be controlled by external criteria.

K elements, representing the symbolic conditions indicating the types of operations at two scales are in fact cyclic separation schemes. S appears as sequences of more or less separated mixtures. Operators U characterize the capability to pass from intended conditions of separation to the reality of separation steps. The possibility P describes the capability of states S to reactivate the separation scheme and to modify the symbolic K description, that is, the separation scheme elements. P shows the activation of some areas of the operations shown in Table 3.37 and the inactivation of others.

## *3.2.4 Perspectives*

### 3.2.4.1 Evolvable Separation Schemes and n-Graphs

The prospective multi-scale scheme shown in Table 3 can be studied in terms of n-graphs (Appendix A5).

The n-graphs are computing tools able to describe asynchronous systems with multiple entrances and exits. Asynchrony allows accelerate processing.

The level n=0 is represented by the 0-graphs or sets. In real separation systems this may be associated to the unit separation devices, columns or beds (Fig. 3.10). They are denoted by f-feed, e-extract, d-desorbent, and r-raffinate.

The $1^{st}$ order evolutionary step corresponds to the transition from sets to the reality level n=1, that is to the 1-graphs. These are directed graphs describing the morphisms that is, relations or connections between beds.

The morphisms are 1-cells. The coupling in the appropriate order allows completing an elementary separation process. This level is associated to 1-categories.

A $2^{nd}$ order evolutionary step delineates the level n=2 corresponding to the 2-graphs. These are graphs plus the so-called 2-cells between paths of same source and target. The 2-cells express the natural grouping of the quadruple elements f, e, d, r beds in just one large bed denoted here by F-feed, or E-extract, or D-desorbent, or R-raffinate, according to its grouped role.

The two compositions of 2-cells, vertical and horizontal correspond to the sequential and to parallel 2-cells. This level, n=2, is associated to 2-categories.

The $3^{rd}$ order evolutionary step outlines the level n=3 corresponding to the 3-graphs.

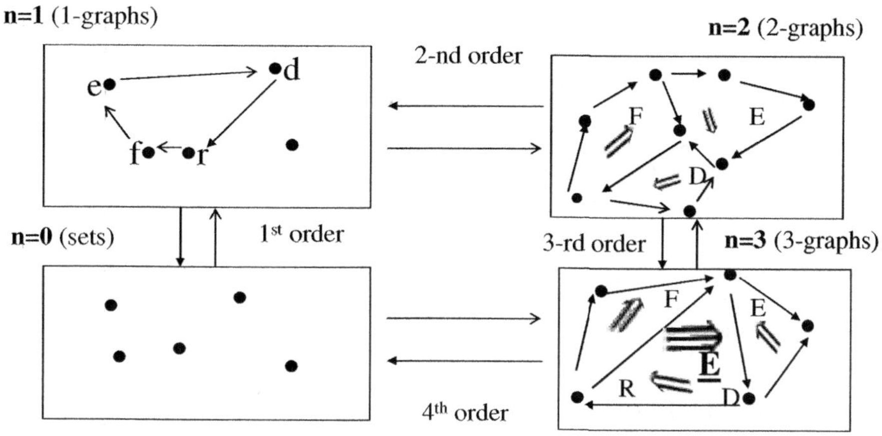


**Fig. 3.10** n-graphs for multiple scales SMB configurations

These are 2-graphs that include 3-cells that is, the cells between the 2-cells. The example shown in Fig. 3.10 corresponds to a complete lumping an extract E. The extract role is assumed to the entire installation.

The 3-graphs represent 2-graphs modification and should be subjected to conditions of natural transformations too. This level is associated to 3-categories.

The higher levels construction, at n=1 n=2 or n=3 following $1^{st}$ order, $2^{nd}$ order and $3^{rd}$ order evolutionary steps should take into account that if the entire system is to be flexible, its constituents need to be far more flexible. This corresponds to the enriching associated to categorification steps (Appendix A4).

Katis et al. (1997) defined symmetric monoidal bicategories with feedback as appropriate modeling frameworks for some concurrent and distribute processes as those shown in Fig. 3.10. In this case the objects in category are the separation devices, beds or columns. Their interconnections represent the relations. The composition should verify axioms concerning associativity, unit and symmetry. In bicategories the objects are 0-cells, the relations between objects, the1-cells and the relations between relations the 2-cells.

There are two different compositions of 2-cells, the vertical and the horizontal.

The vertical composition corresponds to sequential composition, while the horizontal composition corresponds to parallel composition.

It is interesting to observe that the compositions of 2-cells correspond to the two broad classes for integration frameworks in multi-scale systems modeling (Ingram et al. 2004).

The main frameworks are:

- Decoupled frameworks, sequential, are those where one partial model is solved with the data generated by this model used as an input to its integrated models, which is in turn solved.
- Interactive frameworks, parallel, involve the simultaneous solution of the constituent partial models.

These two frameworks are involved in both process operations and modeling.

The bicategories frame allows the theoretical study of process of processes.

A tricategorical approach would be of interest for supplementary understanding of multi-scale processing and modeling.

Interconnecting inputs from different levels of the n-graph and the integrative closure by a $4^{th}$ order step may represent the next generation of configurations.

### 3.2.4.2 Evolvable Control

Current methods of optimal control require computational power which increases exponentially with the state space, with the number of scales.

One potential approach for addressing the issue is to develop weaker but more effective framework than optimal control. For instance, control may seek for evolvability and resilience, or the maintenance of some critical functional properties without demanding the conventional objective of optimal control which is to maximize a function.

For complexity conditions, the evolvability rather than optimization may be a suitable measure of the system potential to be under control. In numerous practical situations, it is impossible to achieve the optimum due to the restrictions for time and resources.

Evolvable systems are those that are more easily modified in accordance to shifting demands and restrictions of the changing environment. Complex systems should be grounded on evolvability rather than optimization criteria. The logic of optimal solutions is often replaced by alternative criteria, such as the level of confidence that options may attract their feasibility, and contributions for the overall evolvability of the system. A change from optimal control to complex multi-scale objectives is required by complexity.

The relation between evolvability and control is obvious. Evolvable arrays may be part of the answer to the difficult control problem for cycling operations in complexity conditions. To ensure evolvability, the control should be evolvable too. It is a need for total multi-scale control of the processes to ensure selectivity and productivity (Charpentier 2007).

Evolvable control is based on the complementarity of "upward" and "downward" control models (Pattee 1997, 2000). The control of cyclic operations is a challenging task due to inherently mixed, discrete and continuous nature of separation operations, due to multiple time scales inducing instability to disturbances and long delays in exhibiting disturbances effects. The need of both "upward" and "downward" control models to ensure evolvable control modeling should be emphasized.

The functionality of the logical or discrete part of the control which incorporates the start–up, the shutdown, the switching policy and the scheme modification represents the "upward" discrete control model. This is significant for the correct operation of the separation array. It affects the separating scheme set-up and allows in an economic way, major improvements of the separation process.

The "upward" control implies that the switching of the inlet or outlet ports for separation beds should be based on observed states.

The "upward" control is sequential and technically easy to model. The model is based on the first-order WE. This model generates the separation schemes. The associated DOE matrices for example Latin squares, represents that kind of cyclic memory space that should be completely crossed over, in a recursive manner, for any new "upward" step. The "upward" control model explains how control can be inherited. The "upward", logical control is heritable and easy to transmit since it may be stored in restricted memory space.

The operators U, from the SKUP, are associated to the "upward" control model.

Some of the implemented SMB valve designs involve "upward" control capabilities Illustrative examples include SMB with several zones, Japan Organo processes, fast startup and shutdown, online decoupled regenerations (Chin and Wang 2004).

Complementing the "upward" discrete control model, the "downward" control model is working with continuous parameters. The appropriate "downward" dynamical control based for instance on solvent flow rates, concentration or temperature measurements may contribute to the closure and finally the entire separation scheme evolvability.

The "downward" control is local, diffuse and parallel. In numerous cases it may be more complicated than the "upward" control. The possibilities P, from the SKUP are associated to the "downward" control model. The "downward" control is not stored in memory but is part of the dynamical mass, heat and momentum transfer processes. The dynamical control model should shows how the large number of component being under logical control can be integrated in the course of process development.

Neither control model, logical or dynamic has much explanatory and predicting value without the other. However, each model considered alone can account for a limited scale of control. There is more to be gained by logical model when the device dynamical model is not well known, since too many parameters are required.

Observe that some conventional control model contains "upward" and "downward" control steps. Conventional control techniques are dynamical models based and focused on the feed-back loop and adaptability. Typically control systems contain controlled, controller, action and perception elements corresponding respectively to s, k, u and p in the general learning or RSCC model (Sect. 2.1.1). The conventional control consists in registering and compensating the deviation from the goal or model. The feedback loop issue having its source in the closure concept is considered also in SKUP schema, but the transformations within the loop, the loop structuring, in many scales are the main objectives here. These multiple scale transformations allow accelerating controls and make the control evolvable. In conventional control, the goals formation that is, for the particular case studied here, the separation schemes set-up is external to the loop. The separation scheme is established too early and is fixed in most cases. Modifications of schemes, if performed, are too late. If the separation scheme configuration is included in the loop, the control may become evolvable, as the controlled system. For instance, in the multi-scale separation scheme, each operation of the scheme is a source of scheme modification by redundancy and a generator of adjacent separation scales, mediated by the U operator and by the wave model, WE. At each scale the separation scheme takes into account results obtained at other scales. A hierarchy of loops is resulting. In this way the separation scheme is not pre-determined and may evolve as its control too. The evolvable control is locally done on the basis of empirical evidence not by reference to a dynamical model.

The existence of several scales in separation scheme, avoid mapping back the results to control structure, too late or too detailed. The domain of results of conventional controls is that of physically measured, real numbers. This numbers serves as inputs for separation scheme modification. This is not the case in real biological control systems where parameter measurement is replaced by pattern recognitions that restrains or favors specific metabolic ways (Pattee 1997). The suggestion from real biological control systems is to use pattern recognition as a fast method to select the acceptable schemes. In SMB technology the chromatograms may be transposed in lookup digital tables adequate to quick pattern recognition. The similarity with a known as good chromatogram may guide the separation scheme evolution.

The technology challenge is to effectively build entirely evolvable separation schemes based on complementarity and continuous back and forth between the wave equation, WE model solutions, that is, a specified separation scheme, and the physical data of separation process itself.

For this, the evolvable device should be able to face hard timing restrictions with respect to measurement analysis. The testing of evolvability capability for such a device would include the accelerated managing of the device start-up and the large disturbance rejection. The uses of evolvable devices and evolvable control lie primarily in exploring beyond the scope of conventional devices and conventional adaptive controls not competing with them. The main concerns are related to robustness, results analysis and scalability.

A number of useful areas of applications have been identified for evolvable separation technologies. Evolvable schemes are necessary in pharmaceutical and food industry, in oil chemistry, in mineral conversion when the mineral composition varies, in air-drying, in hydrogen purification from natural sources, in saline water purification, and so forth.

Evolvable separations schemes offer the prospect of devices to suit a particular individual. Evolvable technologies have the potential to adapt autonomously to changes in their environment. This could be useful for situations where real-time control over systems is not possible such as for space applications. Evolvable control systems are required in such cases. Evolvable separation devices may be of help in the study of crucial concepts as self-repair and self-development.

The organization of the evolvable technology may be considered in the four realms categorical framework from Fig. 3.11 The meta-model or meta-meta-model are supported by the wave equation WE.

Particular solutions are models and they may correspond to implemented flow-sheet schema. In contact with the real environment situation the meta-model application may be changed.

U10, U21, U32 and U30 expresses the decisions that is, the "upward" control, while P01, P12, P23 P03 characterizes "downward" control based on the information coming from real word. This could be mediated but also should be direct as P03 indicates.

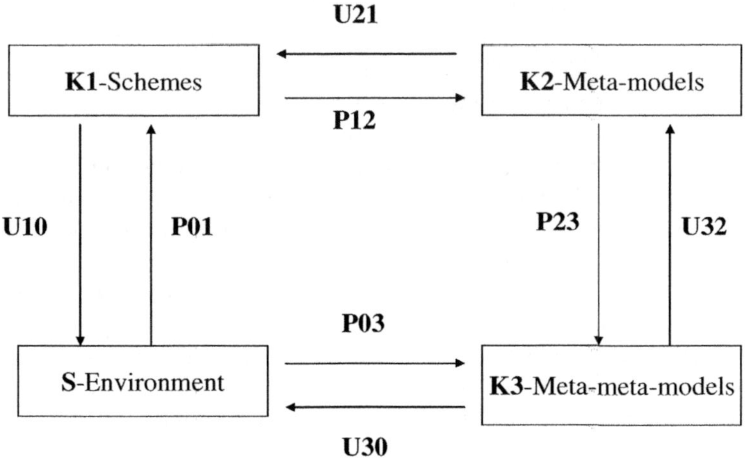


**Fig. 3.11** Evolvable architecture for control

The signification of the functors U and possibilities P is explicit in Fig. 3.11.

U10: K1-Schemes→S-Environment, U21: K2-Meta-models→K1-Schemes, U32: K3-Meta-meta-models→K2-Meta-models, and U30: K3-Meta-meta-models→ S-Environment.

P01: S-Environment → K1-Schemes, P12: K1-Schemes→K2-Meta-models,

P23:K2-Meta-odels→K3-Meta-meta-models, and P03: S-Environment →K3-Meta-meta-models.

Observe that: U30=U10oU21oU32 and P03=P01oP12oP23.

The interaction between S-Environment and K3-Meta-meta-models allow integrative closure and may make the system evolvable and autonomous.

The implementation of autonomic computing concept offers preliminary suggestions for the study of integrative closure problem in evolvable control (Ganek and Corbi, 2003).

At the heart of an autonomic system is a control system, which is a combination of components that act together to maintain actual system attribute close to desired specifications.

An autonomic system embodies one or more closed loops. The autonomic computing system can be modeled in terms of two main control loops, local and global, with sensors for self-monitoring, effectors for self-adjustment, knowledge and planer/adapter for exploiting policies based on self- and environment awareness.

## 3.3 Drug Delivery Systems

### *3.3.1 Complexity of Drug Delivery*

Recent advances in genomics and in the study of complex diseases have shown the necessity for an alternative way of thinking in medicine and pharmacology, a view in which pathology and physiology results from interactions between many processes at different scales.

The new scientific field of systems biology is correlated to this perspective. It focuses on the study of genes, proteins, biochemical reaction networks and cell populations considered as dynamical systems. PSM provides a conceptual framework and effective tools for studying emergent and immergent features from molecules to organisms and contrary.

The method by which a drug is delivered to organism has a significant effect on its efficacy. The traditional approach is to develop drug delivery systems, DDS, which will release the drug at constant rate, to ensure a required concentration level. The drug application method, that may be oral, parenteral, inhalation, transdermal and so on, determines the drug distribution in organism.

This traditional approach is not always suitable or effective. Drugs may induce large metabolic degradation, may lead to adverse effects, may saturate the organisms or may decrease in time the bioavailability. The traditional drug application approach neglects the variability of drug effects and of the organism requirements. Additionally, some of the traditional approaches are associated with the very slow progress in the efficacy of the treatment of complex or severe diseases.

The reason of low efficiency of conventional DDS resides in the complexity of drug-organism interaction process and the complexity of some diseases.

It was observed that most organisms are not passive in their requirements or response to drugs. Organisms manifests as dynamic systems which require different amounts of drug at different times, at different sites for different patients. The biological processes and functions are organized in time, as a specific biological time and space structure, as revealed by physiology and anatomy studies. The time structure is characterized by multiple biological rhythms, while the spatial distribution is characterized by multiple scale heterogeneity. For example, only 2cm $\times$ 2cm of human skin area contains as an average 10 blood vessels, 300 sweat gland, 100 sensors (pain, temperature, and pressure) and 10000 nerve endings. Organisms always involve a wide range of scales both in time (femtoseconds for chemical reactions, seconds for metabolism processes, day to months for cells and years in living organisms) and space (nanometers for molecular structures, micrometers for organelles and cells, centimetres for tissues and organs and meters for organisms).

The drug delivery systems should take into account both the temporal and spatial heterogeneity of the organisms. Despite the fact that some diseases and drug delivery-organism interactions are recognized as highly complex, diseases treatments continue to be implemented by drug delivery regimes and procedures were complexity properties are neglected.

The consideration of different scales, from the macroscopic scale to the atomic and sub-atomic scales is becoming more and more important for drug delivery system design. To illustrate the multiple scales issue we may refer to the design of transdermal delivery systems (Shaeiwitz and Turton 2004).

In terms of multi-scale design, a transdermal delivery system illustrates design from the macroscopic scale through the molecular scale. At the macroscopic scale, the transdermal delivery system must be assembled and be customer friendly. At the mesoscopic scale, between macroscopic and microscopic, the pharmacokinetics of the drug should be modeled. At the colloidal scale, adhesion must be understood in order to facilitate selection of a suitable adhesive. At the microscopic scale, the mechanism of transport of the drug through the skin must be understood. The nano scale study concerns the drug mixing with enhancers and excipients. The former may alter the permeability of the external layer of skin most often acting as the limiting resistance, to facilitate drug delivery. The latter are pharmaceutically inactive ingredients such as skin moisturizers usually found in transdermal patches. A molecular-scale design illustration would be to design the drug based on an understanding of the desired pharmacology.

### 3.3.2 Developments

Innovative DDS able to establish the drug application mapping and timing, in order to be adaptive to the unavoidable variability in living organisms, and to perform an active healing, start to be developed and implemented in the last decade.

New domains as chronopharmaceutics and chronotherapy have emerged (Smolensky and Peppas, 2007). Significant research was devoted to the design and evaluation of DDS that release a therapeutic agent at a rhythm that matches the biological requirements of a given disease therapy.

Constant drug delivery may become ineffectual and needs to be replaced by a pulse of therapeutic concentration in a periodic manner. This imposed the development of modulated, pulsatile DDS (Sershen and West 2002, Anal 2007). In such systems there is rapid and transient release of a certain amount of drug molecules within a short time-period immediately after a predetermined off-release period. Various techniques are available for cyclic delivery like pH, temperature, light, magnetic field, presence or absence of a specific molecule, micro-flora and so on.

To minimize the toxicity associated with the drug delivery, the research focused to the development of modulated drug delivery systems capable of releasing therapeutic agents in response to physiological requirements. These are known as self-regulated or responsive systems.

A major class of modulated drug delivery systems is that in which the pulsatile release of drugs is triggered by external signals. These are known as externally regulated or pulsatile delivery systems. Developments in the field of pulsatile systems based on external triggers such as electrical, ultrasound, magnetic and mechanical have been reported (Sershen and West 2002). Research was done in the field of polymer based temperature-sensitive, pH-responsive, inflammation-responsive and glucose-and other saccharide-sensitive systems; enzyme-based urea-responsive, glucose-responsive and morphine-triggered systems; and systems based on antibody interactions.

Hydrogels are three-dimensional, hydrophilic polymer networks capable of imbibing large amounts of water or biological fluids. The networks are composed of homopolymers or copolymers, and are insoluble due to the presence of crosslinks such as entanglements or crystallites. Hydrogels are capable to swell in aqueous media. They are used to regulate drug release in reservoir-based, controlled release systems or as carriers in swellable and swelling-controlled release devices. Hydrogels as stimuli-sensitive gel systems modulated release in response to pH, temperature, ionic strength, electric field or specific analyte concentration differences (Murdan 2003, Peppas and Leobandung 2004).

In such DDS, release can be designed to occur within specific areas of the body.

The skin patches are useful for drug delivery by iontophoresis. Iontophoresis is a technique used to enhance the absorption of drugs across biological tissues such as skin (Dixit et al., 2007). Patches are either impregnated with drugs or some versions can be used as dispensers of the drugs (Akimoto et al. 2006, Murdan 2003).

To confront the complexity of drug delivery and to go through the frontier of complexity evolvable drug delivery systems, EDDS should be implemented and evaluated.

This approach adopts the point of view according to which the drug delivery system cannot be completely pre-programmed but would be actively constructed by interaction with the organism.

## *3.3.3 Evolvable Systems*

### 3.3.3.1 Scheduling Drug Delivery

Robustness is an important organizing principle of biological systems. Complex diseases appear to be robust systems with some points of fragility. Thus, finding treatments for complex diseases may consists in determining the fragility points of a robust system.

EDDS are designed to be able to release the right amount of drug at the right time and place. EDDS would enable a pulsed release exactly when the chemical or bioactive agent is needed. This means targeting drug delivery both in time as in space.

The EDDS may be useful in completing the biosystems rhythmic release of many endogenous chemicals such as insulin, growth hormones, and so on. It is expected that taking into account the biosystems rhythms will result in enhanced drug treatment of complex diseases as diabetes, dwarfism, autoimmunity or infectious diseases and so on.

Let us consider as an illustrative case the pulsatile or cyclic DDS in iontophoresis. Drug delivery scheduling is of interest. The scheduling is based on the connection with the theory of Latin squares and the designs of experiments (Iordache 2009).

Latin squares resulted as solutions of the wave equation, WE (Sect. 2.2.2).

It is known that the persistent use of a direct current, proportional to time can reduce the iontophoresis flux because of polarisation effect on the skin. This can be overcome by the use of pulsed direct current. It may be a periodical “on”, “off” application of the current. During the “off” stage the skin get depolarized and returns to the initial state.

The analysis of the skin permeation suggests a more complex situation (Akimoto et al., 2006). Transdermal transport of drug molecules during iontophoresis consists of four main phases: absorption, equilibrium, desorption and passive diffusion. These four phases may be indexed by 0, 1, 2, and 3 respectively.

The problem is how to organize and iontophoresis treatment to have any time at least one site in adsorption stage. WE offer a natural solution of this problem (Sect. 3.2.2).

The initial condition for WE means that at the initial degree of separation T=0, the output Y in the Z site is exactly Z. From physical point of view this means, for example, that if there are four drug delivery sites and four drug delivery possible stages, "0"-absorption, A, "1"-equilibrium, E, "2"-desorption, D "3"-passive diffusion, P then, the initial step is as follows: site "0" in step "0", site "1" in step "1", site "2" in step "2", site "3" in step "3".

The solution of WE is shown in Table 3.38 a modified form of Table 3.11.

**Table 3.38** Wave equation solution

| Z\T | 0 | 1 | 2 | 3 |
|---|---|---|---|---|
| #0 | A | E | D | P |
| #1 | E | D | P | A |
| #2 | D | P | A | E |
| #3 | P | A | E | D |

Here Z corresponds to different application areas indexed by #0, #1, #2 and #3 in the DDS scheme. The Table 3.38 offers a description of the stages for different areas along the DDS scheme, as T increases. The characteristic is the condensed presentation of the DDS process and of the cyclic schedule. It may be of interest for continuous delivery of the drug through iontophoresis. This is important for emergence situations, for pain managing.

The connections of wave equation WE, with Latin squares may be of help for the study of configurations with variable number of drug delivery sites or stages, in the study of drug delivery operations coupling in hybrids and in the study of adjustment of drug deliver system confronted with a partial failure.

Results concerning the critical sets for back circulant Latin squares may be applied to DDS (Bate and van Rees 1999). A critical set of order n is a partial Latin square of order n such that it has a unique completion and every entry is necessary.

Table 3.39 shows a minimal critical set containing four elements.

This table suggests that the system may be restricted to two steps, T=0 and T=1 and two locations, Z=#0 and Z=#1. Only three stages are performed, adsorption-A, equilibrium-E and desorption-D while passive diffusion-P is considered as avoidable in this case.

**Table 3.39** Critical set

| Z\T | 0 | 1 | 2 | 3 |
|---|---|---|---|---|
| #0 | A | E | - | - |
| #1 | E | - | - | - |
| #2 | - | - | - | - |
| #3 | - | - | - | D |

Another application of results from the study of Latin square refers to situations when one of the steps needs more time to be performed. In such case the standard Latin square is replaced by a so called frequency square in which each element appears the same number of times in each row and each column but this number of time may be more than once.

Table 3.40 shows an example in which the passive step P was applied for 2 intervals of time.

**Table 3.40** Frequency square

| Z\T | 0 | 1 | 2 | 3 | 4 |
|---|---|---|---|---|---|
| #0 | A | E | D | P | P |
| #1 | D | P | P | A | E |
| #2 | P | A | E | D | P |
| #3 | E | D | P | P | A |
| #4 | P | P | A | E | D |

Gerechte designs may find also applicability for scheduling drug delivery (Bailey et al. 2008). Gerechte designs are specialization of the Latin squares. The n x n grid is partitioned into n regions, each containing n cells of the grid. Such a partition is called a gerechte skeleton. It is required to place the symbols 1, n into the cells of the grid in such a way that each symbol occurs once in each row, once in each column, and once in each region.

The row and columns constraint say that the solution is a Latin square, and the last constraint restrict the possible Latin squares. The purpose of the gerechte designs in experimentation is to ensure that all treatments are fairly exposed to any different conditions in the field. Rows and columns are good for capturing differences as instance distance from a specified area but not for marking out damages or other features that tend to lump in compact areas.

The gerechte skeleton may be considered as a basic application of the DDS.

A problem is to establish the conditions allowing completing the gerechte skeleton to a gerechte design.

The study of gerechte designs of type 4x4 shows that, the steps A, E, D, P of iontophoresis may be operated according to a different solution of WE, complementing that shown Table 3.38. This follows from the fact that there are only two possibilities for the group structure with four elements, the cyclic group C(4) and the Klein 4-group. The different schedule results by replacing in Table 3.38 the cyclic group C(4) by the Klein 4-group.

#### 3.3.3.2 Coupling Delivery Operations

The previously presented scheduling refers to one level, that is, to one time scale.

Cyclic drug delivery systems with evolvability based on coupled delivering systems at different conditioning levels are also of interest.

This type of analysis is important when several delivery rhythms are superposed.

This would be the case if completing the circadian rhythms there are several ultradians ones. Consider the case when 24 hours rhythms are coupled with 12 hours rhythms, for example. Circadian stages are indexed in this case by "0" or "1".

For the two-level cases the elements of WE, are: $Y=y_0y_1$, $T=t_0t_1$, $Z=z_0z_1$ and $F=f_0f_1$.

The method was developed in detail in Sect. 3.2.2.

The solution at the conditioning level m=0, referring to the circadian stage, is given by the Table 3.27 while that at the level m=1 referring to A, E. D, P stages by the Table 3.38.

The Table 3.41 shows the product of these two solutions and contains 16 doublets.

**Table 3.41** Double stages

| | | | | | | | |
|---|---|---|---|---|---|---|---|
| 0A | 0E | 0D | 0P | 1A | 1E | 1D | 1P |
| 0E | 0D | 0P | 0A | 1E | 1D | 1P | 1A |
| 0D | 0P | 0A | 0E | 1D | 1P | 1A | 1E |
| 0P | 0A | 0E | 0D | 1P | 1A | 1E | 1D |
| 1A | 1E | 1D | 1P | 0A | 0E | 0D | 0P |
| 1E | 1D | 1P | 1A | 0E | 0D | 0P | 0A |
| 1D | 1P | 1A | 1E | 0D | 0P | 0A | 0E |
| 1P | 1A | 1E | 1D | 0P | 0A | 0E | 0D |

Coupling cyclic operations with other operations, cyclic or not, in series or in parallel, receives increasing attention in drug delivery study.

Hybrid systems are comprised by two or more delivery operations of different nature that performs significantly different with constituents operating independently. In a hybrid system, the performance of any component process is affected by the performance of all other processes in the system.

Such hybrids may be based on different hydrogels. It is know that hydrogels as stimuli-sensitive gel systems modulated release in response to temperature, mechanical pressure, electric field or specific analytic concentration differences.

Coupling of different cyclic technologies as for instance: pressure, temperature, magnetic or concentration modulated separation with separation by membrane is also of interest. The multiple and superposed periodic separation fields, encountered in forced pressure modulated adsorption with multiple adsorbents, hierarchical arrays of DDS represent interesting developments and adaptations in the field of cyclic drug delivery systems.

## *3.3.4 Perspectives*

### 3.3.4.1 Patterns for Printed Patches

Remarkable potentialities for evolvable drug delivery are offered by the printed patches as for instance the skin patches based on electrophoresis.

A tiny electrical current is passed through the skin by electrodes in the patches typically powered by a small battery laminated in the patch. Some companies use low cost environmentally safe printed paper battery and printer resistor network.

It is possible to consider designed patches and to allow circuitry formation as for the Pask electrochemical device (Cariani 1993).

The Walsh matrices are particular solutions of the wave equation WE (Sect. 2.2.3). This suggests the design patterns for patches electrodes.

Table 3.42 illustrates a solution of the kinetic model in which we suppose the rate Q to be constant in the wave equation WE.

The patch areas are represented here by "o", for negative areas (-1), and "•" for positive areas (+1) in the Walsh type matrices.

The elements of the Table 3.42 may form a patch to be tested in diffusion cells.

**Table 3.42** Kinetic model for patches

| | | | |
|---|---|---|---|
| • | • | • | • |
| • | o | • | o |
| • | • | o | o |
| • | o | o | • |

The Walsh-Hadamard designs allow quick experiments (Iordache 2009).

### 3.3.4.2 Patches Development and n-Graphs

Drug delivery systems may be studied in the general framework of n-graphs.

The n-graphs are appropriate tools for asynchronous regimes modeling.

Fig. 3.12 illustrates a development cycle for patches application for transdermal drug delivery.

At n=0 the patches areas, prepared to be cathods or anods are isolated.

At n=1 interactions appear and electrical cells are resulting. The $1^{st}$ order evolutionary step runs through patch application level n=1 and corresponds to the 1-graphs.

The real phenomenon is driven by patch interaction with the skin.

Electrochemical cells are separated in the n=1 stage but they interact at the n=2 level to form arrays of interacting cells. The $2^{nd}$ order evolutionary step allows transition to the level n=2 corresponding to the 2-graphs.

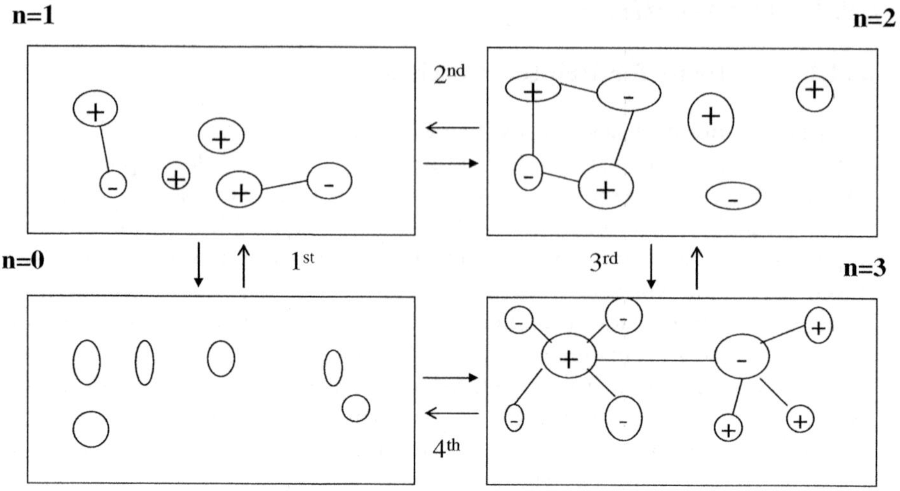


**Fig. 3.12** n-graphs for multiple scale patches

The n=2 level shows the coupling of two or more cells in frames going beyond cells areas isolation. A $3^{rd}$ order evolutionary step allows transition to the level n=3 corresponding to the 3-graphs. The level n=3 is associated to the global action of the whole patch.

The integrative closure, accomplished by a $4^{th}$ order step, connecting levels n=0 and n=3 represents a challenge for such drug delivery systems.

A potential road towards integrative closure is represented by the smart sensory systems based on organic computing (Dressler 2007) or pervasive computing (Estrin et al. 1999) implementation. Wireless sensor networks consists of several battery constrained sensor nodes equipped with sensor detection capabilities as well as some computing and storage resources. Web based architectures may assist health care.

The successive levels and the final results should take into account the biorhythms of the patient. The drug is administrated according to need rather than the time of the day.

This whole patch emerges from the component patches and the interactions with the patient. Its pattern allows identifying disease, reviewing critical cases in the perspective of self-healing. The coalescence of patches areas should be reversible.

#### 3.3.4.3 Drug Targeting and n-Graphs

The n-graph framework has a general applicability for drug delivery systems study.

To illustrate this we will consider the case study of magnetic drug targeting classifications.

Recall that the targeting refers to both time and space aspects of drug delivery.

Lübbe et al., (2001) presented a classification discriminating between $1^{st}$, $2^{nd}$ and $3^{rd}$ order targeting. These are in line with the $1^{st}$, $2^{nd}$ and $3^{rd}$ order steps in n-graphs construction as shown in Fig. 3.12 (Appendix A5).

The $1^{st}$ order refers to delivery to discrete tissues, to localization of the drug at the capillary bed of the target site, a tissue. This is associated to 1-categories.

The $2^{nd}$ order refers to drug delivery to specific cell type within tissue. This is associated to 2-categories.

The $3^{rd}$ order refers to delivery of the drug directly to specified intracellular compartments. It is based on intracellular transport of drug by natural biological processes as cell fusion, endocytosis and pinocytosis.

This is associated to 3-categories. As defined, the higher levels in n-graphs and n-categories take into account that if any system is to be flexible its constituents need to be more flexible. In other words, the $3^{rd}$ order drug targeting would be most difficult to accomplish and requires a solution of the previous $1^{st}$ and $2^{nd}$ order drug targeting.

Recent developments in the field of magnetic targeting drug delivery make use of magnetic enhanced gene therapy (Scherer 2002). The $4^{th}$ order drug delivery achieving the integrative closure and targeting for nucleic acids, is the site specific genomic integration discussed by Scherer (Scherer 2006).

## 3.4 Entropy Production

### *3.4.1 Complexity Issues*

Complexity is omnipresent in biological and cognitive systems and gains increasing importance for physicochemical technological systems.

If complex systems contradict or no the second law of thermodynamics augment or diminish the rate of entropy production towards extreme values, represent questions of high practical and theoretical interest. The second law of thermodynamics establishes that in isolated systems, entropy increases with time and approaches a maximum value. According to the theorem of minimum entropy production, MinEP, systems constrained to remain slightly away from thermodynamic equilibrium, with fixed boundary conditions, will take on the configuration of forces and flows that minimizes the rate of entropy production (Prigogine 1955, 1980).

Numerous complex systems are non-isolated and not near equilibrium. Challenging the MinEP validity for far from equilibrium and variable boundary conditions, numerous studies assert that a non-linear and non-equilibrium system develops so as to maximize entropy production, towards MaxEP, under constraints (Jaynes 1980, Swenson 1989, Dewar 2003). Observe that MaxEP is similar to the maximum entropy principle, a statistical principle.

Instead of finding the most probable state, in the MaxEP one looks for the most typical trajectory or process.

As MinEP, the MaxEP principle is not without controversy. Entropy and entropy production for cognitive system was naturally studied in terms of information theory (Swenson and Turvey 1991, Adami 2002). Numerical simulations involving decrease of informational entropy production have been noticed in the domain of cognitive systems (Kugler and Turvey 1987, Parunak and Brueckner 2001).

Non-equilibrium thermodynamics, finite-time thermodynamics results have been useful in the study of complex technological systems, to evaluate effectiveness of energy and chemical conversion in environment or in thermo-chemical devices (Dewar 2003, Salamon et al. 2001).

The entropy balance for complexity domains is presented here in the framework of polystochastic models, PSM. These allow explaining the emergent properties that make complex systems as difficult as interesting to study.

The new framework is able to explain entropy and entropy production variations and to evaluate both principles MaxEP and MinEP in the same theory.

The PSM framework outlines the possibility that the periods of increasing entropy production, are followed by periods of constant entropy production, and by that of decreasing entropy production in biosystems and in ecosystems.

The PSM framework offers simple explanations for the apparent contradictions of the second law of thermodynamics in cognitive multi-agent systems. Moreover it offers suggestions for fast process synthesis procedure for heat integration within distillation trains and for evolvable technologies.

### *3.4.2 Entropy Balance*

PSM framework suggests correlating the scheme or architecture states, K, and the processes states S.

Entropy balance for both S and K should be established.

In the real field of state space S, the entropy production p(y) is:

$$p(y) = \frac{\partial y}{\partial t} + f(y) \geq 0 \tag{3.40}$$

$$y(0) = g \tag{3.41}$$

Here y is the entropy, f(y) denotes the entropy flow through the surface of the system and g denotes the initial value of entropy (Prigogine 1955).

The entropy balance equation in K should be able to generate schemes, architectures or configurations accommodating the states as described by the entropy balance in S.

The correspondence between Boltzmann's statistical interpretation of physical entropy as disorder and Shannon's formulation of variety as informational entropy signals a deep connection between information and entropy.

To ensure the K and S models correspondence, the entropy balance in K will be similar to equation (3.40) in S but we will replace y, t, f, g with their counterparts in K, Y=$y_0y_1y_2\ldots y_j$, T=$t_0t_1t_2\ldots t_j$, F=$f_0f_1f_2\ldots f_j$, G=$g_0g_1g_2\ldots g_j$ where $y_j$, $t_j$, $f_j$ and $g_j$ are digits.

The succession of digits corresponds to the succession of scales.

In this case:

$$\frac{\partial Y}{\partial T} \oplus F \geq 0 \tag{3.42}$$

$$Y(0) = G \tag{3.43}$$

The solution in C(2) is presented here for illustration purposes (Sect. 2.2.1).

Y, T, F, G are defined in C(2). This means that we refer here to cyclic processes (Iordache et al. 1993b).

In C(2), "0", denotes the null element. The null value "0" may be interpreted as that corresponding to a "non-activated" scale while the unit value "1" corresponds to an "activated" scale of the system.

The real product and the sum were translated to C(2) operations.

The solution of the equality in (3.42) similar to Euler solution for differential equations will be:

$$Y(T) = Y(0) \oplus (F \otimes T) \tag{3.44}$$

Suppose that Y(0) = 1. In this case the solution (3.44) of the balance equation is Y(T, F) as shown in Table 3.43.

**Table 3.43** Y(T, F), Y(0)=1

| F\T | 0 | 1 |
|---|---|---|
| 0 | 1 | 1 |
| 1 | 1 | 0 |

Denote the resulting "0" by the symbol, "-1". This is related to the fact that here we refer to the logical value and we can denote "yes" by "1" and "no" by "0" or "-1".

The Table 3.44 is equivalent to Table 3.43.

**Table 3.44** Y (T, F), Y (0) =1 Equivalent form

| F\T | 0 | 1 |
|---|---|---|
| 0 | 1 | 1 |
| 1 | 1 | -1 |

Suppose now that Y, T, and F are vectors with three components: Y=$y_0y_1y_2$ T=$t_0t_1t_2$ and F=$f_0f_1f_2$. These corresponds to three scales.

The first order entropy balance reduces in fact to three equations, one for each scale:

$$\frac{\partial y_0}{\partial t_0} \oplus f_0 = 0 \quad (3.45)$$

$$y_0(0) = g_0 \quad (3.46)$$

$$\frac{\partial y_1}{\partial t_1} \oplus f_1 = 0 \quad (3.47)$$

$$y_1(0) = g_1 \quad (3.48)$$

$$\frac{\partial y_2}{\partial t_2} \oplus f_2 = 0 \quad (3.49)$$

$$y_2(0) = g_2 \quad (3.50)$$

Consider that the initial condition is:

$$Y(Z,0) = g_0 \times g_1 \times g_2 \quad (3.51)$$

Here "×" denotes the tensor product.

In this case the solution of the model is: $Y(T) = y_0 \times y_1 \times y_2$.

Moreover if $g_0=g_1=g_2=1$ then $y_0=y_1=y_2$ are 2x2 matrices as given by Table 3.44.

Table 3.44 offers the one scale solution $Y=y_0$.

Table 3.45 shows the two scales solution $Y=y_0 \times y_1$ the product of two matrices 2x2.

**Table 3.45** $Y= y_0 \times y_1$. Two scales.

| F\T | 00 | 01 | 10 | 11 |
|---|---|---|---|---|
| 00 | 1 | 1 | 1 | 1 |
| 01 | 1 | -1 | 1 | -1 |
| 10 | 1 | 1 | -1 | -1 |
| 11 | 1 | -1 | -1 | 1 |

**Table 3.46** $Y = y_0 \times y_1 \times y_2$. Three scales.

| F\T | 000 | 001 | 010 | 011 | 100 | 101 | 110 | 111 |
|---|---|---|---|---|---|---|---|---|
| 000 | 1 | 1 | 1 | 1 | 1 | 1 | 1 | 1 |
| 001 | 1 | -1 | 1 | -1 | 1 | -1 | 1 | -1 |
| 010 | 1 | 1 | -1 | -1 | 1 | 1 | -1 | -1 |
| 011 | 1 | -1 | -1 | 1 | 1 | -1 | -1 | 1 |
| 100 | 1 | 1 | 1 | 1 | -1 | -1 | -1 | -1 |
| 101 | 1 | -1 | 1 | -1 | -1 | 1 | -1 | 1 |
| 110 | 1 | 1 | -1 | -1 | -1 | -1 | 1 | 1 |
| 111 | 1 | -1 | -1 | 1 | -1 | 1 | 1 | -1 |

Table 3.46 shows the three scales solution, $Y(T) = y_0 \times y_1 \times y_2$.
Table 3.47 shows the Table 3.45 in an equivalent form.

**Table 3.47** $Y(T) = y_0 \times y_1 \times y_2$. Equivalent form.

| F\T | 000 | 001 | 010 | 011 | 100 | 101 | 110 | 111 |
|---|---|---|---|---|---|---|---|---|
| 000 | 1 | 1 | 1 | 1 | 1 | 1 | 1 | 1 |
| 001 | 1 | 0 | 1 | 0 | 1 | 0 | 1 | 0 |
| 010 | 1 | 1 | 0 | 0 | 1 | 1 | 0 | 0 |
| 011 | 1 | 0 | 0 | 1 | 1 | 0 | 0 | 1 |
| 100 | 1 | 1 | 1 | 1 | 0 | 0 | 0 | 0 |
| 101 | 1 | 0 | 1 | 0 | 0 | 1 | 0 | 1 |
| 110 | 1 | 1 | 0 | 0 | 0 | 0 | 1 | 1 |
| 111 | 1 | 0 | 0 | 1 | 0 | 1 | 1 | 0 |

The elements of the Table 3.44, 3.45 and 3.46 corresponds to well known Walsh functions (Hedayat et al. 1999). The case of two scales but with a three valued digits ("1", "0", "-1") is also of interest.

## *3.4.3 Case Studies*

### 3.4.3.1 Entropy Production for Biosystems

The MinEP theorem of Prigogine has received the status of a principle of organization for biosystems that in some conditions grow toward a state of minimum metabolism per unit mass. The validity of the MinEP theorem is evidently broken in embryogenesis where an initial increase of the heat production more in the spirit of MaxEP is observed. After initial increasing, the entropy production decreases during adult and senescent stages. There are notable exceptions from this pattern as for instance tissue regeneration of malignant growth.

Entropy production in living system consists of multi-stages with time, early increasing, later decreasing and possible with intermediate stages (Aoki 1995). According to Aoki the entropy production in plants leaves oscillates during the period of one day paralleling the daily solar absorbed by leaves. For this case of environment studies the cyclic character of T is obvious. Multi-scale structure of cyclic time may be due to day-night cycle coupled to tidal cycles or other environmental cycles.

Salthe (2003) evaluated the possibility that the periods of increasing entropy production (immature stage), are followed by periods of constant entropy production (mature stage), and by that of decreasing entropy production (senescence stage).

This kind of entropy production behavior may be easily explained in the presented here framework starting from the fact that the living systems are not isolated and not nearly equilibrium.

The entropy results from informational entropy in K coupled with dynamical entropy in S. The total entropy results from informational entropy in K coupled with dynamical entropy in S.

Observe that the entropy in K as calculated in Table 3.44 to Table 3.47 is not a monotone function as second law would require.

Depending on the values of the entropy flow F, different solutions Y(T) results. Y(T) is in this case a two-valued function but it may be considered as a monotone increasing function except in a finite number of points when the evolution starts again from zero with new increasing entropy dependence. The production of entropy varies from large values to zero in a periodic way.

Summing the entropies production in K and S, domains of non-monotonic entropy production dependences may result if the part of K is significant.

Non-monotonic entropy production dependence may describe the evolution of highly non-linear systems, far from equilibrium without fixed boundaries. It may be of interest in the study of self-healing systems, animate or not.

#### 3.4.3.2 Entropy for Bio-cognitive Systems

The fluctuating behavior of entropy as observed for some cognitive systems seems to contradict the second law of thermodynamics that captures the tendency of systems to disorder that is towards increasing entropy. The apparent contradiction was explained in terms of multiple coupled scales of dynamic activity—the Kugler-Turvey model (Kugler and Turvey 1987). Self-organization and the loss of entropy occur at the macro-scale, while the system dynamics on the micro-scale generates increasing disorder. Kugler and Turvey approach refers to rhythmic movements fitting to the cyclical time T from our model.

Significant examples are also described by Parunak and Brueckner (2001) and Gambhir et al. (2004) in the context of pheromone based coordination. Their work defines a way to measure entropy at the macro scale that is in K (agents' behaviors lead to orderly spatiotemporal patterns) and at the micro scale that is in S (chaotic diffusion of pheromone molecules). The micro scale serves as the entropy sink - it permits the overall system entropy to increase, while allowing self- organization to emerge and manifest itself as coordinated multi-agent activity on the macro scale. In summary, macro-scale metrics may capture the quality of the emergent solutions in terms of observable coordination activities, while micro-scale metrics may verify the solution in terms of the multi-agent communications.

Parunak and Brueckner (2001) simulations illustrates how cognitive behavior emerge from a simple entropy increasing behavior and also that the resulting self-organization does not defy the second law of thermodynamics since the price paid for the entropy reduction at the macro system level is the increase in entropy generated by the random process that produces and maintains the gradient.

Notice that probability of violations of the second law of thermodynamics have been predicted and verified experimentally for small systems and short time scales (Dewar 2003).

### 3.4.3.3 Distribution of Heating and Cooling Areas

Heat integration within distillation trains is a problem of engineering interest in which entropy play a significant role (Andersen 2002, Salamon et al. 2001).

How to place few thermally active stages along a distillation column is an issue that may be addressed with the help of solutions presented in Table 3.46 or Table 3.47.

Three scales for cyclic separation have been considered in this case.

They correspond to the entire distillation column, to distances between distributed heating and cooling elements and to the trays.

According to the Table 3.48 the entropy flow F, takes the minimum non-null value, F=001. This corresponds to the situation in which only one-scale is involved in entropy flow process. To this solution correspond the minimum entropy production, MinEP. The Table 3.47 shows an example in which there are 16 tray distillation columns.

They have been lumped in pairs of trays since Table 3.46 or Table 3.47 contain 8 elements.

The minimum entropy production, MinEP corresponds to F=001.

The maximum entropy production, MaxEP corresponds to F=111.

The entropy flow implies in this case the three available scales. The heated trays are in the groups (1, 2) (5, 6) (9, 10) (13, 14) for MinEP. These groups are bolded and underlined in Table 3.48.

This may be compared with the 1, 6, 11, 15 optimal tray locations for a 19 trays example studied by detailed models by Andersen (2002). The distance is of one tray only.

The MinEP distribution corresponds to a kind of equipartition, a situation when local entropy production is uniformly distributed over the space or time variables of the process.

Close to MaxEP solution corresponds to entropy flow F=110 and the heated trays as follows: (1, 2) (3, 4) and (13, 14) (15, 16). The groups are bolded and underlined in Table 3.48. This is the situation with column extremity heated that is similar to adiabatic column.

The maximum production of entropy MaxEP corresponds to F=111 and the heated trays: are in the groups (1, 2) (7, 8) (11, 12) (13, 14). The groups are bolded and underlined in Table 3.48.

A distillation column may have active heating elements on each of its trays.

For this distillation column we may run DOE that is experiments from the Table 3.45 or Table 3.46. Any row in such tables is an experiment that is to

**Table 3.48** Entropy Y(T)

| | | | | | | | | |
|---|---|---|---|---|---|---|---|---|
| F=001 | 1 | -1 | 1 | -1 | 1 | -1 | 1 | -1 |
| MinEP | **1, 2** | 3, 4 | **5,6** | 7,8 | **9,10** | 11,12 | **13,14** | 15,16 |
| F=110 | 1 | 1 | -1 | -1 | -1 | -1 | 1 | 1 |
| HighEP | **1,2** | **3,4** | 5,6 | 7,8 | 9,10 | 11,12 | **13,14** | **15,16** |
| F=111 | 1 | -1 | -1 | 1 | -1 | 1 | 1 | -1 |
| MaxEP | **1,2** | 3,4 | 5,6 | **7,8** | 9,10 | **11,12** | **13,14** | 15,16 |

another variant of heating. They contain the more condensed experiments to be done to clarify the desired heating element positioning.

#### 3.4.3.4 Evolvable Technologies

The presented study allow clarifying the elements of the PSM framework and suggesting how to build and to operate evolvable devices and technologies.

The states S represent micro-scale system. In the case of biosystems S represents the phenotype, enzymes, amino acids, and so forth. For inanimate technological systems S represents the states of the flowing system.

K represents the genotype, the schemes conditions, the scheme devices, and so forth.

Evolvable systems are characterized by evolvable control and by self-organization.

They modify both the scheme that is the configuration and the detailed dynamical processes, that is, K and S.

Evolvability perspectives for technologies start to be evaluated and implemented by engineers. Some aspects related to chemical engineering have been reviewed by Villermaux (Villermaux 1996). Local distributed process control was considered as a key factor for evolvability. This control implies that the energy and chemical supply to be localized to the site were it is required. It should be based on embedded arrays of local sensors and actuators. Micro reactors or multi-sectioned reactors represent possibilities for evolvability. When combined with nonstationary cyclic operation such micro structured programmable devices offer new possibilities for innovative chemical synthesis (Matlosz 1995).

Increased efficiency, productivity and selectivity could be obtained through intelligent operation and multiscale control. The so-called smart design would involve assemblies of structured, modular components and precise computer control based on information transfer between distributed arrays of local sensors and actuators.

An evolvable installation contains several micro-devices in series or in parallel.

The self-building installation is supposed to be intrinsically scaleable and flexible in operation.

Not all the properties of the micro-devices are studied and known. The array of micro-devices itself has a stochastic or deterministic evolution in the sense that they self-organize and grows towards a state of maximum entropy.

There are at least two time scales, one short timescale of the fluids or phenotype level and a large time-scale of self-organization process, for schemes or genotype level. The first is the scale refers to S, the second to K. An analogy can be made with evolution and natural selection in ecosystems where the evolutionary timescale is much larger than the ecological timescale of population dynamics. It should be noted that multiple scales would be beneficial for improved evolvability and autonomy.

Periodic functioning, cyclic rhythms may pass the system from MaxEP regime to MinEP regime. Suppose that the functioning of the distillation column is in periodic regime due for instance to environmental or economic conditions.

If the energy decrease is the main interest, then in the starting period MaxEP will be the regime of interest. Then since the functioning is cyclic, in the relaxation functioning regime the MinEP, that is, low F may be of interest.

The distillation column may be programmable to run the experiment and adapt to different goals.

## *3.4.4 Perspectives*

### 3.4.4.1 Multiple Scales Entropy Production

The study of entropy production of natural or artificial systems at several scales may have significant applications (Bruers 2007, Ingram et al. 2004).

Defined as close association of an abiotic environment and a collection of living organisms, an ecosystem is characterized by a great number of physicochemical factors and biological entities which interact with each other. The multiplicity and diversity of these interactions, the vast range of levels of organization and the broad spectrum of space and temporal scales impose the study of ecosystems by complexity theory methods.

Bruers considered the three levels of description, the fluid, the climate and the ecosystem.

To this may correspond the micro, macro and mega-scale and the categories S, K1 and K2 as shown in Fig. 3.13.

Organizing levels of increasing complexity have been considered also in chemical process engineering (Li and Kwauk 2003). Charpantier (2007) refers to the succession: nano-micro-meso-macro-mega scale. For process engineering systems this corresponds to: molecules-particles-reactors-plants-environments succession. The challenge for chemical engineers is to understand and describe the relationship between events at nano and micro scales, to better convert molecules onto useful products at the process scale.

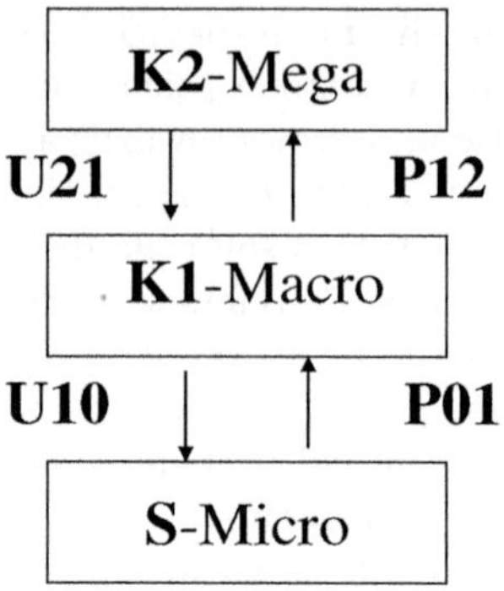


**Fig. 3.13** Three levels framework for environmental studies

The grouping of nano and micro scales and of meso and macro scales may reduces the study to three levels only. A comparison between the three levels behavior is of interest.

An integrative closure of the system shown in Fig. 3.13 would imply the direct interconnection between K2-Mega scale and S-Microscale.

The problem is to evaluate relationship between entropy productions at different levels and the entropy production for the whole system.

Organic computing studies offer some insights (Mnif and Muller-Schloer 2006).

The correspondence between Boltzmann's statistical interpretation of physical entropy as disorder and Shannon's formulation of variety and imprecision, as informational entropy suggests comparing the informational results with those from thermodynamics.

The Table 2.16 and Table 2.17 from Sect. 2.4.2 refer to the increase of informational entropy with the number of experiments that is in time, and is analogous to the $2^{nd}$ law in thermodynamics.

Some two-level calculations proved that it is possible that the information entropy decreases with new experiments apparently contradicting the second law of thermodynamics. Firstly it should be emphasized that the above calculated entropies are for the conditions space, K1 or K2 that is they are informational entropies. The two-level calculus includes in some sense, the condition space K and also the real state system S, with its entropy.

The apparently inconsistent behavior of informational entropy in K1, K2 or in a grouping of both, have been signaled by several authors as for instance in the study of separation sequences (Iordache et al. 1993c), Parunak and Brueckner (2001) in the study of multi-agent systems, Adami (2002) in the study of the physical complexity, and Mnif and Muller-Schloer (2006) in the study of emergence and organic computing.

This discrepancy may be explained in all these cases in terms of the Kugler-Turvey (1987) model. Self-organization and loss of entropy occurs at the macro-level or mega-level that is in K1 or K2, while the system dynamics at the real micro-level that is in S, generates increasing disorder and serves as entropy sink. This permits the overall system total entropy that is in K1, K2 and S, to increases while allowing self-organization to emerge, in K1 or K2 (Parunak and Brueckner 2001). It is an emergence related to the level of observation. A microscopic observation can distinguish between all the microstates, hence there is no emergence at this level, while an observation at the macroscopic or megascopic level notes a degenerate macro state-the microscopic realization of that particular macro or mega state not being accessible to the observer.

The information distance that is the conditional entropy drops observed in the sequencing study (Iordache et al. 1993c) was interpreted as emergence of a new classifications or separation schemes at the macro-level. In that study the micro-level that is the real space S was represented by the random process generating the real valued weights, differing from the exact powers of 0.5. The entropy associated to the random weights in S, was able after few numerical step to support the emergence of new separation schemes, in K.

Adami (2002) simulations show also domains of complexity decreasing in time. Physical complexity as introduced by Adami assigns low complexity to both ordered and random systems and high complexity to those between them. Physical complexity measures the amount of information that a system stores in its space of conditions K1 or K2 about the environment S in which it evolves. Thus, evolution increases the amount of knowledge a population accumulates about its niche. Since entropy is a measure of potential information, biological evolution leads to decrease of entropy. Dynamic environment that is S, continually creates the problem to be solved. To survive in the environment means to solve problems, the solutions being embodied knowledge. By thermodynamic reasoning it is possible to identify differentiation from environment, dissipative structures or other far from equilibrium structures. The basic idea for physical complexity definition was to relate complexity to system's information about its environment. In this way the informational and thermodynamic approaches to complexity may be closely related.

Moreover the study of both types of entropy allows detecting emergence.

Table 2.16 and Table 2.17 from Sect. 2.4.2 show that the increases in entropy from one experiment to another starts with a minimum increase of entropy, for the first time step from n=1 to n=2. Then, the increase of entropy in each step increases with n. This behavior may be explained in terms of the approach to non-equilibrium in statistical thermodynamics. The first experiment activates only one of the possible levels that is, only one scale contributing to the entropy production. This situation parallels the minimum production of entropy theorem (Prigogine 1980). As more experiments are performed and n increases, other scales of entropy production may be activated and the system tends towards the maximum production of entropy (Dewar 2003).

It would be of interest to study the entropy and the corresponding entropy production rules in integrative closed frameworks showing S, K1, K2 and K3 levels.

Observe that in S we are looking for the most probable state while in K1 one looks for the most typical trajectory or process. The study refers to entropy in S and to the entropy production when K1 is included too. Dewar (2003) restricts his study to this level.

If K2 is considered we need to take into account the production of production of entropy. This study is looking for states, trajectories and meta-trajectories that is, trajectory between trajectories.

The production of production of entropy should have extremum properties.

This permits the overall system total entropy that is in K2, K1 and S, to increases while allowing a higher order self-organization to emerge, in K2.

Similar conclusions should be valid if K3 is included as the study of meta-meta trajectories.

Notice that as the level increases we have appropriate criteria for studying not only physical but also other processes as biological, cognitive and intelligent.

## References

Adami, C.: What is complexity? Bioessays 24(12), 1085–1094 (2002)

Anal, A.K.: Time-controlled pulsatile delivery systems for bioactive compounds. Recent Patents on Drug Delivery & Formulation 1, 73–79 (2007)

Andersen, T.R.: Optimal Design and Operation of Process Integrated Distillation. Ph.D Thesis, Insitut for Kemiteknik, DTU, Denmark (2002)

Aoki, I.: Entropy production in living systems: from organisms to ecosystems. Thermochimica Acta 250, 359–370 (1995)

Bailey, R.A., Cameron, P.J., Connelly, R.: Sudoku, gerechte designs, resolutions, affine space, spreads, reguli, and Hamming codes. Amer. Math. Monthly 115, 383–404 (2008)

Bate, J.A., van Rees, G.H.J.: The size of the smallest strong critical set in a Latin square. Ars Combinatoria 53, 73–83 (1999)

Bruers, S.: Energy and Ecology. On entropy production and the analogy between fluid, climate and ecosystems. Ph. D Thesis, Katholieke Universiteit Leuven (2007)

Cariani, P.: To Evolve an Ear: Epistemological Implications of Gordon Pasks's Electrochemical Devices. Systems Research 10(3), 19–33 (1993)

Charpentier, J.C.: Four main objectives for the future of chemical and process engineering mainly concerned by the science and technologies of new materials production. Chemical Engineering Journal (107), 3–17 (2005)

Charpentier, J.C.: Modern Chemical Engineering in the Framework of Globalization, Sustainability, and Technical Innovation. Ind. Eng. Chem. Res. (46), 3465–3485 (2007)

Chiang, A.S.T.: Arithmetic of PSA process scheduling. A.I.Ch.E.J. 34(11), 1910–1912 (1988)

Chin, C.Y., Wang, L.N.H.: Simulated Moving Bed Equipment Designs. Separation and Purification Reviews 33(2), 77–155 (2004)

Denes, J., Keedwell, A.D.: Latin Squares and Their Applications. Academic Press Inc., NY (1974)

Dewar, R.C.: Informational theory explanation of the fluctuation theorem, maximum entropy production, and self-organized criticality in non-equilibrium stationary states. J. Phys. A 36, 631–641 (2003)

Dressler, F.: Self-organization in sensor and actor networks. Wiley, NY (2007)

Estrin, A., Govindan, R., Heidemann, J., Kumar, S.: Next century challenges: Scalable coordination in sensor networks. In: Proceedings of the International Conference on Mobile Computing and Networks, Seattle, WA, pp. 263–270 (1999)

Gambhir, M., Guerin, S., Kauffman, S., Kunkle, D.: Steps toward a possible theory of organization. In: Minai, A. (ed.) Proceeding of the International Conference on Complex Systems, Boston, MA. W. H. Freeman, New York (2004)

Ganek, A.G., Corbi, T.A.: The dawning of the autonomic computing era. IBM Systems Journal 42(1), 5–18 (2003)

Garg, M.K., Douglas, P.L., Linders, J.G.: An expert system for identifying separation processes. Canad. J. of Chem. Engng. 69, 67–75 (1991)

Gomez, M., Seader, J.D.: Synthesis of distillation trains by thermodynamic analysis. Comput. Chem. Engng. 9, 311–341 (1985)

Harmuth, H.F.: Sequency Theory, Foundations and Applications. Academic Press, New York (1977)

Hedayat, A.S., Sloane, N.J.A., Stufken, J.: Orthogonal Arrays. Theory and Applications. Springer, New York (1999)

Heller, M., Schleicher, A., Westfechtel, B.: A Management System for Evolving Development Processes. In: Proceedings of the 7th International Conference on Integrated Design and Process Technology (IDPT 2003), Austin, Texas, USA, p. 10. SDPS (2003)

Henley, E., Seader, J.D.: Equilibrium-Stage Separation Operations in Chemical Engineering. J. Wiley, New York (1981)

Hur, J.S., Wankat, P.C.: New designs of simulated moving bed (SMB) for ternary separations. Ind. Eng. Chem. Res. 44(6), 1906–1913 (2005)

Ingram, G.D., Cameron, L.T., Hangos, K.M.: Classification and analysis of integrating frameworks in multiscale modeling. Chemical Engineering Science 59(11), 2171–2187 (2004)

Iordache, O.: Evolvable Designs of Experiments Applications for Circuits. J. Wiley VCH, Weinheim (2009)

Iordache, O., Corriou, J.P., Garrido-Sanchez, L., Fonteix, C., Tondeur, D.: Neural network frames applied for biochemical kinetic diagnosis. Comp. Chem. Engng. 17(11), 1101–1113 (1993a)

Iordache, O., Valentin, G., Corriou, J.P., Pons, M.N., Pethö, A.: Intermittent Interfacial Transfer. A Dyadic Model. Acta Chimica Hungarica, Models in Chemistry 1(130), 1–18 (1993b)

Iordache, O., Corriou, J.P., Tondeur, D.: Separation Sequencing. Use of Information Distance. Canad. Journ. of Chem. Engng. 71, 955–966 (1993c)

Iosifescu, M., Grigorescu, S.: Dependence with complete connections and applications. Cambridge Univ. Press, Cambridge (1990)

Jaynes, E.T.: The minimum entropy production principle. Ann. Rev. Phys. Chem. 31, 579–601 (1980)

Katis, P., Sabadini, N., Walters, R.F.C.: Bicategories of processes. Journal of Pure and Applied Algebra 115, 141–178 (1997)
Kugler, P.N., Turvey, M.T.: Information natural law, and self-assembly of rhytmic movement. Lawrence Erlbaum, Mahwah (1987)
Kullback, S.: Information Theory and Statistics. J. Wiley, New York (1980)
Li, J., Kwauk, M.: Exploring complex systems in chemical engineering-the multiscale methodology. Chemical Engineering Science 58, 521–535 (2003)
Lübbe, A.S., Alexiou, C., Bergemann, C.: Clinical applications of magnetic drug targeting. J. Surg. Res. 96, 200–206 (2001)
Ludemann-Hombourger, O., Nicoud, R.M., Bailly, M.: The "Varicol" Process: A new multicolumn continuous chromatographic process. Separation Science and Technology 35(12), 1829–1862 (2000)
Matlosz, M.: Electrochemical Engineering Analysis of Multisectioned Porous Electrodes. Journal of the Electrochemical Society 142, 1915–1922 (1995)
Mnif, M., Müller-Schloer, C.: Quantitative Emergence. In: Proceedings of the 2006 IEEE Mountain Workshop on Adaptive and Learning Systems (SMCals 2006), Pacificaway, NJ, USA, pp. 78–84 (2006)
Montani, S., Anglano, C.: Case-Based Reasoning for autonomous service failure diagnosis and remediation in software systems. In: Roth-Berghofer, T.R., Göker, M.H., Güvenir, H.A. (eds.) ECCBR 2006. LNCS (LNAI), vol. 4106, pp. 489–503. Springer, Heidelberg (2006)
Morbach, J., Wiesner, A., Marquardt, W.: OntoCAPE 2.0-a (re) usable ontology for computer-aided process engineering. In: Braunschweig, B., Julia, X. (eds.) 18th European Symposium on Computer Aided Process Engineering-ESCAPE, vol. 18, pp. 991–996 (2008)
Murdan, S.: Electro-responsive drug delivery from hydrogels. Journal of Controlled Release 92, 1–17 (2003)
Pajula, E., Seuranen, T., Koiranen, T., Hurme, M.: Synthesis of separation processes by using case-based reasoning. Comp. Chem. Eng. 25, 775–782 (2001)
Parunak, H.V.D., Brueckner, S.: Entropy and Self-Organization in Multi-agent Systems. In: Muller, J., Andre, E. (eds.) Proceedings of the Fifth International Conference on Autonomous Agents, pp. 124–130. ACM Press, New York (2001)
Pattee, H.H.: The physics of symbols and the evolution of semiotic control. In: Workshop on Control Mechanisms for Complex Systems: Issues of Measurement and Semiotic Analysis, Las Cruces, New Mexico, December 8-12, 1996. Santa Fe Institute Studies in the Sciences of Complexity, Proceedings. Addison-Wesley, Redwood City (1997)
Pattee, H.H.: Causation, control and the evolution of complexity. In: Anderson, P.B., et al. (eds.) Downward Causation, pp. 63–67. Aarhus University Press, Aarhus (2000)
Peppas, N.A., Leobandung, W.: Stimuli-sensitive hydrogels: ideal carriers for chronobiology and chronotherapy. J. Biomater. Sci. Polymer Edn. 15(2), 125–144 (2004)
Piaget, J.: The construction of Reality in the Child. Ballantine Books, New York (1971)
Prigogine, I.: Introduction to thermodynamics of irreversible processes. Interscience, NY (1955)
Prigogine, I.: From Being into Becoming. W. H. Freeman, San Francisco (1980)
Rhee, H.K., Aris, R., Amundson, N.R.: First Order Partial Differential Equations II. In: Theory and applications of hyperbolic systems of quasilinear equations. Prentice-Hall, Englewood Cliffs (1989)

Ruthven, D.M.: Principles of adsorption and adsorption processes. John Wiley, New York (1984)
Ruthven, D.M., Ching, C.B.: Counter current and simulated counter current adsorption separation processes. Chem. Engng. Sci. 44(5), 1011–1038 (1989)
Salamon, P., Nulton, J.D., Siragusa, G., Andersen, T.R., Limon, A.: Principles of Control Thermodynamics. Energy 26, 307–319 (2001)
Salthe, S.N.: Infodynamics, a developmental framework for ecology/economics. Conservation Ecology, 3 (2003)
Scherer, F., Anton, M., Schillinger, U.: Magnetofection: enhancing and targeting gene delivery by magnetic force in vivo and in vitro. Gene Therapy 9, 102–109 (2002)
Scherer, F.: Establishment of magnetofection –a novel method using superparamagnetic nanoparticles and mgnetic force to enhance and to target nucleic acid delivery, Ph D Dissertation, Ludwig Maximilian University, Munich (2006)
Seader, J.D., Westerberg, A.W.: A combined heuristic and evolutionary strategy for synthesis of simple separation sequences. AIChE J. 23, 951–960 (1977)
Sershen, S., West, J.: Implantable, polymeric systems for modulated drug delivery. Advanced Drug Delivery Reviews 54, 1225–1235 (2002)
Shaeiwitz, J.A., Turton, R.: Design of a Transdermal Delivery System: A Case Study in Product Design and Multi-scale Design. In: ASEE 2004 Conference Proceedings, Session 3413 (2004)
Smith, O.J., Westerberg, A.W.: Mixed-integer programming for pressure swing adsorption cycle scheduling. Chem. Engng. Science 45(9), 2833–2842 (1990)
Smolensky, M.H., Peppas, N.A.: Chronobiology, drug delivery, and chronotherapeutics. Advanced Drug Delivery Reviews 59, 828–851 (2007)
Surma, J., Braunschweig, B.: REPRO: Supporting Flow-sheet Design by Case-Based Retrieval. In: Smith, I., Faltings, B.V. (eds.) EWCBR 1996. LNCS, vol. 1168. Springer, Heidelberg (1996)
Sutariya, V.B., Mashru, R.C., Sankalia, M.G., Sankalia, J.M.: Preparation of rapidly disintegrating tablets of ondansetron hydrochloride by direct compression method. Ars Pharm. 47(3), 293–311 (2006)
Swenson, R.: Emergent Attractors and the Law of Maximum Entropy Production: Foundations to a Theory of General Evolution. Systems Research 6(3), 187–197 (1989)
Swenson, R., Turvey, M.T.: Thermodynamic reasons for perception-action cycles. Ecological Psychology 3(4), 317–348 (1991)
Tondeur, D., Migault, G., Wankat, P.C.: Gas purification by pressure-swing adsorption. Process Analysis. Entropie 123, 39–55 (1985)
Varmuzza, K.: Pattern Recognition in Chemistry. Springer, Berlin (1980)
Venkatasubramanian, V., Zhao, C., Joglekar, G., Jain, A., Hailemariam, L., Suresh, P., Akkisetty, P., Morris, K., Reklaitis, G.V.: Ontological informatics infrastructure for pharmaceutical product development and manufacturing. Comput. Chem. Eng. 30(10-12), 1482–1496 (2006)
Villermaux, J.: New horizons in Chemical Engineering. In: Proceeding of the 5th World Congress on Chemical Engineering, San Diego, USA, pp. 16–23 (1996)
Yang, R.T.: Gas Separation by Adsorption Processes. Butherworths, Boston (1987)
Zhang, Z., Mazotti, M., Morbidelli, M.: Continuous Chromatographic Processes with a Small Number of Columns: Comparison of Simulated Moving Bed with Varicol, PowerFeed, and ModiCon. Korean Journal of Chemical Engineering 21, 454–464 (2004)

# Chapter 4
# Biosystems and Bioinspired Systems

**Abstract.** Artificial genetic codes, neural networks and neural codes are presented as theoretical frames for evolutionary computation and biomimetic devices.

Models for genetic code evolution offer suggestions for chemical and biochemical inspired computations as for instance artificial chemistry or chemical programming.

Neural networks architecture issues require evolvability as outlined by growing neural nets or by protein based neural networks.

The significance of neural coding, symbolic connectionist hybrids, neural binding, temporal synchrony studies for unconventional computing and neurocognitive devices is highlighted.

Evolutionary circuits based on electrochemical filaments are proposed. The perspectives of evolvable circuits based on bio-molecules properties, are evaluated.

Case studies show how technological innovation should find the right moment to free the artificial system designer from the detailed experimental data of real systems.

## 4.1 Artificial Genetic Codes

### *4.1.1 Genetic Code Evolution*

The main objective of the genetic code theoretical study in terms of PSM is to understand and make use of genetic code evolution scenarios as suggestions for new computing and information technologies. To re-apply this understanding in developing new ways of study or explanations of biological relevance the for real genetic code, may be considered as a long-term objective only.

Evolutionary computation studies and evolvable devices may make use of biological principles but do not attempt to model or to mimic detailed data or processes from real genomes. Bio-inspired artificial design is not constrained by high fidelity to the original natural complex system. Examples include genetic algorithms calculus inspired by Darwinian evolution and genetics, artificial neural networks and artificial neural codes inspired by neuroscience but not restricted to this.

The expanding code scenario from single-bases nucleotides to doublets and then to triplets that is to codons offer interesting suggestions for evolvability studies and applications.

Several hypothetic scenarios have been advanced to explain the genetic code structure and its origin (Weberndorfer et al. 2003, Koonin and Novozhilov 2009). The main concepts on origin and evolution of the code are the stereochemical theory, the co-evolution theory and the adaptive theory.

The stereo-chemical theories suppose that the specificity of a codon for a particular amino acid is based on a direct interaction of amino acid and nucleotides. Amino acids might have been binding directly to the codons when the code was established and such binding imposed the code. The co-evolution theory explains the non-randomness of the code by the fact that the code system is an imprint of the pre-biotic pathways of amino-acid formation. According to this theory the genetic code evolution reflects the relationship among amino acids and their biosynthesis. In an early code used fewer codons and amino-acids and then expanded to include new amino-acids arising from biosynthesis coded for by new codons, with the resulting code assigning similar codons to amino acids that are related by biosynthesis.

Adaptive codes theories attempted to explain the observed patterns in genetic code and its evolution by postulating optimality of the code. Adaptation theories state that selection pressure resulted in the emergence of a code optimized for some measure, such as for minimization of the physicochemical effects of single mutational or translational errors.

A notable approach in the study of genetic code evolution is the Eigen's work on hypercycles systems of mutually autocatalytic components. It consider the question of under what conditions, the system can self-organize to a dynamic stability (Eigen and Schuster1979). Eigen approach was based on the view that the self-organization including the development of hypercycles is a process that can occur in a homogeneous system by intrinsic necessity.

Eigen and Schuster (1979) considered that the primitive code may use units of less than three bases. During its early evolution, the code would have increased both the number of codons and the coded aminoacids and the present code would reflect the pattern of this historical expansion (Wilhelm and Nikolajewa 2004, Patel 2005, Wu et al. 2005).

In partial contrast with Eigen approach, in the view of H. Kuhn, understanding the origin of living systems is a particular engineering problem: to find a sequence of physicochemical stages, beginning with prebiotically reasonable conditions and leading to self-organization of matter and to systems equipped with a life like genetic apparatus (Kuhn and Waser 1981, 1994, Kuhn H and Kuhn C 2003).

All these theories suggest that the genetic codes are information communication system that should reflect the physico-chemical properties of the amino acids. The different theories are not mutually exclusive and probably the code was shaped by a compromise of several scenarios (Ardell and Sella 2002).

## 4.1.2 Model for Code Evolution

### 4.1.2.1 Genetic Code

The genotype of cells is laid down in a linear sequence of four nucleotides: A-adenine, C-cytosine, U-uracil and G-guanine. The genetic information is transcribed in mRNA used as instructions for protein translation. Translation requires a mapping of the four nucleotides in 20 amino acids. Triplets of the four different RNA bases are read sequentially from mRNA. DNA is transcribed to mRNA that makes use of an RNA adaptor, tRNA to interpret nucleotides in amino acids. The four bases C, G, U and A, might form 64 different simple triplets patterns, the so-called codons. The 20 amino acids and the start and stop signals are coded redundantly by these 64 codons (Alberts et al. 1994).

There are some symmetry elements in genetic code. The symmetry supported the use of algebraic frames to characterize the genetic code. It has been suggested that the overall layout of the code can be accurately described in the algebra of group theory or of fields (Danckwerts and Neubert 1975, Findley et al. 1982, Jimenez-Montano et al. 1996, Jimenez-Montano 1999). These symmetries may be of help in explaining regularities and periodicities as observed in proteins sequences. They have been correlated to the possible evolution scenarios of the genetic code.

The relevant group to describe the symmetries of the bases {C, G, U, A} should be a group of order 4. There are only two possibilities for the group structure, the cyclic group C (4) and the group associated to the Galois field, GF (4). This is the so called Klein 4-group.

Several codes can be associated to the genetic code according to the order of importance for bases and of their positions in codons.

For triplets or codons the ranking: position 2>position 1>position 3 in establishing the amino-acid is acknowledged (Perlwitz et al. 1988).

One of the proposed nucleotide hierarchical ordering is: C>G>U>A. This hierarchy was established starting from the observation that C, in position 2 in codon, is anytime able to be source of a single amino acid. G is able to determine the amino acids in majority of cases, U only in some cases and A never. In other words, C base passes any time a single message while U, and A are credited with at least double message. G passes stronger messages than U or A, concerning the coded amino acid.

It is possible to associate to any base in codon a two-digit vector: (hydrogen bonds, chemical nature). The first digit refers to hydrogen bonds and the second to the chemical nature. We will use first digit “1” for high number of hydrogen bonds that is for G and C and second digit “1” for chemical nature pyrimidines that is for C and U.

We will use first digit “0” for low number of hydrogen bonds that is for A and U and second digit “0” for chemical nature purines that is for A and G.

In this way we may describe the basis by the two-digit vectors:

C: (1, 1), G: (1, 0), U: (0, 1), A: (0, 0)

This corresponds to the hierarchy: C>G>U>A, and to the real numbers 3, 2, 1, and 0 associated to C, G, U, and A respectively. More exactly:

0=(0, 0), 1=(0, 1), 2=(1, 0), 3=(1, 1).

Of course, restricting the nucleotide characterization to only two properties: (hydrogen bonds, chemical nature) is a drastic simplification.

### 4.1.2.2 Expanding Genetic Code

The WE model (Sect. 2.2.2) highlights some particularities of genetic code Tables, the order of amino-acids availability and in this way, it may be of interest for evolutionary molecular devices and evolutionary computation.

Several scenarios for genetic code development from primitive bases will be considered in terms of WE model.

Table 4.1 represents the sum $Y=Z\oplus T$, resulting as a solution of WE, as shown in Chapter 2. It is the sum for GF (4) or Klein 4 group relevant for genetic code.

The cyclic group C (4) offers a different solution. Z and T are identified with their two digit expressions for bases or equivalently with 0, 1, 2 and 3.

The development is supposed to start with all the four bases, C, G, U, and A.

This situation corresponds to the Table 4.1 for one dimensional Z and T.

Table 4.1 with GF (4) Klein-4 group table structure is well known in genetic code study.

Since the Table 4.1 contains 4 nucleotide of each type and the coding accomplished by nucleotides should allow conflict free access to parallel memory we will limit the Table 4.1 to vectors containing only distinct elements. They represent particular solutions of WE at specified levels in development.

An example is the vector $Y(T) = y_0 = (C, G, U, A)$.

**Table 4.1** Matrix of singlets

| | | | |
|---|---|---|---|
| A | U | G | C |
| U | A | C | G |
| G | C | A | U |
| C | G | U | A |

A specific folding operation allows rewriting $y_0$, as the Table 4.2.

There are only four elements and it is possible to represent $y_0$ as a 2x2 matrix like in Table 4.2.

**Table 4.2** Single bases $y_0$

| | |
|---|---|
| G | A |
| C | U |

Other types of folding and other 2x2 matrices may be considered too. The folding with A and U interchanged was also studied. Various folding algorithms and their significance for genetic programming have been described by Banzhaf (1993).

It may be supposed that initially only one nucleotide is able to form doublets, then triplets and that this founding molecule is guanine G (Hartman, 1975).

It results the situation shown in Table 4.3. The codification is accomplished by GC, GG, GA and GU. These doublets corresponds to the amino-acids: Ala, Gly, Asp, Glu and Val.

**Table 4.3** Two-bases matrix

| | | |
|---|---|---|
| GG | | GA |
| | **G** | |
| GC | | GU |

Table 4.4 is the matrix-like presentation of product of two identical vectors that is

$Y(T) = y_0 \times y_1$ with $y_0 = y_1$

This includes the elements in Table 4.2.

**Table 4.4** Doublets $y_0 \times y_1$

| | | | | | | |
|---|---|---|---|---|---|---|
| GG | | AG | | GA | | AA |
| | G | | | | A | |
| CG | | UG | | CA | | UA |
| | | | | | | |
| GC | | AC | | GU | | AU |
| | C | | | | U | |
| CC | | UC | | CU | | UU |

At this stage some amino acids resulted as follows: Pro, Ser, Leu, Thr, Arg coded without any ambiguity by CC, UC, CU, AC and CG.

The remaining amino acids will be coded by triplets showing multiple codifications.

They are: (Phe and Leu), (Ile and Met), (Cis, Trp and Stop), (His and Gln), (Tyr and Stop), (Ser and Arg) and (Lys and Asn).

They are coded by UU, AU, UG, CA, UA, AG and by AA respectively.

In Table 4.4 the eight "strong" double-bases: CC, GC, CG, GG, GU, CU, UC and AC are intertwined with the eight "weak" double-bases: AA, UA, AU, UU, UG, AG, GA and CA. The mirror symmetry G↔A and U↔C relative to median Y-axis is obvious.

The new letters, A, C, U, G have been put adjacent to the first two ones to the left side or right side. In this way the significance order corresponds to the order of letter acquisition.

Table 4.5 is a product of three 2x2-matrices-like tables,

$Y(T) = y_2 \times y_0 \times y_1$ with $y_0=y_1=y_2$

The new letters have been put adjacent to the first two, to the right side.

A version of the genetic table with 64 codons is resulting. Many other artificial genetic code Tables may be obtained by changing the hierarchy for codons, the developmental rules, the initial set of bases, the position of concatenation, and so forth Notice also that there are two groups of order 4, the Klein group and the cyclic group.

The hypothesis concerning the right or left concatenation of solutions is related to other proposed evolution scheme. For instance it was suggested by Wu et al. (2005) that triplet codons gradually evolved from two types of ambiguous doublet codons, those in which the first two bases of each three-base window were read ("prefix" codons) and those is which the last two bases of each window were read ("suffix" codons).

The right or left concatenation of solutions is correlated also to the reverse recognition conjecture of Nikolajewa et al. (2006).

Table 4.5 contains the 64 codons grouped in 4 large quarters each with a common base in the center (C, G, U, and A), each formed by 16 codons. Each large quarter contains 4 new quarters for doublets with the central base in second position and finally 16 codons since each doublet is the center of a new quarter with the doublets in the first two positions. Table 4.5 is similar to the conventional genetic code Table (Alberts et al. 1994). Instead of representing the codons with a central nucleotide on a column the codons associated to a central nucleotide C, G, U or A may be found in a quarter. Similar Tables have been discussed by other authors (Jimenez et al. 1996, Benyo et al. 2004).

For presentation purposes it is easier to portray the codons in plane that is by Tables than by hyper-cubes. However the modes of presentations are equivalent.

According to the above analysis there are several stages in the development of the genetic code. To these corresponds stages of amino-acids availability in the order:

- $1^{st}$ stage: [Ala, Gly, Asp, Glu and Val]
- $2^{nd}$ stage: [Pro, Ser, Leu, Thr, Arg]

**Table 4.5** Triplets. Codons-matrix, $y_2 \times y_0 \times y_1$

| | | | | | | | | | | | | | |
|---|---|---|---|---|---|---|---|---|---|---|---|---|---|
| GGG | | GGA | | AGG | | AGA | GAG | | GAA | | AAG | | AAA |
| | **GG** | | | | **AG** | | | **GA** | | | | **AA** | |
| GGC | | GGU | | AGC | | AGU | GAC | | GAU | | AAC | | AAU |
| | | | **G** | | | | | | | **A** | | | |
| CGG | | CGA | | UGG | | UGA | CAG | | CAA | | UAG | | UAA |
| | **CG** | | | | **UG** | | | **CA** | | | | **UA** | |
| CGC | | CGU | | UGC | | UGU | CAC | | CAU | | UAC | | UAU |
| | | | | | | | | | | | | | |
| GCG | | GCA | | ACG | | ACA | GUG | | GUA | | AUG | | AUA |
| | **GC** | | | | **AC** | | | **GU** | | | | **AU** | |
| GCC | | GCU | | ACC | | ACU | GUC | | GUU | | AUC | | AAU |
| | | | **C** | | | | | | | **U** | | | |
| CCG | | CCA | | UCG | | UCA | CUG | | CUA | | UUG | | UUA |
| | **CC** | | | | **UC** | | | **CU** | | | | **UU** | |
| CCC | | CCU | | UCC | | UCU | CUC | | CUU | | UUC | | UUU |

- $3^{rd}$ stage: [(Phe and Leu), (Ile and Met), (Cis, Trp and Stop), (His and Gln), (Tyr and Stop), (Ser and Arg) and (Lys and Asn)]

This temporal order is not so far from that resulting from co-evolution theory (Wong 1975).

A first stage groups the aminoacids Ala, Gly, Asp and Glu. The connection Ala, Val is not presented in co-evolution theory. The next stage involves Pro, Ser, Thr, Arg in both theories. The difference refers to Leu and is a consequence of the missing connection Ala,Val for the first stage.

According to the theory of Eigen and Winkler-Oswatitsch (1981) the first amino acids were Gly, Ala, Asp and Val a result confirmed by some classical experiments.

Kuhn and Waser (1994) selected as plausible steps in the evolution of genetic code Gly Ala,Val, Asp, Glu followed by a class containing Leu, Ile, Ser, Thr, Lys, then Arg, Gln, Asn, then Pro and so on.

The above results may be compared with consensual chronology of amino acids (Trifonov 2000). Trifonov presents the codon chronology as follows:

Gly, Ala, Val, Asp, Pro, Ser, Glu, Leu, Thr, Asn, Arg and so on.

### *4.1.3 Codons and Amino Acids*

Each codon, codes for the introduction of a specific amino acid into a growing protein, a process that involves recognition of the anti-codon sequence.

There exists a relationship between the codons and the properties of the coded amino acids. To outline this relationship it is possible to associate to any base in codon a two-digit vector. The Table 4.6 contains the vectors associated to codons as well as the coded amino acids.

For the Table 4.6 the associated vector is in the significance order for bases is:

position2>position1>position3.

For example AUC will be replaced by 010011 corresponding in succession to U:(01) (position 2), then A:(00) (position 1) and then C:(11) (position 3). This is in fact the supposed evolutionary pathway for development.

Codons area as outlined by Table 4.6 is again in general agreement with some assertions of the co-evolution theory (Ronneburg et al. 2000). In the evolutionary map of the genetic code based on precursor product pairs the founding precursor is Ala coded by GC. Close to this there are Ser coded by UC and Gly coded by GG. This can be inferred from the Table 4.6 too. Relatively far from the Ala precursor in evolutionary map there are Asn and Lys coded by AA or Leu and Phe coded by UU, Tyr coded by UA, Met coded by AU. This is obvious from Table 4.6 too.

Table 4.6 may be of use in the context of stereo-chemical theory. This theory assumes that the physical and chemical properties of a given amino acid are related to the nature of the codons. If the codons are similar the amino acids will be similar and reverse.

Moreover similar amino acids might replace each other. This assumption is in agreement with those theories that place specific constraints on the assignment of codons to amino acids. For example a significant correlation was observed between hydrophobic ranking of the amino acids and the hydrophobic character of the anti codons.

### 4.1.4 Polypeptides

A graphical illustration of the polystochastic framework is presented here. The Table 4.7 reproduces the genetic code Table and includes free places for amino acids.

Some examples clarify the PSM framework for polypeptide synthesis.

Elements of the SKUP are emphasized by Table 4.7.

Denote the singlet conditioning level by m=0, the doublet level by m=1 and the triplet level by m=2.

The coding for amino acids may be done by doublets or by triplets.

S, K, U and P will be vectors denoted as follows: S = ($s^1$, $s^2$); K = ($k^1$, $k^2$); U = ($u^1$, $u^2$); P = ($p^1$, $p^2$). Upper index refers to levels while lower index will refers to time step.

We start with an m=1 example. It is known that CU coding for Leu may serve as start.

Let CU=$k_0^1$, UC=$k_1^1$, CC=$k_2^1$, GC=$k_3^1$ UG=$k_4^1$. The upper index refers to level while the lower index refers to the time step. The states and the conditions at the level m=1 are represented in Table 4.7 by high thickness border cells.

Correspondingly the states at the second level will be: $s_0^1$ = Leu, $s_1^1$ = Leu.Ser, $s_2^1$ = Leu.Ser.Pro, $s_3^1$ = Leu.Ser.Pro.Ala, $s_4^1$ = Leu.Ser.Pro.Ala.

**Table 4.6** Triplets vectors and amino acids

| | | | | | | | | | | | | | |
|---|---|---|---|---|---|---|---|---|---|---|---|---|---|
| GGG<br>101010<br>Gly | | GGA<br>101000<br>Gly | | AGG<br>100010<br>Arg | | AGA<br>100000<br>Arg | GAG<br>001010<br>Glu | | GAA<br>001000<br>Glu | | AAG<br>000010<br>Lys | | AAA<br>000000<br>Lys |
| | **GG** | | | | **AG** | | | **GA** | | | | **AA** | |
| GGC<br>101011<br>Gly | | GGU<br>101001<br>Gly | | AGC<br>100011<br>Ser | | AGU<br>100001<br>Ser | GAC<br>010011<br>Asp | | GAU<br>010001<br>Asp | | AAC<br>000011<br>Asn | | AAU<br>000001<br>Asn |
| | | | **G** | | | | | | | **A** | | | |
| CGG<br>101110<br>Arg | | CGA<br>101100<br>Arg | | UGG<br>100110<br>Trp | | UGA<br>100100<br>Stop | CAG<br>001110<br>Gln | | CAA<br>001100<br>Gln | | UAG<br>000110<br>Stop | | UAA<br>000100<br>Stop |
| | **CG** | | | | **UG** | | | **CA** | | | | **UA** | |
| CGC<br>101111<br>Arg | | CGU<br>101101<br>Arg | | UGC<br>100111<br>Cys | | UGU<br>100101<br>Cys | CAC<br>001111<br>His | | CAU<br>001101<br>His | | UAC<br>000111<br>Tyr | | UAU<br>000101<br>Tyr |
| | | | | | | | | | | | | | |
| GCG<br>111010<br>Ala | | GCA<br>111000<br>Ala | | ACG<br>110010<br>Thr | | ACA<br>110000<br>Thr | GUG<br>011010<br>Val | | GUA<br>011000<br>Val | | AUG<br>010010<br>Met | | AUA<br>010000<br>Ile |
| | **GC** | | | | **AC** | | | **GU** | | | | **AU** | |
| GCC<br>111011<br>Ala | | GCU<br>111001<br>Ala | | ACC<br>110011<br>Thr | | ACU<br>110001<br>Thr | GUC<br>011011<br>Val | | GUU<br>011000<br>Val | | AUC<br>010011<br>Ile | | AUU<br>000101<br>Ile |
| | | | **C** | | | | | | | **U** | | | |
| CCG<br>111110<br>Pro | | CCA<br>111100<br>Pro | | UCG<br>111001<br>Ser | | UCA<br>110100<br>Ser | CUG<br>011110<br>Leu | | CUA<br>011100<br>Leu | | UUG<br>010110<br>Leu | | UUA<br>010100<br>Leu |
| | **CC** | | | | **UC** | | | **CU** | | | | **UU** | |
| CCC<br>111111<br>Pro | | CCU<br>111101<br>Pro | | UCC<br>110111<br>Ser | | UCU<br>110101<br>Ser | CUC<br>011111<br>Leu | | CUU<br>011101<br>Leu | | UUC<br>010111<br>Phe | | UUU<br>010101<br>Phe |

The operator U, associated to this one level process is: $u^1$ $(s_0^1,k_1^1) = s_1^1$.

Here $s_1^1$ is the amino acid associated to the codon $k_1^1$ coupled to previous chain of amino acids. Observe that: $k_0^1$, $k_1^1$, $k_2^1$, $k_3^1$, $k_4^1$ is a trajectory in the K space, while: $s_0^1$, $s_1^1$, $s_2^1$, $s_3^1$, $s_4^1$ is a trajectory in S-space.

Consider now examples of level m=2 of coding evolution. The states and the conditions at this level are indicated in Table 4.7 by medium thickness border cells. Suppose that the construction starts at AUG. The codon AUG codes for methionine and serves as an initiation site. This is the initial condition $k_0^2$ = AUG. The associated state is methionine that is $s_0^2$ = Met.

Then the trajectory may evolve towards the condition $k_1^2$ = UAC.

This corresponds to the amino acid Tyr. The new state $s_1^2$ is the succession Met.Tyr.

Then the new condition may be $k_2^2$=UAG. This is a terminal codon. Hence $s_2^2 = s_1^2$.

Observe that $s_1^2$ depends on $s_0^2$ and on $k_1^2$, $s_2^2$ depends on $s_1^2$ and on $k_2^2$.

The operator u associated to this one level process is: $u^2$ $(s_0^2,k_1^2) = s_1^2$.

Here $s_1^2$ is the amino acid associated to the codon $k_1^2$ coupled to previous chain of amino acids.

Possibility as p $(k_1^2 \mid s_0^2)$ depends on the fact that it is a genetic transition (purine to purine or pyrimidine to pyrimidine) or a genetic transversion

**Table 4.7** Codons and amino acids schema

| | | | | | | | | | | | | | | | | |
|---|---|---|---|---|---|---|---|---|---|---|---|---|---|---|---|---|
| | | | | | | | | | | | | | | | | |
| | GGG | | GGA | | AGG | | AGA | | GAG | | GAA | | AAG | | AAA | |
| | | | | | | | | | | | | | | | | |
| | GGC | | GGU | | AGC | | AGU | | GAC | | GAU | | AAC | | AAU | |
| | | | | | | | | $s_4^1$ | | | | | | $s_2^2$ | | |
| | CGG | | CGA | | UGG | | UGA | | CAG | | CAA | | UAG | | UAA | |
| | | | | | | UG | | | | | | $s_1^2$ | | | | |
| | CGC | | CGU | | UGC | | UGU | | CAC | | CAU | | UAC | | UAU | |
| | | | | $s_3^1$ | | | | | | | | | | $s_0^2$ | | |
| | GCG | | GCA | | ACG | | ACA | | GUG | | GUA | | AUG | | AUA | |
| | | GC | | | | | | | | | | | | $s_{0'}^2$ | | |
| | GCC | | GCU | | ACC | | ACU | | GUC | | GUU | | AUC | | AUU | |
| $s_2^1$ | | | | | | | | $s_0^1$ | | | | $s_{3'}^2$ | | | | $s_{1'}^2$ |
| | CCG | | CCA | | UCG | | UCA | | CUG | | CUA | | UUG | | UUA | |
| | | CC | | | | UC | | | | CU | | | | $s_{2'}^2$ | | |
| | CCC | | CCU | | UCC | | UCU | | CUC | | CUU | | UUC | | UUU | |
| | | | | $s_1^1$ | | | | | | | | | | | | |

(purine to pyrimidine or vice-versa). Recall that A, G are purines while C, U are pyrimidines. Examples of possibilities are the similarities as defined in Sect. 2.3.1.

Periodicities may arise as for instance in the example at level m=2: $k_{0'}^2$=AUC.

Then $k_{1'}^2$=AUU, $k_{2'}^2$=UUA, $k_{3'}^2$=UUG.

The conditions and states trajectories are outlined by medium thickness border cells. More large excursions in the Table 4.7 may be considered and "junk" steps may be very frequent. Simulation of process as exemplified serves to illustrate the proposed code evolution scenario.

### *4.1.5 Basic Framework Evaluation*

Life involves a semantically closed organization between symbolic records and dynamical constraints (Pattee 1995). Symbols, as discrete functional switching-states, are seen in all evolvable systems in form of genetic codes, and at the core of all neural systems in the form of informational mechanisms that switch behavior DNA molecules represent the symbolic aspects here, that is, the genome. This corresponds to the conditions K, in SKUP framework. The dynamic material aspects are represented by the phenotype that is by proteins, organisms and eventually by the environment. This corresponds to states S. The genome generates different dynamical systems that promotes their stability and survive and in that way serves as seeds of a generally evolvable system.

The genome may be interpreted as a possible solution of the wave model.

It is an apparently "time-less" model since the time T is defined on a finite group and has a cyclic character. As discussed by H. Kuhn this temporal cycling is crucial for genetic code emergence and evolution. Dynamical model with usual real time, characterizing the kinetic equations completes the evolvable system description.

The closure between symbolic that is digital and real aspect of the closure are clearly illustrated in the PSM framework. The operator U may be associated to tRNAs. The tRNAs performs decoding activities. It incorporate two codes: one to read the info from mRNA and a second code that determines the amino acid with which the tRNA is loaded. This outlines the role of operator U in the transition from discrete symbol to real material aspects. U correlates informational and chemical data.

The enzymes as RNA-replicase may be associated to possibilities P. It performs encoding activities. On the primitive genetic code, the tRNAs and RNA replicases could have been involved but other closure possibilities exists.

PSM framework illustrates a minimal closed organism with translation (Webernsdorfer et al. 2003). It has a genome that carries genes for a protein replicase and tRNAs, a translation apparatus, and system loading tRNAs with amino acids.

P and U correspond to the so-called upward and downward causation respectively (Pattee 2000). For the life and artificial life, AL, situations, the semantic genetic control can be viewed as up-down causation, while the dynamics of organism growth controlling the expression of the genes can be viewed as down-up causation. The closure concept is an essential relation of these causations. Semantic closure is limited to two levels, denoted here by K and S.

The interplay between the WE in the so-called sequence space and the more or less similar real valued equations of thermodynamics and chemical kinetics represents the specificity of living systems. Obviously the closure mediated by the operators U and possibilities P is compatible with both co-evolution and with stereo-chemical theories (Webernsdorfer et al. 2003).

Table 4.8 illustrates the SKUP schema associated to hypercycles.

The relation of Table 4.8 to WE is similar to that outlined for Table 4.7.

In this case the conditions K are associated to RNA while the states S to enzymes.

The hypercycle is a self reproducing macromolecular in which RNAs and enzymes cooperate. There are RNA matrices ($I_i$), the i-th RNA codes the i-th enzyme $E_i$. The enzymes cyclically increase RNA's replication rates, namely,

**Table 4.8** Schema for hypercycles

| | | | | | | | | |
|---|---|---|---|---|---|---|---|---|
| | | | | $E_1$ | | | | |
| | | | | | | | | |
| | | $I_1$ | | | | $I_2$ | | |
| | | | | | | | | |
| $E_0$ | | | | | | $e_{31}$ | | $E_2$ |
| | | | | | $i_{31}$ | | $i_{32}$ | |
| | | $I_0$ | | $e_{30}$ | | $I_3$ | | $e_{32}$ |
| | | | | | $i_{30}$ | | $i_{33}$ | |
| | | | | $E_3$ | | $e_{33}$ | | |

$E_0$ increases replication rate of $I_1$, $E_1$ increases replication rate of $I_2$, $E_2$ increases replication rate of $I_3$, and $E_3$ increases replication rate of $I_0$. The hypercycle is represented in Table 4.8 by high thickness border cells.

The mentioned macromolecules cooperate to provide primitive translation abilities, so the information, coded in RNA-sequences, is translated into enzymes analogous to the usual translation processes in biosystems.

The cyclic organization of the hypercycle ensures its structure stability. For effective competition, the different hypercycles should be placed in separate compartments.

Table 4.8 shows that some RNA may induce the reproduction of other metabolites in cyclic manner. Supposing that $I_3$ is in this situation, $e_{30}$ increases replication rate of $i_{31}$, $e_{31}$ increases replication rate of $i_{32}$, $e_{32}$ increases replication rate of $i_{33}$, and $e_{33}$ increases replication rate of $i_{30}$. The secondary cycle is represented in Table 4.8, by high thickness border cells.

The number of RNAs in each cycle may vary.

The wave model WE, characterizes the genetic bio-chemical reactor in a discrete space. It includes the "convection" or "drift" term $V \otimes \frac{\partial Y}{\partial Z}$ and the "kinetic" term $Q \otimes Y$.

Observe that just one wave equation replaces the entire system of differential equations for quasispecies (Eigen and Schuster 1979).

This WE is adequate for highly non-linear processes modeling. The time T is a more natural expression for time to record qualitative developments than the usual linear time. The cyclic and diversified characters of environment, as described by Kuhn are accounted for by T and Z introduced here. The different values of T correspond to the developmental or pattern recognition stages.

Q takes into account the mutations and selections. The velocity V takes into account the "convection". It could happen that the convection contribution is more significant than that of mutations for evolution.

## 4.1.6 Perspectives

### 4.1.6.1 Three Realms Framework

Research within evolutionary computation has identified properties of biological coding that may be significant to evolutionary algorithms (Rocha 1997, Kargupta 2001, Suzuki and Sawai 2002). Applying computation results back to biology suggests that the genetic basis of life may enhance the power of natural mechanisms as selection as a search algorithm.

This approach offers a partial answer to the present need to elaborate common mathematical frames for evolvable systems.

Fig. 4.1 shows the categorical framework for the three levels of the refined central dogma of biology. According to the central dogma proteins are not made directly from genes but require an intermediary, and this intermediary is RNA.

Here S denotes the proteins level. For computing purposes, K1 and K2 are the two conditions levels. K1 is associated to RNA K2-is in this case the meta-level representing the DNA.

The strategies are defined at the level K2 since the information is transmitted from DNA to protein through RNA.

U10: K1 $\rightarrow$ S describes the translation. U21: K2 $\rightarrow$ K1 describes the transcription.

Implicit in the central dogma view is the idea of a unique mapping from gene to protein in which RNA plays only a mediatory role.

Some refinements of central dogma refer to the possibilities P01 and P12. P01: S $\rightarrow$ K1, and P12: K1 $\rightarrow$ K2 effects may be associated to the regulation processes.

For the operon model the DNA encodes two classes of proteins, structural and regulatory. It refers to a splitting of S in two non-interacting realms S1 and S2. Structural proteins play a functional role in the cell's metabolism. Regulatory proteins interact with DNA to control the rates of transcription of other genes. This links the proteins S to K2 realm (Jacob and Monod 1961).

Geard and Wiles (2003) evaluated the possibility of splitting of K1 in two functional modules corresponding to small RNA and standard RNA. The small RNA molecules may provide a kind of meta-level of evolution allowing for the evolution of new and complex functions by modulating the control

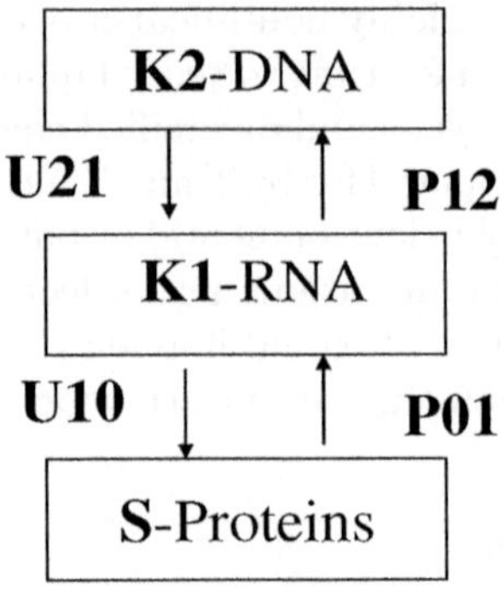


**Fig. 4.1** Refined central dogma

architecture of a stable proteome. In this case the DNA should offer the meta-information.

Fig. 4.2 shows a theoretical model in which S-Proteins are supposed to play a regulatory role for both K1-RNA and K2-DNA levels.

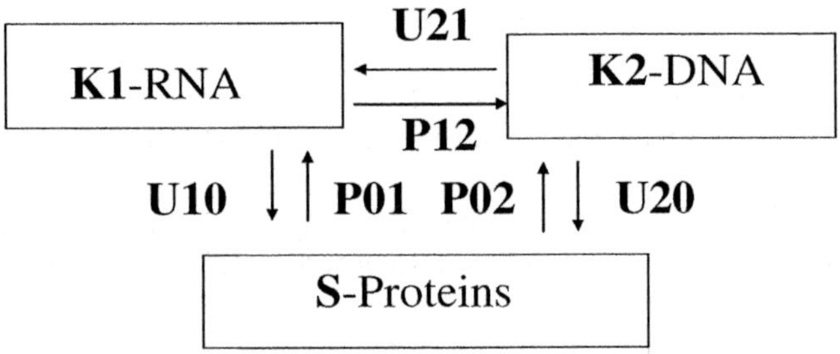


**Fig. 4.2** Regulatory models

The categorical framework describes hypothetical interactions as follows: U10: RNA → Proteins, U20: DNA → Proteins, P01: Proteins → RNA, P02: Proteins → DNA. The DNA is a code for RNA level.

The framework shown in Fig. 4.2 is an integrative closure and should correspond to a degree of evolvability and autonomy of the system.

#### 4.1.6.2 Higher Order Genetic Code Hypothesis

It was observed that the genome may contain more information than it has been anticipated. This redundant information suggested investigating if there is a higher level genetic code that directs evolution (Caporale 1984).

An examination of the rate of codon substitution during gene evolution reveals that not all so-called silent mutations, that is, the mutations to another codon that signifies the same amino acid, behave as if they are neutral.

There appear to be constraints of codons selection so that in a given context two codons, although thought to be synonymous are not in reality equivalent. Another issue is the evidence that the so-called neutral third position in codons may also carry a message may be as important as specifying the amino acid.

Rejecting purely probabilistic mechanism of genetic variation is not a refutation but rather a higher understanding of the Darwinian theory of selection. Genomes that evolve efficient biochemical systems to navigate through the space of possible future genomes would be favored by natural selection and would allow adapting more quickly when confronted by environmental challenges.

Several other facts allowed the hypothesis of a fourth realm, controlling the significant modifications of codes, or the transitions between several codes.

A four realms framework may be considered for code evolution.

Fig. 4.3 illustrates this higher order genetic code hypothesis.

The signification of the functors U and possibilities P is explicit in Fig. 4.3.

U10: K1-RNA→S-Proteins, U21: K2-DNA→K1-RNA, U32: K3-Metacode →K2-DNA, U30: K3-Metacode→S-Proteins, P01: S-Proteins→K1-RNA, P12: K1-RNA→K2-DNA, P23:K2-DNA→K3-Metacode and P03: S-Proteins→ K3-Metacode.

Observe that: U30=U10oU21oU32 and P03=P01oP12oP23.

The DNA is a code for RNA level.

The new realm, K3-Metacode, is a kind of meta-meta-model that completes and closes the frameworks shown in Fig. 4.1.

K3 organizes the multiplicities of codes and their overlapping (Trifonov 1999).

K3 represents an efficient way to explore codes and may favor the strategies that increase the rate of adaptation.

The architecture shown in Fig. 4.3 offers suggestions for modeling genetic control hierarchies. Biological progress may be accelerated if models are formulated and applied for global genetic control structures.

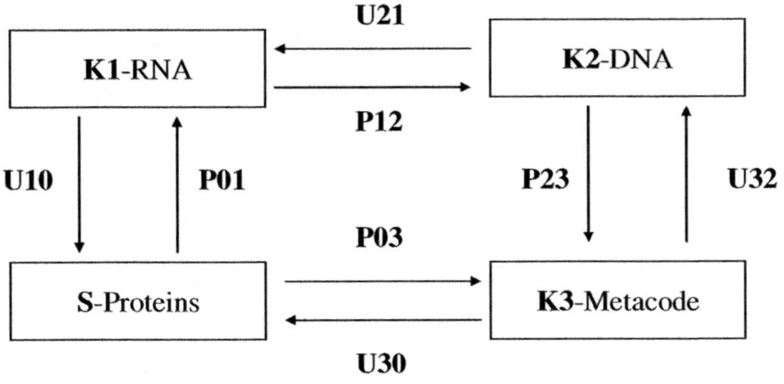


**Fig. 4.3** Higher order genetic code hypothesis

The concept of evolution as a complex hierarchical process was illustrated by a framework similar to that shown in Fig. 4.3. Gould identified three levels or "tiers" of evolution (Gould 1985). The first tier selection is the conventional Darwinian selection of individual organisms and corresponds to K1. The second-tier selection corresponding to K2 emerges from differential speciation and extinction among lineages. The third tier selection reflects the infrequent catastrophes which may eliminate forms of life without respect to their adaptive or competitive advantage. It corresponds to K3 in Fig. 4.3.

Confirmed at least in part by the study of real biosystems such unconventional models offer in turn interesting suggestions for chemical and biochemical inspired computations and devices as for example in genetic or chemical programming (Keller and Banzhaf 1999, Matsumaru and Dittrich 2006).

#### 4.1.6.3 Chemical Programming

The architecture of the chemical programming developed in organic computing studies outlines four realms similar to that shown in Fig. 4.3.

In this case the realms shown in Fig. 4.4 are as follows: S-Hardware, sensor, actuator, K1-ChemOS, chemical operating system, K2-ChemVM, chemical virtual machine, and K3-Compiler.

The compiler takes a high level description of a chemical program as input. The chemical program consists of a list of molecules and reaction rules including kinetic laws. The compiler generates chemical byte code which can be processed by the chemical virtual machine that is able to run a chemical program. Communication between the chemical program and other hardware such as sensors or actuators is handled by the chemical operating system.

The integrative closure refers to compiler and hardware relationship.

Taking bio-chemical information processing as an inspiration for organic computing is attractive since biochemical systems possess a number of desirable properties. However chemical programming does not aims to replace the current computing systems. For example, implementing a word

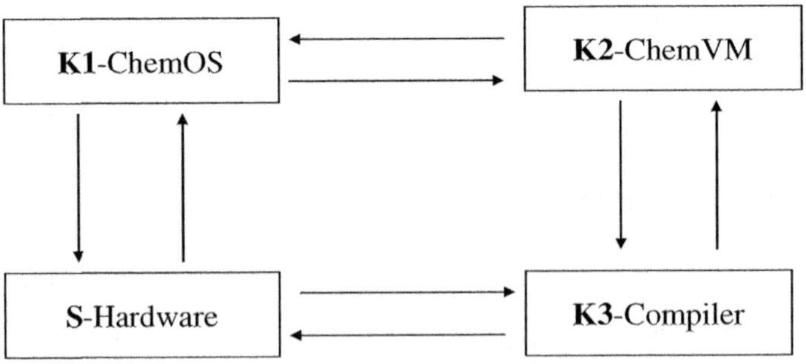


**Fig. 4.4** Chemical programming framework

processor on a chemical basis is not feasible, and techniques for programming chemical-like technical systems are still missing. It is more likely that "artificial chemistry" or "artificial chemical engineering" will be integrated as subsystems together with other high-level computing concepts and methodologies (Matsumaru and Dittrich 2006). There exists applications were the molecular level analog computer may have distinct advantages.

## 4.2 Artificial Neural Networks

### *4.2.1 Architecture Problem*

The domain of evolutionary computation is important for challenging applications as visual pattern classification, failure diagnosis, signal detection, sensor fusion, identification and control, planning and robotics, trading and so forth (Baeck et al. 1997).

The success and speed of training for neural networks NN is based on the initial parameter settings, weights, learning rates and architecture. NN simulation studies show that many complex problems cannot be solved by a learning algorithm in conventional fully connected layered NNs.

In spite of much research activity in the area of neural networks, NN, the design of architecture of NN was considered as less significant that the learning rules.

Evolutionary artificial neural network, EANN, refer to a class of artificial neural networks NNs, in which evolution is another fundamental form of adaptation in addition to learning. Evolutionary neural networks, EANN make use of evolutionary algorithm, EA, to improve NN architecture. EA are used to perform various tasks such as connection weight training, architecture design, learning rule extraction and adaptation (Yao 1999, Balakrishnan and Honavar 1995).

Evolutionary behaviour is considered as a preliminary method to confront complexity advent in computation. As a condition for evolutionary behaviour, EANN systems should be characterized by closure between the dynamics that is the phenotypic physical process of the material aspects of NN and the constraints that is the genotypic, syntactic rules or schemes of the symbolic aspects of the NN organization.

Evolutionary computational systems should be able to change both architectures and learning rules for automatic implementation without operator intervention.

The evolutionary artificial NN, the development models algorithms, are naturally represented in the PSM framework.

Innovative is the architecture indirect encoding based on the wave equation, WE. For genotypic, that is symbolic aspects of evolutionary systems the WE, generating the computation schemes or architectures is operational.

The novelty of the PSM framework lies also in hierarchical multiple scaled structure, in specificity of operations of decoding, from genotype to phenotype, and encoding, from phenotype to genotype.

The existing algorithms for evolutionary computations have many elements in common.

This follows the fact that they are inspired by the evolution theory in biology.

It would be beneficial to outline the common elements, to standardize the algorithms by applying the same generic framework to different complex problems.

The elements of a generic framework are that of the basic SKUP, that is the states S, the conditions K, the operators U and the possibilities P.

The state space S-represents the material aspects, the phenotypes, the realized experiments. For biological systems S corresponds to proteins, amino-acids, neurons.

The states changes are governed by dynamical laws and represent the factual aspects of the closure concept. The semantic closure concept refers to two levels architecture.

The conditions space K, describes genotypes, schemes, architectures.

For real biological system K is associated to DNA or to genome.

DNA contains the information needed by a biological organism to carry out its functions.

DNA represents the logic or informational part of the closure.

K is the space of genotypes. K elements may be indirectly specified by the wave model, WE. The WE solutions pertain to the set of conditions, K.

The genotypes are elements of a high dimensional search space. In conventional genetic algorithms, GA, the genotypes are binary strings of some fixed length, say n, that code for points in an n-dimensional Boolean search space. More generally a genotype can be considered as a string of genes.

Each genotype encodes for one or for a set of phenotypes. Such encodings employ genes that take on numerical values.

For SKUP framework U-is the operator associating to any element of K elements from S.

It is a decoding operator that produces the phenotypes corresponding to genotype and to previous phenotypes. This operator associates results, in S, to experiments designed in K.

Operators U may be stochastic or deterministic.

In real biological systems the genetic information written in DNA is translated in amino acids by means of a set of molecules known as amino acyl-tRNAs. These represent the biological interpretation of operators U.

The possibilities P-are related to the procedure to learn, to express information, to fitness evaluation. P facilitates the encoding process. This suggests that to the complementarity between K and S correspond a complementarity relation between U and P.

The phenotypes may be equipped with learning algorithms that train using evaluations on a given task. This evaluation of the phenotype determines the fitness of the corresponding genotype. In real biological systems P may be associated to RNA-replicase an enzyme that catalyzes the self-replication of RNA.

The generic SKUP framework elements for some of the existing evolutionary computation algorithms will be emphasized in what follows.

### *4.2.2 Graph Generation Grammar*

Classical learning algorithm for NN aim at finding weights for an NN whose architecture is frozen. Considerable performances are resulting by modifying NN architectures and the learning rules. The application of evolutionary algorithms, EA, to neural network, NN illustrates the increasing interest in combining evolution and learning (Yao 1999).

The graph generation grammar, developed by Kitano (1990) combines a genetic algorithm, GA with an L-system (Lindenmayer 1968). The L-system is a rewriting formalism introduced to model the growth of plants and a neural net, NN to enable modeling of the development process (Appendix A7). The GA is used to acquire graph rewriting rules, for the graph L-system, instead of directly acquiring the NN network topology. The introduction of developmental stages is considered more plausible biologically and computationally efficient.

The developed graph generation grammar (Kitano 1990) contains rules of the form:

$$A \rightarrow \begin{matrix} B & C \\ D & E \end{matrix} \tag{4.1}$$

The left-hand-side LHS of the production rule is a symbol and the right hand side RHS is a matrix of symbols from an alphabet.

The terminal symbols are constant rules as for instance:

$$0 \rightarrow \begin{matrix} 0 & 0 \\ 0 & 0 \end{matrix} \quad 1 \rightarrow \begin{matrix} 1 & 1 \\ 1 & 1 \end{matrix} \tag{4.2}$$

After a specified number of steps in L-system the matrix symbols restrict to 0 or 1. The resulting matrix is considered as a connectivity matrix and as in direct encoding methods an NN graph is associated to this.

Subsequently the NN weights are modified by learning methods as the back-propagation, for example.

A new graph generation model is presented in what follows, outlining the generic PSM framework elements and based on WE.

In this generic PSM framework, the genome corresponds to the elements of K.

Starting from an alphabet the WE generates more elements of K.
Consider the WE model with Y=$y_0y_1$ and T=$t_0t_1$.
The wave equation WE model has among its convection type solutions:

$$y_0 = \begin{matrix} 0 & 1 \\ 1 & 0 \end{matrix} \tag{4.3}$$

$$y_1 = \begin{matrix} 0 & 1 & 2 & 3 \\ 1 & 2 & 3 & 0 \\ 2 & 3 & 0 & 1 \\ 3 & 0 & 1 & 2 \end{matrix} \tag{4.4}$$

The general solution may be the categorical product Y=$y_0 \times y_1$:

$$Y = y_0 \times y_1 = \begin{matrix} 00 & 01 & 02 & 03 & 10 & 11 & 12 & 13 \\ 01 & 02 & 03 & 00 & 11 & 12 & 13 & 10 \\ 02 & 03 & 00 & 01 & 12 & 13 & 10 & 11 \\ 03 & 00 & 01 & 02 & 13 & 10 & 11 & 12 \\ 10 & 11 & 12 & 13 & 00 & 01 & 02 & 03 \\ 11 & 12 & 13 & 10 & 01 & 02 & 03 & 00 \\ 12 & 13 & 10 & 11 & 02 & 03 & 00 & 01 \\ 13 & 10 & 11 & 12 & 03 & 00 & 01 & 02 \end{matrix} \tag{4.5}$$

The matrix Y contains elements of K. To this matrix, a connectivity matrix CM and an NN is associated in different ways.

Suppose for instance, that instead of the double digit vectors (ij) in Y we put 0 if the difference between digits is equal or higher than 2 and 1 if the difference is lower or equal to 1.

The rule may be:

$$u(ij) = \begin{matrix} 1 & i - j \leq \pm 1 \\ 0 & i - j \geq \pm 2 \end{matrix} \tag{4.6}$$

Using this rule, the connectivity matrix CM is resulting instead of Y and an NN may be associated to this matrix. In this particular case the neuron 1 is connected with 2, 5, 6 and 7, the neuron 2 is connected with 1, 4, 5, 6 and 8 and so on.

$$CM = \begin{matrix} 1 & 1 & 0 & 0 & 1 & 1 & 1 & 0 \\ 1 & 0 & 0 & 1 & 1 & 1 & 0 & 1 \\ 0 & 0 & 1 & 1 & 1 & 0 & 1 & 1 \\ 0 & 1 & 1 & 0 & 0 & 1 & 1 & 1 \\ 1 & 1 & 1 & 0 & 1 & 1 & 0 & 0 \\ 1 & 1 & 0 & 1 & 1 & 0 & 0 & 1 \\ 1 & 0 & 1 & 1 & 0 & 0 & 1 & 1 \\ 0 & 1 & 1 & 1 & 0 & 1 & 1 & 0 \end{matrix} \tag{4.7}$$

Elements of generic PSM framework are identifiable and easy to be compared with those of Kitano model.

The conditions K may include the initial alphabet and the matrices resulting as WE solutions.

The grammar rules are based on concatenation as for instance:

$$0 \rightarrow \begin{matrix} 00 & 01 & 02 & 03 \\ 01 & 02 & 03 & 00 \\ 02 & 03 & 00 & 01 \\ 03 & 00 & 01 & 02 \end{matrix} \qquad 1 \rightarrow \begin{matrix} 10 & 11 & 12 & 13 \\ 11 & 12 & 13 & 10 \\ 12 & 13 & 10 & 11 \\ 13 & 10 & 11 & 12 \end{matrix} \tag{4.8}$$

These are resulting by categorical product "×" operations used in WE solutions.

The WE model includes grammar as that used in Kitano model.

The operator U-determine the rule translates a matrix like Y in the connectivity matrix CM. It is a specific rule u, associated to this translation.

The states S- may be identified with the NN weight of the connections associated to NN.

The possibilities P, give the learning rules for weights.

Table 4.9 contains the solutions $Y = y_0$ and $Y = y_0 \times y_1$ presented together (equations 4.3 and 4.5). Selected states are bolded and underlined in Table 4.9.

**Table 4.9** WE solutions at m=0 and m=1. Selected states

| | | | | | | | | | |
|---|---|---|---|---|---|---|---|---|---|
| 00 | 01 | | 02 | 03 | **10** | **11** | | **12** | **13** |
| 01 | 02 | | 03 | 00 | **11** | **12** | | **13** | **10** |
| | | 0 | | | | | **1** | | |
| 02 | 03 | | 00 | 01 | **12** | **13** | | **10** | **11** |
| 03 | 00 | | 01 | 02 | **13** | **10** | | **11** | **12** |
| 10 | 11 | | 12 | 13 | 00 | 01 | | 02 | 03 |
| 11 | **12** | | **13** | 10 | 01 | 02 | | 03 | 00 |
| | | **1** | | | | | 0 | | |
| 12 | **13** | | **10** | 11 | 02 | 03 | | 00 | 01 |
| 13 | 10 | | 11 | 12 | 03 | 00 | | 01 | 02 |

Applying for the area selected the rule u, the Table 4.10 results.

It is supposed that only "1" represents a large node "•" at the level m=0 and a smaller node "•" at the level m=1. Table 4.10 shows the cells resulting after the application of the rule u and area selection.

High border thickness outlines the level m=0 while medium border thickness outlines the level m=1. Next steps in NN development should establish connections between different nodes. The superposed arrows show an oriented NN.

**Table 4.10** Associated NN

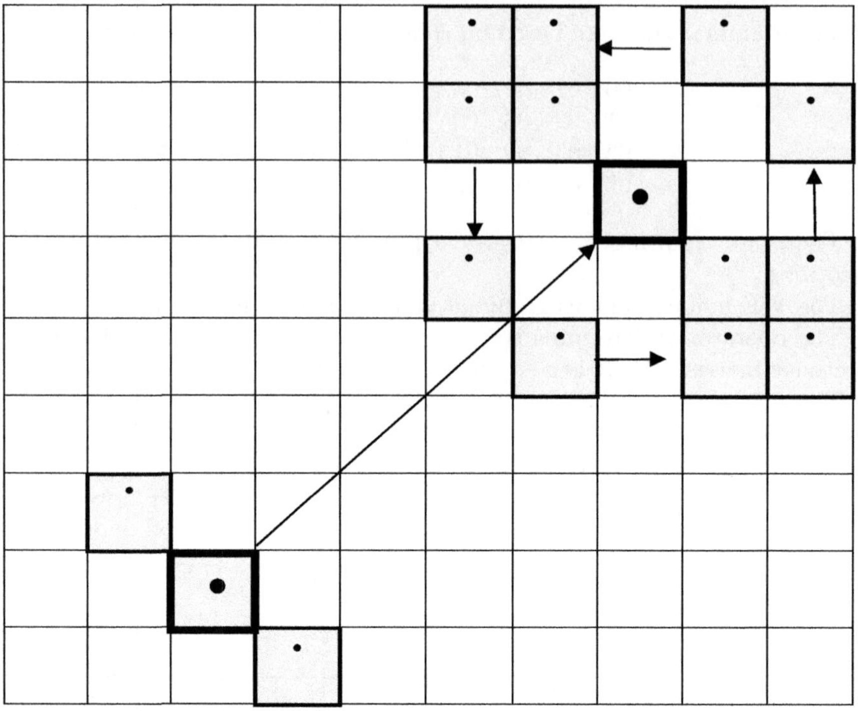

## 4.2.3 Cell Space Encoding

### 4.2.3.1 Growing Neural Networks

Several indirect encoding methods are inspired by the development and morphogenesis biological processes. An illustrative example is the algorithm due to Nolfi and Parisi (Nolfi and Parisi 1995a, 1995b). This algorithm presents a method for encoding NN architecture into a generic string, suggested by the real neural development. Inherited genetic material specifies developmental instructions that control the axonal grow and the branching process of a set of neurons. The neurons are encoded with coordinates in a two-dimensional space.

The neurons located in a 2D space are associated to the space K of conditions in generic PSM framework. K is the set of initially established neurons or cells.

The dendrite growing is part of the states S-process. For different neurons or cells, the GA mechanism induces a tree as a random walk in the space S. If the growing axonal branch of a particular neuron reaches another neuron a connection is established.

The resulting NN contains only the completely connected neurons and the branches.

The temporal dimension of developmental process was taken into account in the Nolfi and Parisi models. Several time scales may be naturally considered in WE study.

The NN architecture was tested for specific problem as for instance for pattern recognition.

Then according to fitness criteria S is modified. This is the learning process associated to possibilities P.

#### 4.2.3.2 Schema for Cell Encoding Algorithm

The SKUP framework for new cell space encoding model is presented in what follows. The hierarchy of neurons will be considered. Here the large dots "•" denotes the level m=0, while smaller dots "•" denotes the level m=1 (Table 4.11).

For two level evolution, m=0, m=1 the quadruple SKUP consists of the vectors

$$S = (s^0, s^1); K = (k^0, k^1); U = (u^0, u^1); P = (p^0, p^1)$$

**Table 4.11** Schema for cell space encodings

| | | | | | | | | |
|---|---|---|---|---|---|---|---|---|
| | | | | $s_2^0$ | | | | |
| | | | | | | | | |
| | | $\bullet k_1^0$ | | | | $\bullet k_2^0$ | | |
| | | | | | | | | |
| $s_1^0$ | | | | | | $s_2^1$ | | $s_3^0$ |
| | | | | | $\bullet k_1^1$ | | $\bullet k_2^1$ | |
| | | $\bullet k_0^0$ | | $s_1^1$ | | $\bullet k_3^0$ | | $s_3^1$ |
| | | | | | $\bullet k_0^1$ | | $\bullet k_3^1$ | |
| | | | | $s_0^0$ | | $s_0^1$ | | |

Let: $k_0^0$, $k_1^0$, $k_2^0$ and $k_3^0$ denotes the possible conditions at the level m=0. The upper index refers to levels while the lower index refers to the time step. It should be emphasized that the time steps at different levels may be different and this is a key feature for evolutionary behavior. The states and the conditions at the level m=0 are represented by high thickness border cells in Table 4.11.

The system initial state is $s_0^0$. With possibility $p^0(k_0^0|s_0^0)$ the condition $k_0^0$ is selected. Based on this, the operator $s_1^0=u^0(k_0^0,s_0^0)$ allows the transition to the new state $s_1^0$. Then with possibility $p^0(k_1^0|\, s_1^0)$ the new condition, $k_1^0$ arises. In the new condition, the operator $u^0(k_1^0,\, s_1^0) = s_2^0$ allows the system to reach the state $s_2^0$.

Observe that: $s_1^0=u^0(k_0^0,\, s_0^0)$ implies $s_2^0=u^0\ (k_1^0,\, u^0(k_0^0,\, s_0^0))$.

With possibility $p^0(k_2^0|s_2^0)$, the condition $k_2^0$ is selected and finally the new state results $s_3^0=u^0(k_2^0,\, s_2^0)$ results. Observe that $s_3^0=u^0(k_2^0,\, u^0(k_1^0,\, u^0(k_0^0,\, s_0^0)))$.

The states are resulting not necessarily in a recursive way since, in practical cases the operators may varies with the step.

The possible states at the level m=0 are: $s_0^0$, $s_1^0$, $s_2^0$, $s_3^0$, $s_4^0$. The interpretation of the higher-thickness border cells trajectory is the process description as follows: from the state $s_0^0$ through condition $k_0^0$ towards the state $s_1^0$, then through condition $k_1^0$ towards the state $s_2^0$, and so on.

The net development may be continued at the level m=1 for different new conditions $k_0^1, k_1^1, k_2^1, k_3^1$.

The states and the conditions at the level m=1 are represented by medium thickness border cells. The system initial state at the level m=1 is $s_0^1$. With possibility $p^1(k_0^1|s_0^1)$ the condition $k_0^1$ arises. Based on this, the operator $u^1(k_0^1,\, s_0^1) = s_1^1$ describes the transition to the new state $s_1^1$. Then with possibility $p^1(k_1^1|\, s_1^1)$ the new condition, $k_1^1$ arises. In the new condition, the operator $u^1(k_1^1,\, s_1^1)=s_2^1$ allows the system to reach the state $s_2^1$.

Observe that: $s_2^1=u^1(k_1^1,\, u^1(k_0^1,\, s_0^1))$ and $s_3^1=u^1(k_2^1,\, u^1(k_1^1,\, u^1(k_0^1,\, s_0^1)))$.

The states at the level m=1 are: $s_0^1$, $s_1^1$, $s_2^1$, $s_3^1$. The conditioning at the level m=1 is represented by the loop: $k_0^1$, $k_1^1$, $k_2^1$, $k_3^1$.

The interpretation of the standard border thickness trajectory is as follows: from the initial state $s_0^1$ through condition $k_0^1$ to the state $s_1^1$, then through condition $k_1^1$ to the state $s_2^1$, and so on.

The elements of generic PSM framework are clearly indicated here.

The conditions are $K = (k^0, k^1)$ were $k^0$ corresponds to large neurons "●" while $k^1$ corresponds to smaller neurons "•"

The states $S = (s^0, s^1)$ are the NN. There are two types of NN here and this allows evolutionary computation. This corresponds to a decomposition of the problem in sub-problems more easy to solve.

Rules for $U = (u^0, u^1)$ and $P = (p^0, p^1)$ depend on the studied case.

## 4.2.4 Perspectives

### 4.2.4.1 Hierarchy of Prediction Layers

From the point of view of a natural agent the external environment does not provide any direct indication on how the agent should act to attain a given goal. The environment provides a large number of data, the sensory states. The system should be able to extract regularities from time series through prediction learning.

Nolfi and Tani (1999) shows that the ability to extract regularities from data can be enhanced if we use a hierarchical architecture in which higher layers are trained to predict the internal state of lower layers when such states change significantly.

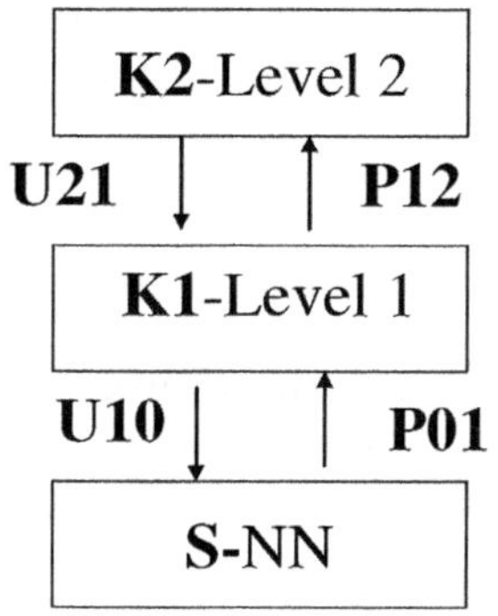


**Fig. 4.5** Hierarchy of prediction layers

Fig. 4.5 shows the architecture with three levels one corresponding to sensor states and two prediction levels.

For this type of architecture the sensory information will be progressively transformed going from lower to higher levels.

S denotes the environment that is the NN nodes.

K1 and K2 are the two prediction levels. K1-represents the basic level while K2-is the meta-level.

A change in K2 has higher leverage because it represents multiple changes at lower level K1. U10: K1→S describes the decoding and actions, P01: S→K1, describe the fitness evaluation for S. The information change between the basic level and the meta-level is characterized by the operator U21 and the possibilities P12.

The first prediction level K1 predicts the states of the sensors by receiving as input their state in the previous time step and the planned action. The higher level K2 predicts the internal state of the lower level K1 by receiving as input their previous state.

By training an architecture of this type to predict the next sensory state of an animate navigating in an environment divided into two rooms Nolfi and Tani (1999) showed how the first level prediction layer extracts low level regularities such as walls and corners while the second level prediction layer extracts higher level regularities such as the left side wall of the large room.

That prediction learning can extract higher level regularities from time series was shown by Elman studies of languages (Elman 1990, 1993). He showed how by training a simple recurrent NN to predict the next word in sentences of a language the network was able to extract high level regularities for words such as nouns or verbs.

The architecture shown in Fig. 4.5 lacks some elements of a potential top level K3 and also the integrative links between the level, K2 or K3 and environment S.

The missing links induce limitations and prevents this architecture to become evolvable and to detect regularities for changing environments as for instance the presence of a new object for animates or of new words for language.

#### 4.2.4.2 Structured GA

It was recognized that genetic algorithms work well in some cases and not in others, but it is not yet clear why this happens.

To address some of the difficulties encountered by the traditional GA in problem solving Dasgupta and Mc Gregor (1992) introduced a two-level structure for the chromosome in genetic algorithms.

The structured GA is based on hierarchical genomic structure and a gene activation mechanism in its chromosome. Genes at different level can be either active or passive. Higher level genes activate or deactivate sets of lower level genes. Thus the dynamic behaviour of genes at any level is governed by the high level genes.

In biological systems, there appear to be many possible strands of evidence supporting this model.

Fig. 4.6 outlines the categorical framework of evolutionary computation algorithm for the three levels. S represents the phenotype that is the NN nodes.

K1 and K2 are the two conditions levels. K1-represents the basic GA chromosome.

K2-is the meta-level representing the structured GA, sGA chromosomes. The strategies are defined at the meta-level since a change in K2 has higher leverage because it represents multiple changes at lower level K1.

U10: K1→S describes the decoding P01: S→K1 describes the fitness evaluations. The information change between the basic level and the meta-level of GA is characterized by the operator U21 and the possibilities P12. They describe sGA rules.

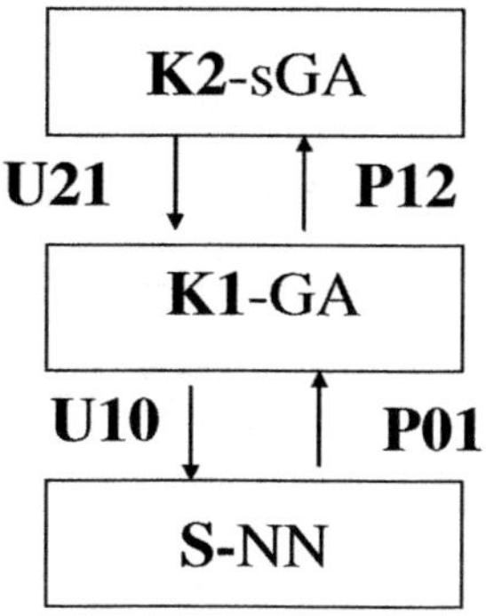


**Fig. 4.6** Evolutionary designs for artificial neural networks

U21 corresponds to transcription processes.

The gene activation mechanism may be expressed by the categorical product.

It is possible to make use of different types of categorical product in K1 and K2.

Observe that the architecture shown in Fig. 4.6 may act like a complex network in which the genes corresponding to K1 and K2 regulates one another activity either directly or through their products, from S. This makes the architecture applicable to the GA-deceptive problems.

Other three level hierarchical architectures useful for modeling are the contextual GA (Rocha 1997) and the chemical GA (Suzuki and Sawai 2002).

#### 4.2.4.3 n-Graphs for Growing Neural Networks

The Nolfi and Parisi growing neural networks (Nolfi and Parisi 1995a, 1995b) may be studied in the n-graphs framework (Appendix A5).

Fig. 4.7 illustrates a potential development for NN dendrites architecture using n-graphs.

For the stage n=0 there are isolated neurons. The $1^{st}$ order evolutionary step is allowed by interactions with the substrate. At n=1 interactions and branches appear. It is the 1-categories level

Branches are separated for the n=1 stage but they interact as the $2^{nd}$ evolutionary step shows. The n=2 level corresponds to 2-categories and allows arrays of interacting branches, the coupling of two or more branches in macro-branches or trees.

The $3^{rd}$ order evolutionary step outlines the final stage, n=3 corresponding to a kind of single tree. The single tree pattern is specific to the growing. Some neurons and some branches remain undeveloped. Unconnected branches or neurons are removed. It is the 3-categories level.

The integrative closure, connecting also the n=0 and n=3 levels represents the challenge for such NN systems. A possibility shown in Fig. 4.7 is the

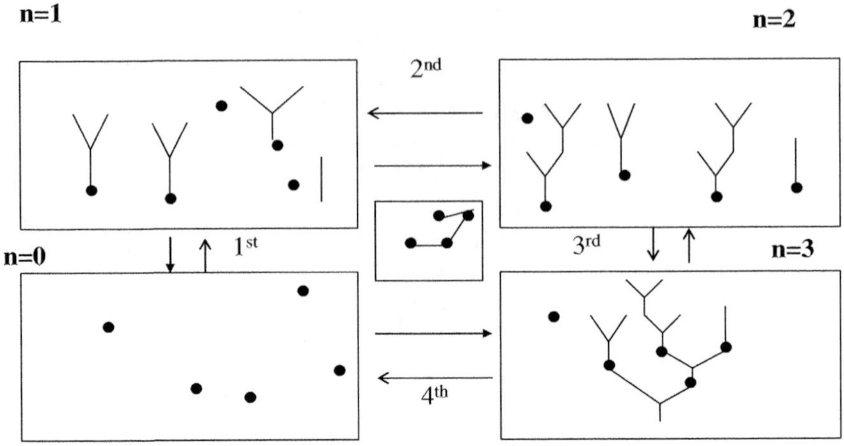


**Fig. 4.7** n-graphs for growing neural networks

restriction to the central structure retaining only the four neurons connected at the levels 1, 2 or 3. Growing net may restart from this central structure and follow the same steps on a higher methodological plane that is at a new dimension in modeling. The branching should be reversible.

The integrative closure between cognitive levels and environment was studied in the so-called artificial life NN, ALNN (Nolfi and Parisi 1977). This study models an NN that lives in a physical environment. An active, embodied knowledge and knowledge acquisition makes ALNN closer to Piaget schema rather than to the classical NN.

Networks that adapt or self-organize structurally to the environment by adding and removing neurons and connections in the system exploit mechanisms that are similar to those used in the growth of an organism (Fritzke 1994). A developmental mechanism implies the presence of a mechanism for cell creation, a requirement for structural adaptation and thus can simplify the design of structure adaptable systems.

### 4.2.4.4 Protein Based Neural Networks

Many proteins in living cells appear to have as a primary function the transfer and processing of information accomplished by the physical or chemical transformation of metabolic intermediates or the building of cellular structures (Bray 1995, Spiro et al. 1997).

Cells perform calculations as a mean of monitoring and responding to their internal and external environment.

It was observed that the highly interconnected network of protein based pathways in living cells share the properties of neural nets allowing cognitive capacities. This refers to memory capacity, pattern recognition, handling fuzzy data, multifunctionality, signal amplification, integration and crosstalk and signal amplification (Paton and Toh 2004). Moreover the mathematical formalism of artificial neural networks is a more accurate approximation for networks of protein molecules than for networks of real neutrons (Bray 1995). Ideas from the category theory can be used to illustrate this point. The internal organization of a protein can be modeled by a diagram of domains that is cooperating objects in which links represents functional relations. A colimit glues a pattern into a single unity in which the degrees of freedom of the parts are constrained by the whole. A limit represents the relationship between the whole that is the single unity and its components. It is possible to reason about functions with regard to how a whole is integrated and coheres out of its parts. Part-whole relations may be described as emergent cohesion reflecting the internal synergy in which interactions and local measurement generate cohesion. Cohesion concerned with part-whole relations is correlated to colimits in the sense that the whole keeps the parts together.

Proteins molecules can act as logical elements and assemblies of proteins can be artificially coupled to perform computations.

Bacterial chemotaxis illustrates the computation potentialities of protein networks in living cells. Chemotaxis is the process by which a cell alters its speed or frequency of turning in response to an extracellular chemical signal.

A four realms framework may describe the circuit mediating the chemotactic response of bacteria.

Fig. 4.8 shows illustrates the organization of protein based neural network for *E. Coli* chemotaxis. Here S denotes the receptors. These include chemoreceptors and amino-acids as ligands. The proteins network includes also a hierarchy of signaling proteins as: CheA, CheB, CheR, CheW, CheY and CheZ.

It may be assumed that the categories of signaling proteins are: K1={CheA, CheW}, K2={CheB, CheY} and K3={CheR, CheZ}.

The signification of the functors U and possibilities P is explicit in Fig. 4.8.

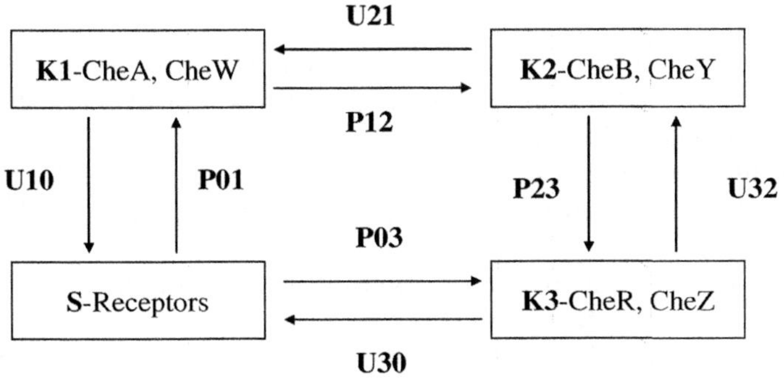


**Fig. 4.8** Protein network mediating the chemotaxis

For instance, the operators U10 describes the interaction {CheA, CheW} → Receptors while the operator U30 describes the interaction {CheR, CheZ} → Receptors.

The mechanism is based on three inter-correlated realms for signaling proteins (Bray 1995). Other chelating proteins and other hierarchical order have been considered in the vast literature dedicated to chemotaxis.

The framework shown in Fig. 4.8 doesn't aims to reproduce the details of the *E. coli* chemotaxis but to retain the basic pattern and to make suggestions how to design artificial neural networks.

It is known that despite its relatively simple structure, *E. Coli* is capable of embodiment and highly autonomous behavior (Quick et al. 1999).

Viewed as signal processing systems, cell signalling networks like that shown in Fig 4.8 can be considered as special purposes computers. In contrast to conventional silicon-based computers, the computation is not realized by electronic circuits but by chemically reacting molecules in the cell.

Such system may achieve the essential properties of integrative closure. A critical feature of this closure is that the steady-state values reached after a changed input should still ensure the autonomous core functioning of the entire system.

As computational devices the proteins networks can be compared to analog computers. Several analog computers have long been displaced by the digital computers due to programming and stability issues. However there are situations were it is required to interface computation with chemical interactions when artificial protein based neural networks may be used to implement special computation and signal processing tasks. This may have direct applications to the so called smart drugs or for other bio-medical interventions.

## 4.3 Artificial Neural Codes

### *4.3.1 Neural Coding*

The neural coding problem in perception of signals involves the interpretation of the neural correlations of sensory registrations (von der Malsburg 1994, Freeman 2000, Cariani 2001).

Sensed information can be encoded in patterns of neurons that respond, the so-called channel codes or in temporal relations between spikes that is temporal codes.

Progresses in the domain of understanding neural codes have been achieved starting from neural activity modeling.

In the firing rate model it is speculated that the information is conveyed by being encoded in the rate at which action potential are generated by neurons. Over short times the network structure of the brain is commonly regarded as fixed. Brain states may be regarded as semantic symbols. They are lacking hierarchical and syntactical structure.

The correlation theory of brain function (von der Malsburg 1986, 1994) challenged the semantic symbol system sketched above and proposed a different interpretation of data in terms of semantic symbols with a richer structure. The correlation theory suggests that information is conveyed in the brain through correlations of neural firing patterns.

This theory received support from a model developed by Damasio which holds that entities and events are represented in the brain by time-locked synchronous neural firing patterns (Damasio 1989).

An approach based on the PSM method and the wave equation, WE, may generate solutions looking like the temporal response patterns registered in brain studies.

The contact with existing symbolic neural architectures may be established and on this basis potentially neurocognitive architectures are resulting.

Neural-symbolic systems are hybrid systems that integrate symbolic logic and neural networks. The goal of neural-symbolic integration is to benefit from the combination of features of the symbolic and connectionist paradigms of artificial intelligence, AI.

In the basic SKUP framework K is associated to symbols while S to neural networks.

An open problem is how to put together in the same framework K and S.

It may be assumed that a compositionality principle would allow computing the meaning of complex formulas using the meaning of the corresponding sub-formulas.

On the other side, it was assumed that NN are non-compositional from principle, making them unable to represent complex data structure like formulas, lists, tables, and so forth.

Two aspects can be distinguished, the representation and the inference problem.

The first problem states that complex data structures can only implicitly be used and the representation of structured objects is a challenge for connectionist networks. This is correlated to possibilities P in the basic SKUP framework.

The second problem, of inference, tries to model inferences of logical systems with neural account. This is correlated to operators U in the SKUP.

## *4.3.2 Symbolic Connectionist Hybrids*

Some authors claimed that connectionist models as the neural networks, NN, did not support symbolic processing and were incapable of adequately representing evolving neurocognitive structures (Fodor and Pylyshyn 1988).

Symbolic connectionist models, implemented as hybrid devices allowed a rebuttal of Fodor and Pylyshyn criticism in both theory and practice.

Hybrid symbolic connectionist techniques allow vectors representing the constituents of a symbol structure to be combined into a single vector representing the whole structure, and for this vector to be decoded into the vectors representing the original constituents. In this manner representations for compositional structures can be built up and, then processed by NN.

The recursive auto-associative memory (RAAM) was among the NN models developed to address the question of how compositional structures may be stored within a connectionist framework (Pollack 1990). The data for a RAAM network consists of a collection of trees and a representation that is a pattern of "0", "1" and so on for each terminal symbol occurring in those trees. The task for the network is to provide a means of compressing each tree into a representation, an activation vector, and reconstructing the tree from its representation. The elements of the SKUP are naturally associated to the RAAM elements. The input and output units may be associated to the set of conditions, K. The RAAM architecture contains encoding or compressor networks, associated to possibilities P in the SKUP framework. The RAAM contains also decoding or reconstruction networks associated to operators U. The hidden units are associated to the states S. As shown in Sect. 2.3.1, to any matrices containing discrete information as "0", "1" and so on, classification trees may be associated based on similarity calculations. Another promising NN architecture is the distributed associative memory developed by Austin (Austin 1996). Associative memories operate differently from the memories typical for current computer architectures. This type of architecture take a set of data often in the form of an image, and scan the entire set of data in memory until it finds a set that matches it, as much as possible.

Fig. 4.9 shows a typical neurocognitive architecture making use of two categorical frames for conditions, K1 and K2 with two types of tensorial product, the coproduct "∪" for K1 and the product "×" for K2.

The architecture proposed by Austin makes use for symbolic processing, the component-wise operations in GF (2). The categorical product "×" is in this case a vectorial outer product, while the categorical coproduct, "∪" is a concatenation followed by superimposed coding.

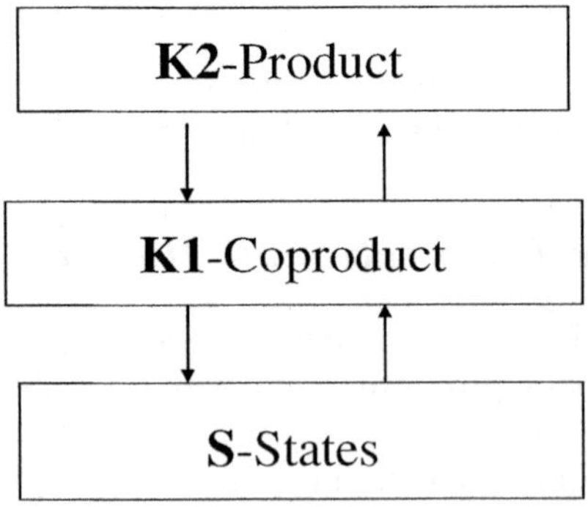


**Fig. 4.9** Typical three levels framework

Another strategy to meet the challenges posed by connectionism critiques for both models and devices is offered by the so-called holographic reduced representations HRR, (Plate 1995). Associative memories are conventionally used to represent sets of pairs of vectors. Plate describes a method for representing complex compositional structures in distributed representations. The method uses circular convolution to associate items which are represented by vectors. The representation of an association is a vector of the same dimensionality as the vectors which are associated. The method allows encoding relational structures in fixed width vector representation but it should be noted that this increases the risk of missing the emergent structures.

Plate calls his models, holographic reduced representations, since convolution and correlation based memory mechanisms are close related to holographic storage. The circular convolution may be associated to the categorical product, "$\times$", while the superposition may be associated to categorical coproduct, "$\cup$" (Fig. 4.9).

The properties of higher neurocognitive processes and how they can be modelled by NN have been extensively studied by Halford and collaborators (Wilson and Halford 1994, Halford et al. 1998). They proposed the so-called STAR model of analogical problem solving.

The rank of tensor used by Halford is linked to the arity of relation, that is, to the number of attributes to the relation, and in the end, to the Piaget stages of neurocognitive development. The STAR model uses a tensor of rank-3 to represents a predicate of two arguments.

Halford studies suggests that for early Piaget stages in neurocognitive development, the categorical coproduct, "$\cup$", prevails allowing the associative knowledge. This is a fast and parallel process. During the higher Piaget stages the categorical product, "$\times$" seems preponderant, allowing the relational knowledge. It is a slow, sequential, effortful, higher neurocognitive process (Fig. 4.9).

The study of tensor product networks using distributed representations outlined the significant role of Hadamard matrices (Wilson and Halford 1994). As shown in Sect. 2.2.3 these matrices are special solutions of the WE equations.

Notice that Halford and associates evaluated the significance of Klein-4 group and of Latin squares for learning transfer in NN and in neurocognitive systems. Such structures correspond to the INRC group studied by Piaget (Inhelder and Piaget 1958) as well as to standard solutions of the WE model.

### 4.3.3 Temporal Synchrony

A promising way of dealing with variable binding in connectionist systems is to use the temporal aspects of nodes or neurons activation. Phase synchronization can be used since it allows different phases in an activation cycle to represent different objects involved in reasoning, and representing variable binding by the in-phase firing of nodes (von der Malsburg 1986, Hummel and Biederman 1992).

Based on temporal synchrony, SHRUTI system (Shastri and Ajjanagade 1993) provides a connectionist architecture performing reflexive reasoning. SHRUTI shows how synchronous activation can be harnessed to solve problems in the representation and processing of high level conceptual knowledge. LISA system (Hummel and Holyoak 1997, Hummel and Choplin 2000) used the synchronous activation approach to model analogical inference. Both computational systems demonstrates that temporal synchrony in conjunction with structured neural representations suffices to support complex forms of relational information processing specific to neurocognitive systems.

The problem for such systems is their suitability for reflexive or reflective neurocognitive processes. Reflexive processes are linked to categorical coproduct while reflective processes, are linked to the categorical product (Fig. 4.9).

While reflexive and reflective processes follow different kinds of computational constraints, in most cases, the two types of processes interact and need to be integrated in the performance of a single task.

SHRUTI represents a restricted number of rules with multiple place predicates. There are several types of nodes or neurons in the architecture, denoted for example by circles, triangles and pentagons.

Relational structures as frames and schemas are represented in SHRUTI by focal clusters of cells, and inference in SHRUTI corresponds to a transient propagation of rhythmic activity over such cell-clusters. Dynamic bindings between roles and entities are represented within such a rhythmic activity by the synchronous firing of appropriate role and entity cells. Rules correspond to high-efficacy links between cell-clusters, and long-term facts correspond to coincidence and coincidence-failure detector circuits.

SHRUTI was designed for reflexive reasoning tasks and the model is not suited to account for reflective processes.

To ensure applicability to complex situations, SHRUTI was coupled with systems activating the reflective component of problem solving. Such systems are capable of attention shifting, making and testing assumptions, evaluating uncertainty. The resulting neurocognitive systems presented both reflexive and reflective capabilities and has been used to model decision making in imposed time frames.

LISA is a computational model based on temporal synchrony and designed for analogical inference and for schemes induction. The data for LISA network consists of a collection of trees and a representation that is a pattern of "0", "1" and so on for each terminal symbol occurring in those trees.

The LISA system is shown in Fig. 4.10.

The basic level includes semantic units, s, the next includes the so-called localist units, L, (predicate/object or object/roles), the next level includes the sub-problems and the higher level the problems.

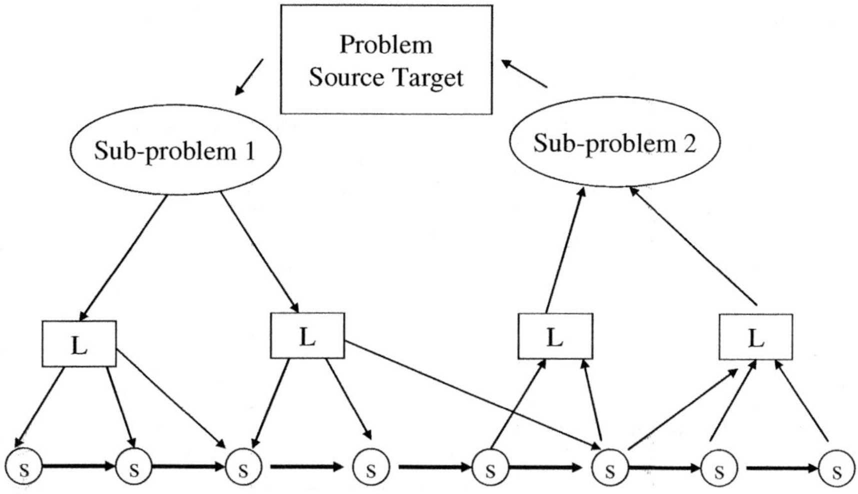


**Fig. 4.10** LISA neurogonitive frameworks

## *4.3.4 Perspectives*

### 4.3.4.1 Three Realms Neurocognitive Architectures

Three realms multi-agent architectures may achieve integrative closure, in this case cognitive evolvability and autonomy (Zachary and Le Mentec 2000, Di Marzo Serugendo et al. 2007).

An agent architecture grounded in models of human reasoning such as Cognet is shown in Fig. 4.11 (Zachary and Le Mentec 2000).

Cognet is a research framework concerning the analysis and modeling of human behavioral and neurocognitive processes in real-time, multi-tasking environments.

Meta-cognition refers to cognition about cognition and in this case to the ability to explicitly and strategically think about and control an agent's own neurogonitive processes. The Cognet architecture allows a meta-cognitive control of neurocognitive processing. An emphasized aspect is that of self-awareness of resources and processes.

The categorical framework is shown in Fig. 4.12. It shows the architecture of conditioning levels with two-sided dependence.

The elements of the categorical framework are as follows:

S-Environment interface allowing action and perception

K1-Cognition processes, K2-Metacognitive processes

The neurocognitive level is structured in K1 and K2 to allow performing integrated cognitive/behavioral tasks.

U10: K1→S actions physical or verbal

P01: S→K1 sense of visual and auditory cues

U20-motor action resources, P02-perception resources

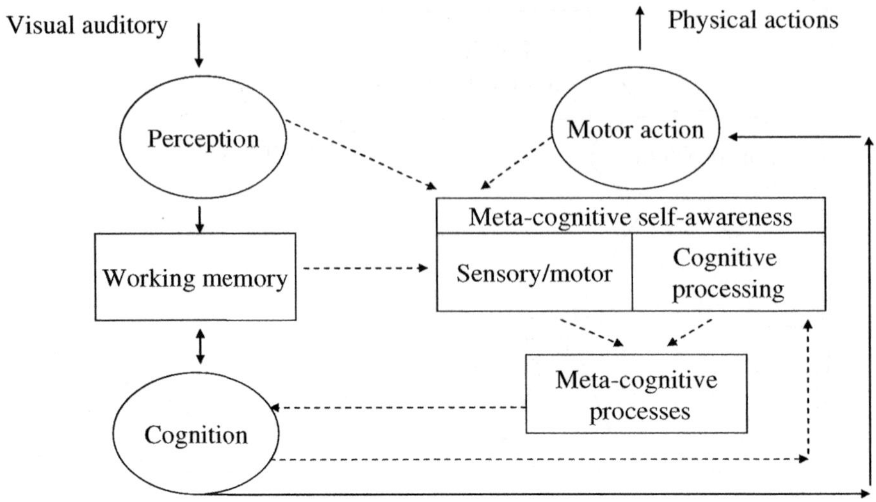


**Fig. 4.11** Cognet information processing framework

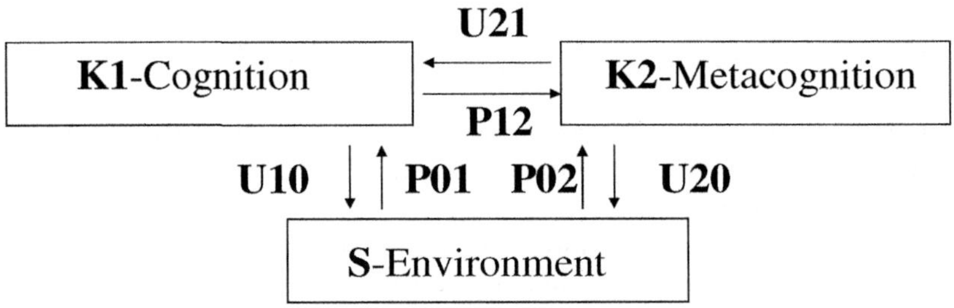


**Fig. 4.12** Three realms neurogonitive framework

A similar framework is shown by the self-adaptive and self-organizing SASO architectures (Di Marzo Serugendo et al. 2007).

In this case, the elements of the categorical framework are as follows:

S-Application components, services
K1-Metadata, K2-Policies
U10: K1→S defines acting
P01: S→K1 defines sensing
U20 and P02 describes the application and acquisition of policies

Elements of the cognitive architecture shown in Fig. 4.12 may be correlated to the adaptive resonance theory ART (Carpenter and Grossberg 1987).

In this case, the elements of the categorical framework are, S-receiver of the input signals.

K1-classifier of the afferent input patterns and K2-attention/orienting subsystem.

Viewed abstractly, the ART classifier network meets the definition of an algebraic structure known as grupoid. Formally a grupoid is any mathematical structure consisting of a set of inputs and an operation on this set possessing the property of closure. A grupoid may be seen as a category in which any morphism is an isomorphism.

#### 4.3.4.2 Four Levels or Realms Neurocognitive Architectures

A four level architecture for LISA approach is presented in Fig. 4.13.

This architecture takes into account that the working capacity of human is typically limited at four relations (Halford et al. 1998). Hummel and Holyoak (1997) correlate the four levels of memory in the LISA neurocognitive framework to the limits in mental storage capacity. Probably, this fact is related to the four modular architecture of the neurocognitive system and to cerebral rhythms (Freeman 2000).

The signification of the functors U and possibilities P is explicit in Fig. 4.13.

S-Semantic units, K1-Localist units, K2-Sub-problems, K3-Problems

U10, U21, U32 corresponds to implementation operations

Observe that: U10:K1-Localist→S-Semantic, U21: K2-Sub-problems→K1-Localist, and U32: K3-Problems→K2-Sub-problems.

P01, P12, P23 and P03 are synthesis steps.

P01: S-Semantic→ K1-Localist, P12: K1-Localist→K2-Sub-problems, and P23: K2-Sub-problems→K3-Problems.

The four realms approach emphasizes the need of contact between the problem and the ground semantics units.

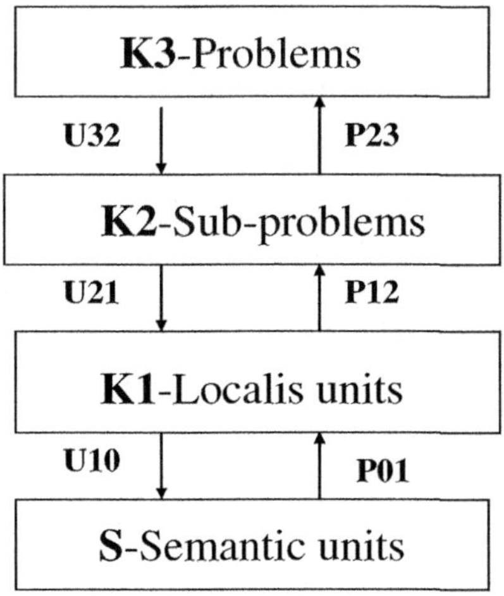


**Fig. 4.13** Four levels neurogonitive framework

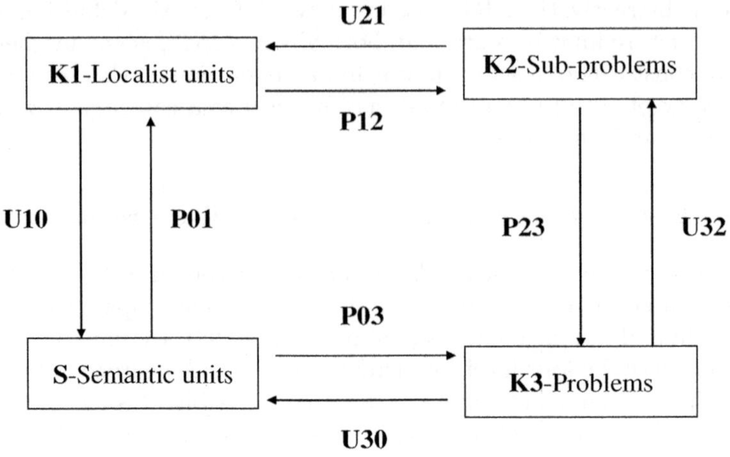


**Fig. 4.14** Four realms neurogonitive framework

Fig. 4.14 shows a prospective four realms architecture developing the LISA approach towards integrative closure and evolvability. It is expected to facilitate the interaction between S-Semantic units and K3-Problems.

U30 correspond to implementation operations. In this case U30: K3-Problems→S-Semantic units and U30=U10oU21oU32.

P03 is a synthesis step. In this case P03: S-Semantic units →K3-Problems and P03=P01oP12oP23

The basic realm n=0 includes semantic units, the next realm n=1 includes the so-called localist units, the realm n=2 includes sub-problems and the realm n=3 the problems to solve.

The task for the LISA network is to provide a means of compressing each tree into a representation, the so-called activation vector, and reconstructing the tree from its representation. SKUP elements are naturally associated to the LISA elements. The problems to solve may be associated to the set of conditions K. LISA contains a driver network associated to operators U, and to the reflective reasoning. They are U10, U21, U32 and U30. As a difference from SHRUTI, the initial LISA model was not developed to account for the reflexive processes. However the representational structure of LISA provides at least a starting point for reflexive reasoning capabilities. LISA propositions are retrieved into memory via guided pattern matching. During retrieval and comparisons the proposition are divided into two mutually exclusive sets: a driver and one or more recipients or receivers. The receiver network is associated to possibilities P. The elements are P01, P12 and P23. The calculus of possibilities for LISA model was studied by Taylor and Hummel (2009).

The switch between reflexive and refractive reasoning passes trough the semantics. The LISA semantics elements are associated to the states S in SKUP.

The activation of semantic units is controlled by time. Often the analysts do not have the time to allow runaway activation of semantics because they needs make inferences quickly. Notice that in contrast to reflexive inferences which are fast, the reflective inferences may require more effort. An open problem is to establish, for imposed time frames, the number of switching from reflexive to refractive and the order in which the switching should be performed.

#### 4.3.4.3 Spatial Cognition

A four level hierarchical architecture was operated in the study of complexity of behavior for spatial cognition (Mallot 1999) (Fig. 4.15).

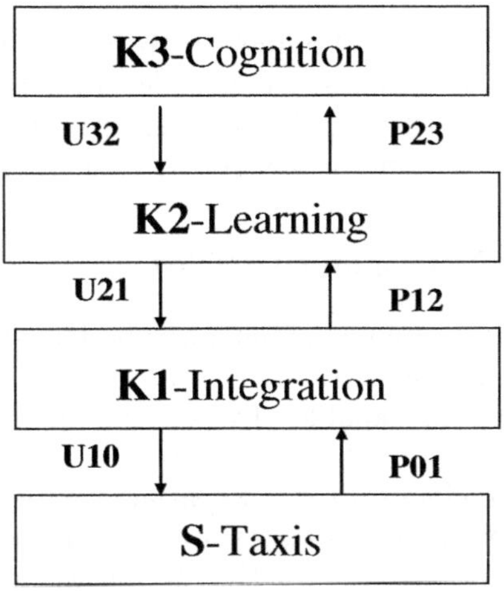


**Fig. 4.15** Four levels for complexity of behavior

The basic level S-Taxis, describes the reflex-life behaviors.

The level K1-Integration, requires spatiotemporal combination of data on the basis of a simple working memory. The level K2-Learning, requires long-term memory for procedures.

The level K3-Cognition, requires declarative memory consisting of neuro-gonitive maps allowing changing behavior according to current goals. Cognitive behavior is characterized by goal-dependent flexibility.

It is difference in time scale for the four levels in the sense that higher levels are slower.

The spatial cognition is important in the study of autonomous vehicles (Trullier and Meyer 1997). The functional model of the role of the hippocampus in navigation was implemented as a multi-level feed forward neural-network (Burgess et al. 1994). The first layer identified as K1 in hierarchy consists of sensory neurons that discharge selectively when the environment-S elements are sensed. The place cells represent the main elements of the K2 layer while the goal cells may be associated to the level K3.

A model of animat navigation based on several neurogonitive modules was proposed by Schmajuk and Thieme (1992). One module encodes topological representation and the other selects movements on the basis of predictions generated by the first module.

The module K2 is the comparator and elaborates next place predictions based on slow changing signals. A neurogonitive map is part of this. The module K1 allows goal predictions based on fast changing signals. The module K3 corresponds to goals.

Another model for spatial cognition was based on mesoscopic dynamics of brain activity (Freeman 2000). Freeman proposed a hierarchy of models that have capacity to show aperiodic behavior similar to that shown by electroencephalograms.

For the K-sets computational model due to Freeman (Freeman 2000) the basic level S is linked to columns, the next level K1 to bulbs, the next level K2 to cortex and the next level K3 to hemispheres of the brain. Notice that the corresponding Freeman notations are KI, KII, KIII and KIV.

#### 4.3.4.4 Arrays of Neurogonitive Tiles

In an attempt to develop a model of memory von Foerster (von Foerster1969) proposed a tessellation of neurogonitive tiles. The idea was to advance a conceptual minimum element capable of observing the desired neurocognitive characteristics of memory (Rocha 1995). The SKUP are considered as neurogonitive tiles. SKUP includes a self-referential level of interactions, in which an internal meaning of measured states of the memory empowered organization is generated, a closed system, with external meaning is obtained.

The SKUP is seen as a through-put system. Sensory information S is compared in a feed-forward fashion and altered in respect to the existing scheme in K. The feedback loop incorporates the delay of the system, that is, the associated time-scales.

We can think at these neurogonitive tiles as a suggested conceptualization of the necessary connections between symbol and matter in order to obtain closure.

The SKUPs may be correlated in an array to obtain an autonomous classification function that is autonomous neurogonitive architectures.

It seems to be of interest to use such arrays not only in higher level modeling approaches to evolvability and cognition as in traditional artificial intelligence, AI, models but also at the presumable lower level of artificial life, AL, models. In a cell we can find different processes with different time-scales. If these processes can be organized into semantically closed groups, then they can be represented by neurogonitive tiles and a functioning of the cell by a tessellation. The network of SKUPs includes processes that affect the time-scales of other processes. With an array of SKUPs we may be able to recognize the true temporal pattern recognition and not simply sequence recognition as in artificial NN. We can thus start to consider a tessellation of neurogonitive tiles as a proper measurement device which becomes dynamically coherent.

Implemented by a network of interacting SKUPs such arrays will be responsible for the recognition of temporal rather than spatial patterns of inputs. The recognition of appropriate temporal patterns will then dictate the neurogonitive system response. Clearly time plays the important role on the functioning of these arrays, unless the time-less switches of conventional artificial NNs.

Complex networks of tessellations can be organized as blocks of tiles dynamically closing higher semantic loops based on other semantic loops. This may correspond to categorification process (Appendix A4).

#### 4.3.4.5 n-Graphs for Neural Symbolic Computation

In neurodynamics studies the entities are embodied in the network's nodes and activated by associations. Logical systems define symbols that can be composed in a generative way but do not posses a microstructure appropriate for perception and learning tasks.

An illustration of the neural symbolic frames is based on the representation of the NN multi-scale evolution in term of n-graphs (Appendix A5).

The n-graphs characterize asynchronous systems with multiple entrances and exits.

Fig. 4.16 describes the process of self-structuring in a neural network and the emergence of symbols. The model is biomimetic.

The role of neocortical self-structuring as a basis for learning in neurodynamics was emphasized by von der Malsburg (1986, 1994), Doursat and Bienenstock (2006), Doursat (2007). It should be emphasized that Doursat (2007) approach is limited to a three level hierarchy.

The level n=0 represents to the 0-graphs or sets. This is associated in this case to the unit isolated neurons. The level n=1 correspond the 1-graphs. These are directed graphs including the morphisms that is, the connections between neurons.

Here the connected neurons are denoted by A, B, C and so on.The morphisms are 1-cells, describing relations. Their coupling in the right order allows the complete signal transfer process.

The level n=2 corresponds to the 2-graphs. These are graphs plus the so-called 2-cells between paths of same source and target. These 2-cells describe relations between relations and express the natural lumping of the simplex A, B, C of neurons in just one loop with specific role.

The level n=3 corresponds to the 3-graphs. These are 2-graphs that include 3-cells that is, the cells between 2-cells. The 2-graphs and 3-graphs represent graphs modification and should be subjected to conditions of

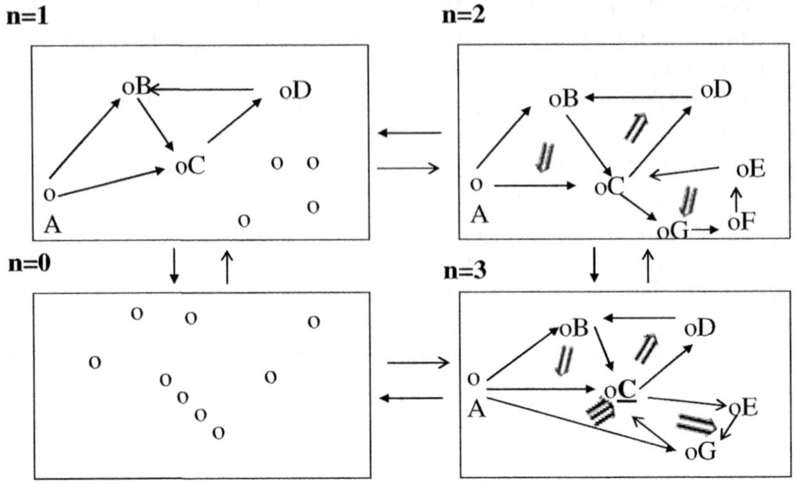


**Fig. 4.16** n-graphs for neural symbolic computation

natural transformations. These are related to travels time between neurons and association of neurons.

Denote by $\tau_{AB}$ the time necessary to travel from node A to node B (Doursat 1991) Conditions as $\tau_{AB} + \tau_{BC} = \tau_{AC}$ or $\tau_{CG} + \tau_{GF} + \tau_{FE} = \tau_{EC}$ should be imposed at the reality level n=2 to ensure the signal route equivalence. Double arrows emphasize these equivalences. There exist two different compositions of 2-cells. The vertical composition corresponds to sequential 2-cells, while the horizontal composition corresponds to parallel 2-cells.

The level n=3 corresponds to the interaction between two component networks of interactions.

In that case coherence conditions are: $\tau_{AG} + \tau_{GC} = \tau_{AC}$.

Triple arrows characterize equivalences between routes including the two component nets.

The integrative closure, connecting n=0 and n=3 represents the challenge for such systems.

The n-graphs for successive stages n, outlines a hierarchy of correlations of successive orders for neural patterns, a hierarchy of binding levels (von der Malsburg 1994, 2004). The mechanism is formally similar to that of concentration zones, CZ, as discussed by Damasio (1989). Damasio modelled the brain as an interacting system of hierarchical sub-networks for different major neural computing tasks. Healy and Caudell (2006) elaborated a category theory approach to NN emphasizing the potential role of Damasio approach. The categorical model, with functors from a category of concepts to a category of NN components and natural transformations between these functors, provides a mathematical model for neural structures consistent with concept-subconcepts relationship. Colimits of diagrams show how concepts can be combined and how a concept can be re-used many times in forming more complex concepts. The functors map commutative diagrams to commutative diagrams capturing this aspect of the colimit structure. Natural transformations express the fusion of single mode sensor representations of concepts in the same neural architecture.

This categorical model is compatible with the model of binding proposed by von der Malsburg or of concentration zones proposed by Damasio.

The architecture shown in Fig. 4.16 suggests considering three stages of binding or of concentration followed by the $4^{th}$ order stage of integrative closure and embodiment.

The stages may be considered in terms of categorical colimits. Categorical limit is the emergent concept summing up in itself the properties of its constituents. This generates the n-graphs hierarchy where at any stage the objects are the colimits of objects of the previous stage. This means that higher level of binding needs n-categorical models.

This idea may be linked to the attempts to make computers more close to natural brain-body system. It was observed that models suited for "off-line" computation such as Turing machine should be replaced with frameworks that are more readily to accommodate "on-line" and "real-time" processing of environment input streams.

An approach in line with "artificial chemical engineering" was proposed by Maass (2007). Maass proposed a framework calling it the "liquid computing" which is a generalization of classical finite states machine to continuous input streams and state space. The "reservoir computing" paradigm (Schrauwen et al. 2007) develops the main idea of "liquid computing" paradigm, that is the separation of the producing output stream from processing the input stream. Finding out what is a good reservoir represents an active research area. The n-categorical point of view may introduce a right structuring in levels of the "reservoir computing".

## 4.4 Evolvable Circuits

### *4.4.1 Evolutionary Circuits*

#### 4.4.1.1 Evolutionary Behavior for Circuits

Evolutionary or proactive circuits have the capability to change the preemptively embedded circuitry elements in order to keep on and to accomplish un-programmed tasks. Evolutionary behavior is a forward-looking perspective from engineering point of view, enabling to the designed circuit to modify faster in the anticipation of the future constraints of the environment.

Evolutionary circuits make use of self-construction elements offered by the basic generic framework, and by the environment. The evolutionary devices and sensors developed by the cybernetician Gordon Pask (Cariani 1989,

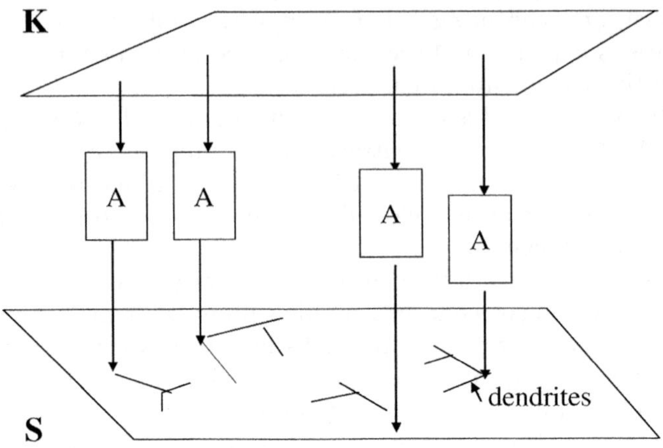


**Fig. 4.17** Pask's evolutionary device

1993), the "evolved radio" described by Bird and Layzell (2002), some developments of "evolutionary hardware" (Thompson 1998) may be regarded as a kind of proactive circuitry implementations.

One way to achieve to the circuit a degree of autonomy is to have sensors constructed by the system itself instead of sensors specified by the designer. Cariani refers to "Pask's Ear" as a first example of such evolutionary circuits (Fig. 4.17). The Pask's system is an electrochemical device consisting of a set of platinum electrodes in an aqueous ferrous sulfate/sulfuric acid solution. When current is fed through the electrodes iron dendrites tends to grow between the electrodes. If no or low current passes through a thread, it dissolves back into the acidic solutions.

The threads that follow the path of maximum current develop the best. In the complex growth and decay of threads, the system mimics a living system that responds to rewards that is more current and penalty that is less current. The system itself is able to discover the most favorable forms for the condition, which may embed information concerning other factors of the environment such as magnetic fields, auditory vibrations, temperatures cycles. The system resembles a model of ants leaving pheromone to reach a target (Virgo and Harvey 2008).

This circuit was trained to discriminate between two frequencies of sound by rewarding structures whose conductivity varied in some way with the environmental perturbation. The Pask's evolutionary device created a set of sensory distinctions that it did not previously have, proving that emergence of new relevance criteria and new circuits is possible in devices. The Pask's device may be considered as an analogous realization of the SKUP and of the categorical framework. The dendrite structures forming in malleable materials correspond to the category S, the resistance, capacitance or ionic resistance linkage to energy is linked to the category K. The evaluation of the signal network developed in malleable material is part of the possibilities P. Amplifying servomechanisms, A, may be linked to the operators U.

Following similar ideas, Bird and Layzell (2002) built an "evolved radio". Like Pask's ear the evolved radio determined the nature of its relation to environment and the knowledge of a part of the environment.

Bird and Layzell emphasized that novel sensors are constructed when the device itself rather than the experimenter determines which of the very large number of environmental perturbations act as useful stimuli.

The relation with von Uexküll *Umwelt* concept is obvious. Both of these devices, the Pask's ear and the evolved radio show epistemic autonomy that is, they alter their relationship with the environment depending on whether a particular configuration generates rewarded behavior.

Evolutionary systems include the four basic parts of the von Uexküll functional cycle: object, sensors, the command generator and the actuator (von Uexküll 1973). These parts are associated to the SKUP with the states S, the possibilities P, the conditions K, and the operators U, respectively. In the categorical framework the perception is associated to the functor P and the action is associated to the functor U. They can be viewed as adjoint connections of the different categories K and S.

Moreover, as in the functional cycle the evolutionary systems includes two levels for K, one related to the control K1, and a higher one K2 related to the coordination.

#### 4.4.1.2 Evolutionary Hardware

Research for evolutionary devices, is associated to the domain of MEMS, micro-electro-mechanical systems, MECS, micro-energy-chemical systems and to evolutionary hardware (Thompson 1998, Mahalik 2005).

MEMS represent the integration of mechanical elements, sensors, actuators and electronic circuits on a substrate through the utilization of microfabrication technology.

MECS focus on energy transfer, microfluids and chemical reactions. The evolutionary circuit is the candidate for the "brain" part of the systems while MEMS or MECS would allows to the micro-systems to sense and to control the environment.

In the associated SKUP quadruple, the environment corresponds to the states S, the evolutionary circuit itself to conditions K, the MEMS or MECS control part is linked to the operators U and MEMS or MECS sense part to possibilities P.

For embedded EDOE, the MEMS or MECS and ultimately, the prined circuits board, PCB may be the physical support material. Coupling evolutionary circuits with PCB and MEMS or MECS may ensure robustness and autonomy (Cheung et al. 1997, Mahalik 2005).

Evolutionary hardware represents an emerging field that applies evolutionary computations to automate adaptation of reconfigurable and polymorphic structures such as electronic systems and robots (Thompson 1998).

Evolutionary computation methods in designs that take the performance of a scheme as prediction for the performance of a modified scheme are suitable for evolutionary circuit development (Koza 1992). Koza elaborated genetic programs that could design band-pass filters that are electrical circuits able to separate signals of one frequency versus another. There is no explicit procedure for conventional designing these circuits due to the large number of optimization criteria. The algorithms work by starting with simple circuits and evolving them. The program, then created different variations, tested them, select the best and used them for the next generations. Implemented on silicon such programs may result in a circuitry that has attributes of novelty. The program may be used to generate evolutionary circuit schemes.

An interesting suggestion for evolutionary hardware architecture is the CellMatrix (Macias 1999). CellMatrix develops self-modifying, self-assembling and self-organizing circuits. These circuits are designed for, and constructed out of, a unique hardware substrate. The Cell Matrix may modify circuit architecture in the direction of locally connected; re-configurable hardware meshes that merge processing and memory.

#### 4.4.1.3 Electrochemical Filaments Circuits, ECFC

Based on electrochemical filaments development, a new type of evolutionary circuits, ECFC became possible.

ECFC construction starts with a generic framework representing the elements of the set of conditions K.

The K-framework elements may be that generated by wave equation, WE. The process in K generates successive K-frames, K0, K1,..., Km, at different levels.

The generic circuitry represented by K-frames is completed by additional circuitry, traces, dendrites, filaments, and supplementary matter, corrosion or degradation products. The processing for these additional circuits is an S-process. S-denotes the physical circuit based on filaments, threads, or micro-channels for fluids allowing the electrical contact or interaction. The K steps and the real environment S-steps have complementary contributions in circuit building. ECFC are expected to be at least partially autonomous. The autonomy includes the capability in building, assembly, modifying, organizing, repairing and destroying. As a difference, if compared to adaptive and self-adaptive devices based mainly on feedback, ECFC make use of the pre-emptively embedded multi-scale K frames. The appropriate K designs and the selective addition and the subtraction of appropriate elements from environment are the processes allowing both self-functionality and evolutionary behavior.

- **ECFC design**

Suggestions of evolutionary behavior may be detected in conventional multi-purpose circuit designs. These circuits have only holes and conductors, which have to be connected or interrupted according to the specific assembly needs. Often components are assembled directly on the components side so that it is possible to make small changes and to keep the whole circuit under control. This technique permits easy adjustment and trials of different components to modify the circuit from design stage. Adjacent to multi-purpose design methodology is the existing design re-use. The need to decrease time to market imposes to make use of known good sub-circuits or known good blocks as building elements. The sub-circuits may be developed individually as component DOE in a design centered EDOE.

The basic elements of ECFC technology are the K-valued generic framework, linked to class of tasks, the environment media for self-construction in non-stationary or oscillatory fields and the self-learning capability by exposure to environmental complexity and to variable tasks.

The ECFC that results by coupling the electrochemical filaments, ECF of different orders m, ECFm, over a pre-existing K-frames, K0, K1, K2, ..., Km is considered here. The circuit may be described using the categorical tensor "$*$" that links different levels in circuitry: ECFC=K0$*$ K1$*\ldots*$Km$*$ECF0$*$ ECF2$*\ldots*$ ECFm.

The tensorial product "$*$" may be the categorical product "$\times$", the co-product "$\cup$" and so on.

The categorical presentation of this architecture is shown in Fig. 4.9.

The K-framework should be a quasi complete printed circuit, with several opens. These opens allows the ECFC versatility and multiple potentialities. The environment is able to fill the opens sequentially in a way that ensure functionality. Potential geometrical variants and architectures for ECFC are: dots, cells, hexagons, triangles, squares, circles arrays, circular crowns, dyadic structure, labyrinths and mazes, high density circuitry, self-similar nested structures, tiles, pre-fractals and multi-fractals used for evolving antennas.

It was established that the wave equation, WE, is able to generate fractal structures making use of categorical product "$\times$". For example Hadamard-Sylvester matrices, reduce to Sierpinski triangles if only the "1"s are considered while the "-1"'s or with other notations the "0"s are neglected since they breaks the circuit (Barnsley 1993).

The switch from categorical product to categorical coproduct determines the size and the shape of the circuit. The switch is determined by the oscillatory fields that accompanies the ECFC development.

- **Materials for ECFC**

The materials should offer opportunities for wet chemistry and for solid physics to play significantly. ECFC's make use of composites and multi-phase media. The materials should be as rich as possible in structural possibilities, for example in phase transitions, on the edge of chaos, in non-linear regimes.

Interesting options are the existing self-adaptive or smart materials that allow phase transition, such as the piezoelectric, thermoelectric, electrorheological, electro active polymers and so forth.

Laminate known as filaments non-resistant as polyester rigid woven glass, paper phenol, or specially contaminated laminates represents valid opportunities since they allow the electrochemical filament fast formation.

Possible K-frames conductor lines make use of materials like Cu, Ag, Sn, Sn/Pb, Zn, Al, Mg and Fe. Metallic inorganic salts for conduction may be: sulfates, chlorides or nitrates of Fe, Cu, Ag, Sn, Pd, Pt, Zn, Al, Mg and catalysts. Metallic oxides may be useful as potential dielectrics. Damaged or fatigued printed or integrated circuits represent new potentialities for proactiveness.

- **ECFC processing**

ECFC should be processed in the environment that is in real field conditions in which the circuit should be functional such as:

- Mechanical (vibration, pressure)
- TRB (temperature, relative-humidity, bias) with direct current, alternative current or pulse plating of variable frequency
- Light, radiation
- Cyclical operation of various types
- Superposed oscillatory fields

These kinds of fields are the usual field of stresses for reliability tests. This suggests that evolutionary circuits may results from some over-tested circuits still able to show new capabilities.

An example of flow chart for ECFC fabrication is based on the following steps:

- Build K-frames-based on the wave equation, WE, solutions
- Select appropriate environment
- Introduce the K-frames and media in field conditions and allows periodic signals, stress field
- Develop the first level of filaments, ECF0 during training for signal that needs to be sensed or for any encountered new signals
- Build the circuit ECFC=K0*ECF0
- Repeat ECF0 procedure and allows ECF1 corresponding to another signal and so on
- Build the circuit ECFC=K0*K1*…*Km*ECF1*ECF2*…*ECFm
- Resulting circuits may be covered with gel, organic coating, photo coating or lacquer to ensure protection and robustness.

The operators U from the associated SKUP describe the evolutionary circuit at different levels of its construction.

The ECFC would be a circuit useful and stable in its building conditions. For any new level another frequency domain of oscillatory field is associated. As much as the oscillatory field still exists the new level would be developed.

If the structured dendrite structures were located in a specific field, the resulting structure would be able to recognize the patterns of that field. Learning and removal of information is possible if any dendrite may continually be formed broken and regenerated. Training to discriminate signals may be accomplished with the help of WH waves. The similarity associated to WH waves as defined in Sect. 2.2.3 is associated to the potentialities P of the SKUP.

- **n-graphs for dendrite circuits development**

The growing of tiny gold wires circuits in voltage controlled colloids is an example of ECFC (Miller and Downing 2002).

Fig. 4.18 illustrates a potential development for dendrites architecture using n-graphs (Appendix A5).

For the stage n=0 the filaments, are isolated wires. At n=1 interactions and dendrites may appear. This is allowed by interaction within the substrate. Filaments are separated in the n=1 stage but they interact in the n=2 stage to form arrays of interacting filaments. The n=2 stage shows the coupling of two or more dendrites in macro-wires.

The final stage, n=3 corresponds to a kind of single dendrite. The single dendrite pattern is specific.

The integrative closure, connecting also n=0 and n=3 is still an open problem for such systems. The dendrites development should be reversible.

A process like this may be compared with the operadic development (Appendix A6). The transition from 2-graphs to 3-graphs may be described as an operad.

The relation between higher categories, n-folds operads and dendrite circuit growth was investigated by Forcey (2008).

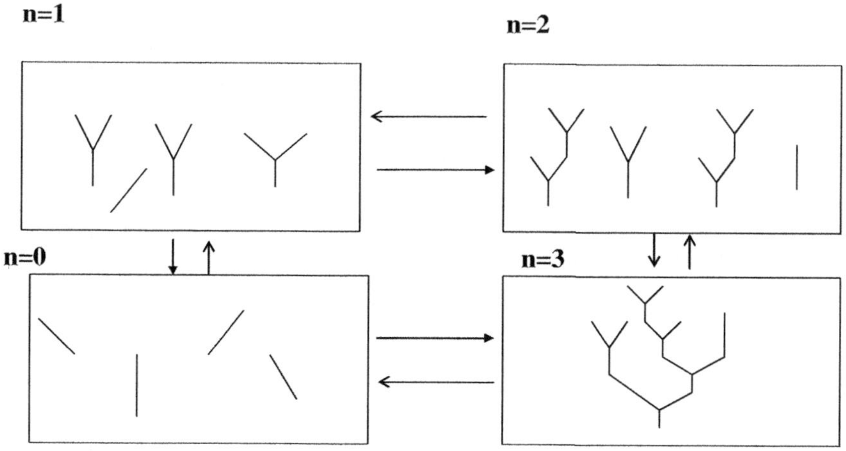


**Fig. 4.18** n-graphs for dendrites framework

## *4.4.2 Evolvable Circuits*

### 4.4.2.1 Evolvability Challenges

Evolvable circuits, EC, and evolutionary circuits are closely related. The difference between them corresponds to the degree of achievement from the point of view of full evolvability. This refers to the number of involved categorical levels and to closure. Evolutionary circuits show hierarchically architecture while evolvable ones suppose the integrative closure. This is also correlated to embodiment degree and to the scales. Transition from fixed circuits to evolvable circuits implies a change of scale, an increasing of the number of correlated scales. This is related to the categorical level too. Evolutionary circuits refer mainly to micro and meso-structured circuitry components while EC focuses also on molecular and nano-molecular structures facilitating the integrative closure. The evolutionary circuit is based on largely extrinsic designed and built circuits, while EC is expected to self-construct and to modify most part of their circuitry based on a genotype-phenotype-like scheme inherent to evolvability. Evolutionary circuit design is mainly from exterior while EC should be autonomous and self-programmed from the interior of the devices in a complex interaction with their environment. There is a threshold below which evolutionary circuit tends towards fixed circuits and above which they may progress towards fully EC.

Evolvable designs of experiments, EDOE was presented as a new modeling and simulation framework for complex problems solving (Iordache 2009). Additionally, EDOE may support the neurogonitive architecture for fully evolvable circuits, EC, practical implementation. The challenge is to build circuits that take advantage and control of their environment in increasingly complex ways. EC is supposed to be an embodied EDOE, able to run EDOE intrinsically, with emergent, behavior.

Unconventional principles, design of configurations, materials, fabrication methods, testing and applications have to be evaluated for evolvable circuitry (Bedau et al. 2000, Miller 2008, Rasmussen et al. 2004, Zauner 2005, Mahalik 2005).

Cellular automata suggest interesting architectures for EC soft. An example is the EvoCA cellular automata system (Taylor 2002). The EC's are supposed to be organizationally closed for matter but informational open. In order to realize evolvable systems, an important representational distinction should be between genotype and phenotype plus a biotic structure. As illustrated by EvoCA, semantically closed constructions may lead to novelty. EvoCA is a system where the environment is represented by a layer made of cellular automata, the physical or dynamical part, S and the genotypes represented by a second genome layer the inert or symbolic part, K. Each genotype controls a given cell in the first layer and evolves through a genetic algorithm. EvoCA-like constructions lead to operationally closed evolvable circuits, embedded in a dynamic environment, having metabolic-like potential, and being capable of self-replication and self-maintenance.

#### 4.4.2.2 Molecular Electronics

Molecular and nano-molecular systems represent the promising domain able to ensure the objectives for EC that is to add evolution capability to devices, to self-construct systems going beyond learning and being capable to act completely autonomous in an indeterminate environment. The circuits may be electronic, optical, molecular, micro-fluidic, and so forth.

As expected, bio-molecules provided potential substrates to build technical information processes systems as EC. For example biologically available conjugated polymers, such as carotene, can conduct electricity and can be assembled into circuits.

Among the bio-molecules, the *Bacterio-rhodopsin*, BR, and the deoxyribonucleic acid, DNA received extensive attention. Hybrid systems that combine the best features of bio-molecular architecture, with optic, electronic, micro-fluidic circuits represent a necessary step in EC development. The hybrid character refers to both formal models and practical devices.

These hybrids are digital-analog devices. The analog aspects are related to rate-dependent processes, and the digital aspects are related to macro-states and to macro-state transition rules. The issue of digital-analog or symbolic connectionist complementarity is closely related to the closure concept and to evolutionary behavior for devices (Pattee 1995, Cariani, 2001).The potential of the hybrid devices and hybrid models remains to be developed, but by all indications, such representational method can provide strategies of unifying low-level connectionist and high-level symbolic models of neurogonitive processes.

#### 4.4.2.3 $Bacterio-rhodopsin$ for Optoelectronic Circuitry

Early use of molecules in information processing has been in the field of optical computing. This suggested as candidate for EC base material, the Bacterio-rhodopsin, BR, which can serve as computer switch (Birge 1995, Mann 1996, Vsevolodov 1998).

BR has two useful properties for molecular level calculation. It exhibits photo chromic switching and shows photoelectric effect also.

The photo-cycle of BR, the sequence of structural changes induced by light-allows the storage of data in memory. Green, red and blue light induce structural changes of BR. Green light transforms BR in an intermediate denoted by "k" that relaxes to the "o" state. Red light transforms "o" state in "p" state that relaxes to "q" state. Blue light converts "q" state back to BR (Birge 1995). Any long lasting states can be assigned to digital values making possible to story information as a series of BR molecules in one or another state.

Discrete states as "0", "1" and so on, allows operating the EC devices. With these identifications the BR substrate may be the source for the symbolic language such as pixels and strings.

The photoelectric effect is another BR property useful for EC realization. Practical use of this property is exigent because it requires the preparation of BR films with highly oriented molecules. The possibility to interface BR electrically is the basis for several applications. The light of a specific wavelength range can be used to change the BR conformational state and the conformation change is accompanied by a color change that can be detected by optical means. It should be observed that the circuits are in this case, at least in part, of optical type.

A significant step in the development of the optoelectronic circuitry and computing was the study of all-light-modulated transmission mechanism of BR films. When a yellow beam and a blue beam illuminate the BR film, the two transmitted beams suppress mutually. Based on this mechanism, an all-optical operating device in which all 16 kinds of double-variable binary logic operations was implemented. The intensity of an incident yellow or blue beam acts as the input to the logic gate and the transmission bears the output of the gate. It is possible to turn this all-optical device into different states using different wavelengths and different intensity illuminations.

Full evolvability of hybrid symbolic connectionist models and associated circuitry that may be based on the unique properties of BR will be evaluated in the following.

#### 4.4.2.4 Embedded EDOE

The perspectives of a hybrid optoelectronic device based on BR molecules properties, in which conventional electronics is used to implement DOE analysis, are evaluated in the following.

Photo-cycle and photoelectric effects allows a direct writing DOE embedding in the BR based substrate. BR memorizing digits should be complemented by standard electronics able to perform the real valued operations.

As shown in Sect. 2.2.4, the DOE are resulting as particular solutions of the wave equation. Consider the solution:

$$Y(T, Z) = Z \oplus (V \otimes T) \tag{4.9}$$

A computing "cell" with three BR molecules is retained here for illustration purposes. Table 4.12 shows the wave solution for V=1.

**Table 4.12** Convection model: Y (T, Z)

| Z\T | 0 | 1 | 2 |
|---|---|---|---|
| #0 | 0 | 1 | 2 |
| #1 | 1 | 2 | 0 |
| #2 | 2 | 0 | 1 |

**Table 4.13** DOE associated to three molecules cell

| Exp | Molec. | Time | Operation |
|---|---|---|---|
| 1 | #0 | 0 | g |
| 2 | #0 | 1 | r |
| 3 | #0 | 2 | b |
| 4 | #1 | 0 | r |
| 5 | #1 | 1 | b |
| 6 | #1 | 2 | g |
| 7 | #2 | 0 | b |
| 8 | #2 | 1 | g |
| 9 | #2 | 2 | r |

The Table 4.12 is resulting by Galois field, GF (3) calculations and is a 3x3 Latin-square. The factors are the time steps 0, 1, 2, the molecules #0, #1, #2 and the operations 0=g, 1=r, 2=b corresponding to the three colors green, red, blue able to induce transitions. The time is multiple of the same time-step.

Standard DOE table may be developed by indicating the conditions associated to any element of the 3x3 Latin square (Table 4.13). Experimental results of DOE application may be the resolution, or any other value or data to be memorized. The DOE selects the significant results and also the significant factors by standard ANOVA calculations done by an external computer. This is Fourier analysis over the real field, for the device functioning parameters.

Successive steps will continue the experiment in the direction of beneficial results. The new experiment means a new DOE based on GF(m) algebra calculation and the wave equation. Following the EDOE suggestion, hardware may be achievable in 2-D or 3-D structures with concentric hierarchically located levels or planes. Light sources should be placed externally (Birge 1995).

Based on special BR properties, new classes of evolvable circuits, embedding and evolving DOE became possible. The evolvability, for the proposed architectures is the challenged result. Among the possible sets of DOE matrices, for n runs, m factors, and s settings we select the Walsh matrix of design, Wn,m,s, or Latin square matrices Ln,m,s generated by first order wave equation, WE (Iordache 2009). As for the EDOE structures, after the implementation of the DOE matrix of the type Wn,m,s or Ln,m,s it is required to perform at least two steps: factor evaluation, on columns in DOE, and experiment classification, on rows in DOE. It is necessary to define thresholds as degrees of acceptability for results. This help to decide when to recognize a pattern to be classified, as new. Various areas throughout the chosen EC layers may be written and addressed simultaneously. It is conceivable to embed Wn, m, s or Ln,m,s matrices in any active areas with memory. EC would be built using in succession similar additive and subtractive steps as for printed circuits and integrated circuits fabrication. Matrices such as Wn,m,s or Ln,m,s play the role of masks in printed or integrated circuits fabrication. These evolvable circuits should be able to drive the input signal and to decode the signal in a manner similar to logical thinking processes. As a difference, if compared to conventional circuits, this kind of EC will be continuously formed and erased, allowing the operation to be in succession forward and backward. The parallel search may be organized to achieve amplification, resonance and coherency. The EC works associatively as well as serially. By parallel processing the experiments would be performed at once, and the recorded results can be presented simultaneously to the center DOE. The EC should be able to record data from different areas to analyze and to give rise to a decision. This means that EC need to have monitoring functions, that is sensors, and executive functions, that is actuators, since the long term technological challenge is to get results by EC, independent of any external analyst or "operator". The EC should be a system that confronts the environment having the ability to evolve autonomously. New environmental conditions for EC may be materialized by a new row in the existing, embodied, component DOE matrices. This is the discrete symbolic step of the EC. Then follows the step in which real field values are associated to discrete DOE. This real valued step goes after data expansion and precedes data compression. With a learned degree of acceptability the sensor information goes backward and is classified in inner levels or layers and finally come back in the center. In this way the material embodiment may regenerate the symbolic description represented by DOE.

#### 4.4.2.5 Hybrid Controlled Micro-fluidic Circuits

In micro-fluidic devices the circuitry from printed or integrated circuits is replaced or completed by micro-channels for fluids. The MEMS became in fact MECS (Mahalik 2005). The transport of molecules in complex biological or chemical process may be programmed as the electric current in standard electronic circuits (van Noort et al. 2002, Verpoorte and de Rooij 2003, Erickson and Li 2004).

The micro-fluidic devices supposes the existence and the development of sensors, able to monitor changing environment, of actuators able to influence environment, coupled with computing and control capabilities for communication and data processing, all physically wired together. Tangen et al. (2006) presented elements of an interesting development in this direction. It focuses on the application of on-line programmable micro-fluidic bio-processing as a vehicle towards the design of artificial cells.

The electronically controlled collection, separation and channel transfer of the bio-molecules are monitored by sensitive fluorescence setups. This makes combinatorial fluidic circuitry and biochemical reactions circuitry feasible.

The basic elements of the SKUP quadruple may be identified for the "biomolecular console" described by Tangen et al. (2006). The reconfigurable electronic interface is linked to the space of conditions, K. The micro-fluidic network represents the states-S. This includes chemicals reservoirs and products. The parallel actuator network is related to operators U, while the monitoring system is linked to possibilities P. An electronic computer guides and controls the molecular circuits and ensures the cyclic functioning.

Another promising micro-fluidic technology consisting of a fluidic layer with a network of micro-channels superposed on layer with external computer programmable electrodes and actuators controlling the flow, has been proposed by Goranovic et al. (2006).

The project applies micro-fluidic nano-techniques to programming molecular reactions and priming an evolution of artificial cells and cell assemblies,

The basic elements of the SKUP are obvious for this technology. The genetic channel is linked to the space of conditions K. The temperature cycles ensures the gene replication. This fits with the cyclic character of the time T.

The metabolic channel is naturally linked to states S. The replication of selected proto-cells is linked to operators U, while the metabolism of selected proto-cell step is related to the possibilities P. The switch from categorical product to coproduct is determined by the oscillatory temperature fields and is able to control the proto-cell replication.

A categorical presentation of the architecture is shown in Fig. 4.9.

An important specificity of this micro-fluidic device is the realization of closed or loop operations, essential for the transition from fixed circuits to evolutionary and then to fully evolvable circuits.

#### 4.4.2.6 Self-constructed Molecular Circuits and Computing

Self-construction and separation in classes may be considered as computational processes and may be utilized to build information processors. Observe that the basic elements of the SKUP quadruple are naturally associated to any self-construction or separation processes. Suppose that from an unstructured environment S, some molecules considered as symbols are able to assembly in a supra-molecular structure linked to the conditions space K. These K structures should be recognized by a receptor and possibly amplified to provide an action U, redirected towards the unstructured environment S. The selection of specific symbols from the environment is done according to possibilities P. This may be a process driven by an optimization criterion as for instance energy or entropy production minimization or maximization (Prigogine 1980, Dewar 1993).

The self-construction may be described by the WE, too. According to the interpretation of the tensor product two main types of configurations are resulting. The tree-like forms are resulting for if the tensor product is a categorical product and a multiple cells stacked configuration are to be expected if the tensorial product is a coproduct. The transition between the two configurations is mediated by the environment conditions.

A categorical presentation of the architecture is shown in Fig. 4.9.

Elements of the general scheme of self-constructed computing are present in different DNA experiments (Adleman 1994, Winfree 2000).

Adleman proposed an approach to information processing with bioprocesses that allowed solving combinatorial problems by making use of specific set of DNA molecules.

DNA-based computing consists of four basic operations: encoding, hybridization, ligation and extraction. Problem solutions are obtained through an exhaustive parallel search by means of the pattern recognition intrinsic to DNA hybridization that is to self construction of complementary DNA strands. Involved chemical reactions such as the activity of restriction enzymes, ligases polymerases or simple hybridization can operate in parallel and this explains the possibility to solve complex problems.

Following similar ideas, cellular automata architectures describing DNA self-constructed circuit patterns for various forms of DNA tiles have been studied by Winfree (2000). Cook et al. (2004) showed how several common digital circuits, including de-multiplexers, random access memory, and Walsh transforms, could be built in a bottom-up manner using biologically inspired self-construction.

The Walsh-Hadamard matrices may be obtained as particular solutions of the wave equation WE. Table 4.14 shows a solution of the kinetic model in which we suppose the rate Q to be constant in the wave equation WE.

It is in fact the so-called Hadamard-Sylvester matrix, similar to Sierpinski triangle as presented by Cook et al. (2004). To highlight this parallelism the bolding and underlining is used for the wired "1" cells. It was assumed that "-1" breaks the circuitry. The non-wired digits are italicized. Notice that only two digits "0" (replaced here by "-1") and "1" need to be present in this case.

Based on operations in GF (4) Sierpinski square like fractals may be generated (Carbone and Seeman 2002 a, 2002 b).

At the present stage, a number of researchers are rather skeptical whether existing DNA based computation strategies will ever follow Bacterio-rhodopsin, BR, on its path in information processing. Critical problems with DNA

**Table 4.14** Kinetic model, Y (T)

| Q\T | 000 | 001 | 010 | 011 | 100 | 101 | 110 | 111 |
|---|---|---|---|---|---|---|---|---|
| 000 | **1** | **1** | **1** | **1** | **1** | **1** | **1** | **1** |
| 001 | **1** | *-1* | **1** | *-1* | **1** | *-1* | **1** | *-1* |
| 010 | **1** | **1** | *-1* | *-1* | **1** | **1** | *-1* | *-1* |
| 011 | **1** | *-1* | *-1* | **1** | **1** | *-1* | *-1* | *1* |
| 100 | **1** | **1** | **1** | **1** | *-1* | *-1* | *-1* | *-1* |
| 101 | **1** | *-1* | **1** | *-1* | *-1* | *1* | *-1* | *1* |
| 110 | **1** | **1** | *-1* | *-1* | *-1* | *-1* | *1* | *1* |
| 111 | **1** | *-1* | *-1* | *1* | *-1* | *1* | *1* | *-1* |

circuits and DNA-bio-computers are related to their inflexibility and to the ineffective accommodation to the variety of computation requests in real conditions. The assembly of DNA molecules of "tiles" has been designed to simulate the operation of any Turing machine. The self-construction DNA structures may be mapped naturally onto the grammars of the Chomsky hierarchy (Chomsky 1966, Winfree 2000). However for strictly algorithmic operations the DNA tiling computer can't compete with silicon machines. It was expected that the DNA computers may be eventually advantageous for the complementary domain of computation, beyond Turing machine capability. Again, it is not the case for the self-assembled DNA circuits as much as they map the Chomsky hierarchy of grammars.

The 1-D chain of DNA or the 2-D crystal tiles represent only the informational that is the K part of the SKUP. Major parts of the actual DNA computation have been accomplished with human operator involvement. To solve complex problems the K structure should be part of SKUP quadruple. K elements should be recognized by a receptor and amplified to provide an action U towards the external non-assembled environment S. The selection of specific symbols from the environment S would be done according to the possibilities P. For this reason the tilling need to be flexible and the tiles could be cycled through alternating assembly and disassembly stages. The self-construction and reconstruction operation may be programmable using glued and un-glued tiles (Carbone and Seeman 2002 b). According to the signification of the tensor product in WE solution, two main types of configurations are resulting. The tree-like forms are resulting if the tensor product is a categorical product and a multiple tiles stacked configuration is to be expected if the tensorial product is a categorical coproduct. The switching from one tensor interpretation to another is induced by environment changes.

A categorical presentation of the architecture is shown in Fig. 4.9.

Interactions between tiles and between tiles and their environment are mandatory to challenge Turing machines.

#### 4.4.2.7 Conventional Circuits versus Evolvable Circuits

For the forthcoming evolutionary and evolvable circuits fabrication a natural query is why do not use traditional methods, such as the physical and chemical study, followed by the modeling and extrinsic implementation of the models in the usual computer based control of circuit fabrication. The answer is that the envisaged control and computing task are impossibly to be extrinsically operated for evolvable systems of high complexity. In conventional circuits design the majority or non-linear interactions that could possibly contribute to the problem are deliberately excluded. The properties characterizing EC constructions should be, at least in part, the consequences of their own dynamic of the computational environment, not of the decision

**Table 4.15** Comparison of conventional circuits and evolvable circuits

| Conventional circuits-PC | Evolutionary or evolvable circuits |
|---|---|
| Single objective for any fabrication step | More general classes of objectives |
| Defined-based on previous learning | Undefined-open for learning, innovative |
| Top-down, linear | Top-down, bottom-up, cyclic, multi-scale |
| Aims for best solution, optimal | Makes workable, evolvable, active |
| Looks for perfect elements | Accepts elements with small defects |
| Conventional design-detailed models | Generic design-based on wave equation |
| Clear processing steps, complete data | Incomplete data and variable ad-lib steps |
| Independent on previous designs | Use everything at hand, if useful |
| Insulate the elements, serial or sequential | Combine elements, distributed, parallel |
| Builds | Builds, disbands, embeds and reorganizes |
| Divide and conquer | Divide and integrates, opportunistic |
| Maintain functionality in different media | Sensitive to environment, multifunctional |
| Restricted, static | Less restricted, rich, dynamic |
| Isolate from medium-protection | Medium, opportunistic exploitation |
| Avoid variability, interactions, transitions | Accept, use variability, interactions |
| Reliable | Robust, multi-reliable |
| High maintenance | Low and proactive-maintenance |
| Catastrophic degradation | Degradation in steps, hindered |

of the designer who is anyway unable to predict the evolution of its construction. EC are supposed to work for their evolution more efficiently than an external computer or operator can do. EC has the potentiality to be developed towards an autonomous system allowing survivability in completely unforeseen situations. It was observed that the more an autonomous system is optimized to perform a task the less capability it has to deal with unexpected changes in an uncertain environment. This implies that in complex environments, evolvability rather than adaptability or versatility may be an appropriate measure of a circuit's potential to carry out tasks.

Complex systems, natural or artificial, seem to opt for evolvability rather than for optimization and adaptability. This may be because in a complex environment it is impossible to achieve the optimum particularly when there are strong interactions between conditions K and states S. One way to proceed is to diversify several acceptable circuit options in a given environment and to let them evolve. This means that evolvable circuits may have several possible non-optimal but acceptable and useful architectures. This implies the discovery of environment properties that can be utilized to solve the imposed tasks.

Table 4.15 summarizes some of the differences between conventional circuits and unconventional ones such as evolutionary circuits and evolvable circuits, EC.

At the present technological level, a project grouping in a manufacture and in a product all the described faculties of EC is unrealistic but it is expected to manufacture EC of increasing capability in small steps.

### 4.4.3 Perspectives

#### 4.4.3.1 Evolutionary Devices

The categorical architectures shown in Fig. 4.19 may be considered for the study of evolutionary devices. Similar frames have been described by Cariani (Cariani 1989, 1998). The three levels are outlined in Fig. 4.19.

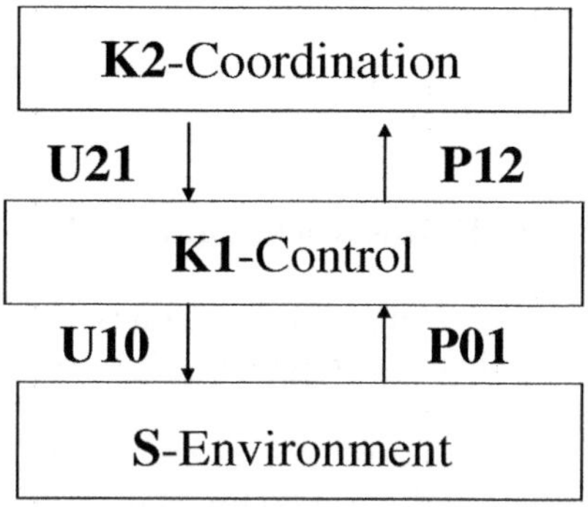


**Fig. 4.19** Three levels framework for evolutionary devices

Cariani shows that a hierarchical framework similar to that from Fig. 4.19, are embedded in the internal structure of individual organisms and evolutionary devices. The elements of the categorical framework are as follows:

S corresponds to the environment; K is structured in two levels to allow performing integrated neurogonitive/behavioral tasks.

K1-Evaluation and control level
K2-Coordination and decision level
U10: K1 → S control actions, decoding
P01: S → K sensor and measuring devices, encoding
U21-effectors resources
P12-evaluation/selection capability

The resulting device is able to evolve new goals to have a creative direction.

The same general framework corresponds to the scientific or engineering methodology. While this method includes only measurements and computations, organisms and evolvable device may act directly on the environment through effector organs of devices. The effectors convert signs into action on the material world. This corresponds to the control U. The basic informational operations of signs (semantic functionalities) present in organisms and devices can be described in terms of measurement (sensing), computation (coordination) and effecting (action). For living organisms, capable of neurogonitive behavior this was described by von Uexküll (1973).

It would be interesting to compare Fig. 1.5 showing the functional cycle with Fig. 4.19 and Fig. 4.9 showing two categorial frames for conditions, K1 and K2 with two basic interpretations of tensorial product, the coproduct "$\cup$" for K1 and the product "$\times$" for K2.

The interaction between S and K1 corresponds to a $1^{st}$ order evolutionary step, while that between K1 and K2 to a $2^{nd}$ order evolutionary step.

Many architectures proposed as evolutionary designs are based on less than four levels and may be considered as still incompletely developed.

The Pask's evolutionary device and the evolved radio appear to lack some elements of the top levels and also a link between coordination level K2 and environment S.

The missing levels and links may induce severe limitations and prevent this device to become evolvable.

#### 4.4.3.2 Three Realms Frameworks and Molecular Computation

There are several molecular computation studies suggesting how to design synthetic chemical or biochemical circuitry able to perform specified algorithms (Miller, 2008).

A method to make use of molecules in computing architectures was by reproduction of computer solid-state components with molecular structure. This is the usual approach taken in molecular electronics research. Typical objectives are the molecular wires, rectifiers or transistors (Siegmund et al. 1990). Another research direction was the chemical computing based on the fact that chemical reaction networks are able to process information in parallel. Kuhnert et al. (1989) demonstrated contour detection, contrast enhancement and same image processing operations on images projected onto a thin layer of a light-sensitive variant of chemical waves reaction medium. This system is a chemical realization of an associative memory and suggests the potential to implement learning networks by chemical means. The research into parallel chemical information processors led to artificial neural network, NN design based on mass-coupled chemical reactors (Hjelmfelt et al. 1992). Real chemical computing employs real chemical processes to compute. For example the simplest nonlinear function XOR can be implemented with reaction-diffusion behavior of palladium chloride (Adamatzky and Costello 2002).

Studies in molecular and supra-molecular design and engineering opened the perspectives for the realization of electronic, ionic and photonic circuits (Lehn 2004). Orchestrated, supra-molecular architectures deliberately designed to carry information allow to accelerate and to direct molecular interactions.

Artificial chemistry and organic computing suggests innovative ways to go beyond the chemical kinetic level and encompass supra-molecular interactions (Dittrich et al. 2001, Dittrich 2005). Interesting projects involve genetic programming (Harding 2005, Miller 2008).

Genetic programming GP, introduced by Koza (1992) is a development of genetic algorithms, GA methods. In GP the operations are as in GA but

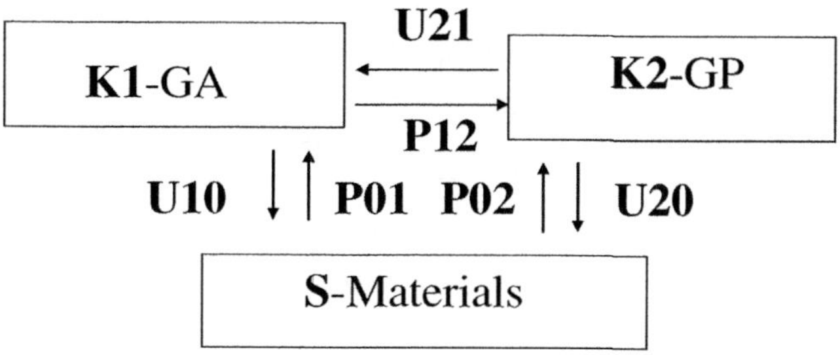


**Fig. 4.20** Genetic programming framework

on populations of programs not on strings. The behavior of each program in population is evaluated using a fitness function. Programs that do better are copied into next generation.

Fig. 4.20 outlines a three realms categorical framework for evolutionary computation.

The notations are as follows: S-Materials, K1-GA, genetic algorithm, K2-GP, genetic programming. The frame is useful for the study of evolution in *materio* (Miller and Downing 2002, Harding 2005).

S represents the material substrate. K1 and K2 are conditioning levels. K1-represents the basic genetic algorithm GA. K2-is the meta-level representing the genetic programming, GP. A change in K2 has higher impact because it represents multiple changes at previous realm K1.

The categorical framework describes interactions as: U10: GA → Materials, U20: GP → Materials, P01: Materials → GA, P02: Materials → GP.

What is interesting in this three realm architecture is the connection between K2 and S that is, between meta-model level and materials. This computation in *materio*, can ensure evolvability and autonomy. For this reason such frameworks were proposed for extracting computation from physical systems, for autonomous experimentation (Lovell et al. 2009).

#### 4.4.3.3 Four Level Frameworks

Driven by the continuously changing environment, living beings developed hierarchical self-repair and self-replicating mechanisms. Embryonics project brings the worlds of biology and electronics closer, by implementing in silicon these features.

Progresses have been reported in the construction of multi-cellular self-replicating systems (Mange et al. 2004). This is significant since one of the characteristic of evolvability is self-reproduction. Mange and coworkers proposed the "Tom thumb" algorithms that made possible to design self-replicating loops with universal construction and universal computation that can be easily embeddable into silicon.

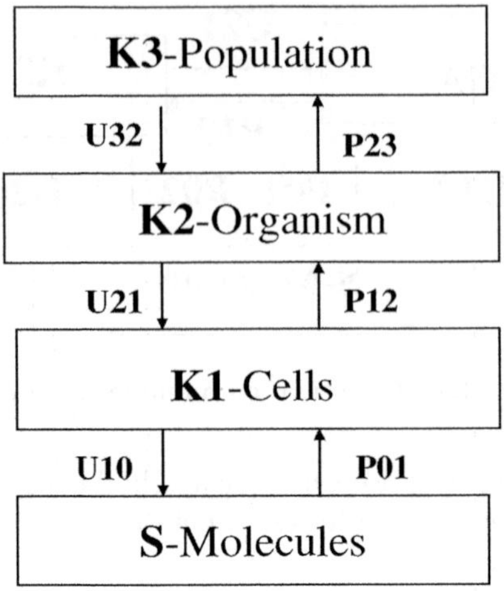


**Fig. 4.21** Four level organization of embryonics system

Mange et al. (1998) proposed a bio-inspired architecture for evolutionary electronic devices.

Embryonics bio-inspired devices are made up of four hierarchical levels (Fig. 4.21).

The multi-scale structure in embryonics project was correlated to the four levels of organization analogous to molecules, cells, organisms and population. The molecular level is represented by the basic field programmable gate array, FPGA elements.

The FPGA is the molecule of the devices. The FPGAs can be put together through a set of programmable connections to realize different types of digital circuits.

Each cell is a simple processor for instance a binary processor realizing a unique function within the organism, defined by a set of instructions.

The organism level is an association of cells while the population level is an association of organisms. The functionality of the organism is obtained by the parallel operation of all the cells. The size that is, the number of cells of an organism is also programmable and given enough space the organisms replicate automatically. Since the functionality of an organism is identical in each replicated copy, this mechanism provides an intrinsic fault tolerance. Given an appropriate cell structure the organisms are capable of learning.

In living cells, the genetic information is processed sequentially. Designing a memory that is inspired by biology suggests a different type of memory, called cyclic memory. Cyclic memory does not require any addressing mechanisms. Instead it consists of a simple storage structure that circulates synchronously its data in a closed circle, much as the ribosome processes the genome inside a living cell.

Observe that for the embryonics project, an architecture showing the integrative closure mappings between the top level K3 and the lower level denoted by S is still missing.

The embryonics architecture is restricted to the $3^{rd}$ order evolutionary step. For this reason the embryonics project may be considered only as evolutionary devices rather then fully evolvable, autonomous devices.

Critical for EC autonomy is the embodiment of the computing capacity that is the interconnection between K3 and S realms. This is the $4^{th}$ order evolutionary step.

Strategies to correlates K3 to S both in programming as in fabrication are suggested by organic computing studies (Müller-Schloer et al. 2004). In this general frame Pietzowski et al. (2006) proposed a system that use the paradigm of antibodies and developed the organic computing middleware system.

For the organic computing middleware architecture, the four levels may be identified as: S-Transport connector interface, K1-Event dispatcher, K2-Service interface, K3-Organic manager. Existing computational systems can be redesigned and redeveloped to engineer evolvable capabilities into them. Evolvability capabilities have to be added gradually and incrementally as organic computation studies suggests. Complete evolvability may be attained only step by step.

#### 4.4.3.4 n-Graphs Organization of Immuno-embryonic Systems

In order to create technological systems that are autonomous robust and evolvable, new engineering approaches must draw inspiration from natural complex systems.

For example in computer security, systems able to mimic the biological immune system can provide solutions against attacks on computer networks.

The immune system has been a major source of inspiration in the design of pattern recognition applications including computer security and virus protection.

Inmunotronics is another bio-inspired concept that has been successfully implemented in evolutionary hardware (Bradley et al. 2000).

Immunity is a multi-layered and multi-scale architecture starting with physical barriers, through physiological barriers, through cellular interactions.

Antibody mediated immunity protects the body from bacteria using B cells to generate antibodies and helper T cells to activate the production of antibodies.

The geometrical shape plays a crucial role for this type of immunity.

The embryonic cells proposed in embryonics lack a real-time method of verifying that each is performing the correct operation with respect to neighboring cells.

Bradley et al. (2000) proposed to incorporate emrbryionic and immune cells.

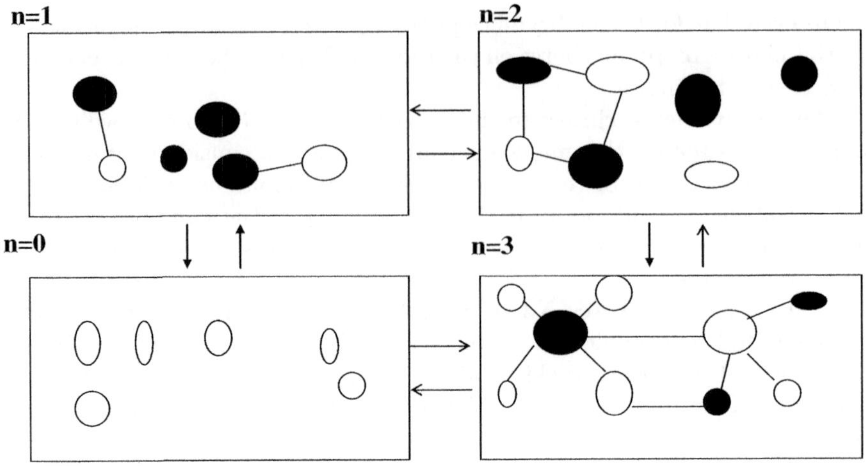


**Fig. 4.22** n-graphs for immuno-embryonics framework

Fig. 4.22 illustrates a potential development cycle for architecture using n-graphs (Appendix A5). It is an appropriate tool for multi-scale systems study.

The immune cells are black and the embryonic cells are white.

At n=0 the cells, prepared to be immune or embryonic are isolated. At n=1 interactions and couples cells appear. This is allowed by interaction within the body. Cells are separated in the stage n=1 but they interact in the stage n=2 to form arrays of interacting cells. The stage n=2 shows the coupling of two or more cells in frames going beyond cells areas isolation. It describes interactions of interactions.

The final stage, n=3 corresponds to a kind of global action of the whole immuno-embryonics architecture. Whole architecture pattern allow to identify faults and to review critical cases. The integrative closure including the interconnection of stages n=0 and n=3 represents the challenge for such systems.

The n-graphs are naturally correlated to n-categories (Appendix A5).

Katis et al. (2000) proposed symmetric monoidal categories with feedback as appropriate modeling frameworks for concurrent and distribute processes as those shown in Fig. 4.22.

In this case the objects in category are the cells, immune or embryonic. Their interconnections represent the relations. In bicategories the objects are cells, the relations between cells corresponds to 1-graphs, and the relations between relations to the 2-graphs. There are two different compositions of 2-cells, the vertical and the horizontal.

Notice that Katis et al. (2000) study is restricted to the 2-graphs that is, to 2-categories.

The tricategorical development would include the 3-graphs as a step towards the integrative closure.

## References

Adamatzky, A., Costello, B.D.L.: Experimental logical gates in a reaction-diffusion medium: The XOR gate and beyond. Phys. Rev. E 197(22), 344–352 (2002)

Adleman, L.M.: Molecular computation of solutions to combinatorial problems. Science 266(5187), 1021–1024 (1994)

Alberts, B., Bray, D., Lewis, J., Raff, M., Roberts, K., Watson, J.D.: Molecular Biology of the Cell. Garland Publishing, NY (1994)

Ardell, D.H., Sella, G.: No accident: genetic codes freeze in error-correcting patterns of the standard genetic code. Philos. Trans. R Soc. London B Biol. Sci. 357(1427), 1625–1642 (2002)

Austin, J.: Distributed associative memories for high speed symbolic reasoning. International Journal on Fuzzy Sets and System 82(2), 223–233 (1996)

Baeck, T., Hammel, U., Schwefel, H.P.: Evolutionary computation: Comments of the history and current state. IEEE Trans. Evolutionary Computation 1, 3–17 (1997)

Balakrishnan, K., Honavar, V.: Evolutionary design of neural architectures-A preliminary taxonomy and guide to literature. Department of Computer Science Technical Report 95-01 (1995)

Banzhaf, W.: Self-replicating sequences of binary numbers, foundations I. General, Biological Cybernetics 69, 269–274 (1993)

Barnsley, M.F.: Fractals Everywhere. Academic Press, New York (1993)

Bedau, M., McCaskill, J.S., Packard, N.H., Rasmussen, S., Adami, C., Green, D.G., Ikegami, T., Kaneko, K., Ray, T.S.: Open problems in artificial life. Artificial Life 6, 363–376 (2000)

Benyo, B., Biro, J.C., Benyo, Z.: Codes in the codons: Construction of a codon/amino acid periodic Table and a study of the Nature of specific nucleic acid-protein interactions. In: Proceedings of the 26th Annual International Conference of the IEEE EMBS, San Francisco, pp. 2860–2863 (2004)

Bird, J., Layzell, P.: The Evolved Radio and its Implications for Modeling the Evolution of Novel Sensors. In: Proceedings of the Congress on Evolutionary Computation, CEC 2002, pp. 1836–1841 (2002)

Birge, R.: Protein-based computers. Sci. Am. 272, 90–95 (1995)

Bradley, D.W., Ortega-Sanchez, C., Tyrrel, A.M.: Embryonics+Immunotronics: A bioinspired approach to fault tolerances. In: Proceedings of the 2nd NASA/DoD Workshop on Evolvable Hardware (2000)

Bray, D.: Protein molecules as computational elements in living cells. Nature 376, 307–312 (1995)

Burgess, N., Recce, M., O'Keefe, J.: A model of hippocampal function. Neural Networks 7, 1065–1081 (1994)

Caporale, L.H.: Is there a higher level genetic code that directs evolution? Molecular and Cellular Biochemistry 64, 5–13 (1984)

Carbone, A., Seeman, N.C.: Circuits and programmable self-assembling DNA structures. PNAS 99, 12577–12582 (2002a)

Carbone, A., Seeman, N.C.: A route to fractal DNA assembly. Natural Computing 1, 469–480 (2002b)

Cariani, P.: On the Design of Devices with Emergent Semantic Functions. Ph D Dissertation, Binghamton University (1989)
Cariani, P.: To evolve an ear: epistemological implications of Gordon Pask's electrochemical devices. Systems Research 10, 19–33 (1993)
Cariani, P.: Towards an evolutionary semiotics: the emergence of new sign functions in organisms and devices. In: Van de Vijver, G., Salthe, S., Delpos, M. (eds.) Evolutionary Systems, pp. 359–377. Kluwer, Dordrecht (1998)
Cariani, P.: Symbols and dynamics in the brain. Biosystems 60, 59–83 (2001)
Carpenter, G., Grossberg, S.: ART2: Selforganization of stable category recognition codes for analog input patterns. Applied Optics 26, 4919–4930 (1987)
Cheung, P., Berlin, A., Biegelsen, D., Jackson, W.: Batch fabrication of pneumatic valve arrays by combining MEMS with PCB technology. In: Symposium on Micro-Mechanical Systems, ASME, Dallas, TX, 62, pp. 39–46 (1997)
Chomsky, N.: Cartesian Linguistics. Harper & Row, New York (1966)
Cook, M., Rothemund, P.W.K., Winfree, E.: Self-assembled circuit patterns. In: Chen, J., Reif, J.H. (eds.) DNA 2003. LNCS, vol. 2943, pp. 91–107. Springer, Heidelberg (2004)
Damasio, A.R.: The brain binds entities and events by multiregional activation from convergence zones. Neural Comput. 1, 123–132 (1989)
Danckwerts, H.J., Neubert, D.: Symmetries of genetic code-doublets. J. Mol. Evol. 5, 327–332 (1975)
Dasgupta, D., McGregor, D.R.: Designing Neural Networks using the Structured Genetic Algorithm. In: Proceedings of the International Conference on Artificial Neural Networks (ICANN), Brighton, UK (1992)
Dewar, R.C.: Informational theory explanation of the fluctuation theorem, maximum entropy production, and self-organized criticality in non-equilibrium stationary states. J. Phys. A 36, 631–641 (2003)
Di Marzo Serugendo, G., Fitzgerald, J., Romanovsky, A., Guelfi, N.: A Generic Framework for the Engineering of Self-Adaptive and Self-Organising Systems. Technical Report CS-TR-1018, School of Computing Science, University of Newcastle (2007)
Dittrich, P.: The Bio-Chemical Information Processing Metaphor as a Programming Paradigm for Organic Computing. In: ARCS Workshops, pp. 95–99 (2005)
Dittrich, P., Ziegler, J., Banzhaf, W.: Artificial chemistries-a review. Artificial Life 7(3), 225–275 (2001)
Doursat, R.: A Contribution to the Study of Representations in the Nervous System and in Artificial Neural Networks. Ph D Thesis, University Paris (1991)
Doursat, R.: Of tapestries, ponds and RAIN. Toward fine-grain mesoscopic neurodynamics in excitable media. In: International Workshop on nonlinear brain dynamics for computational intelligence. Salt Lake City, USA (2007)
Doursat, R., Bienenstock, E.: Neocortical self-structuration as a basis for learning. In: 5th International Conference on Development and Learning, ICDL 2006, Indiana University, Bloomington, IN (2006)
Eigen, M., Schuster, P.: The hypercycle a principle of natural self-organization. Springer, Berlin (1979)
Eigen, M., Winkler-Oswatitsch, R.: Transfer-RNA: The early adaptor. Naturwiss 68, 217–228 (1981)
Elman, J.L.: Finding structure in time. Cognitive Science 14, 179–211 (1990)
Elman, J.L.: Learning and development in neural networks:The importance of starting small. Cognition 48, 71–99 (1993)

Erickson, D., Li, D.: Integrated microfluidic devices. Analytica Chimica Aca. 507, 11–26 (2004)

Findley, G.L., Findley, A.M., McGlynn, S.P.: Symmetry characteristics of the genetic code. Proc. Natl. Acad. Sci. USA 79, 7061–7065 (1982)

Fodor, J.A., Pylyshyn, Z.: Connectionism and cognitive architecture: a critical analysis. Cognition 28(1/2), 3–71 (1988)

von Foerster, H.: What is memory that it may have hindsight and foresight as well? In: Bogoch, S. (ed.) The Future of the Brain Sciences, pp. 19–65, 89–95. Plenum Press, New York (1969)

Forcey, S.: 2-fold operads Young diagrams and dendritic growth, University Research Symposium, Tennessee State University (2008)

Freeman, W.J.: Neurodynamics. An exploration of mesoscopic brain dynamics. Springer, London (2000)

Fritzke, B.: Growing cell structure-a self organizing network for unsupervised and supervised learning. Neural Networks 7(9), 1441–1460 (1994)

Geard, N., Wiles, J.: Structure and dynamics of a gene network model incorporating small RNAs. In: IEEE Congress on Evolutionary Computation, pp. 199–206 (2003)

Goranovic, G., Rasmussen, S., Nielsen, P.E.: Artificial life forms in microfluidic systems. In: Proceedings microTAS 2006, Tokyo, Japan, vol. 2, p. 1408 (2006)

Gould, S.J.: The paradox of the first tier: an agenda for paleobiology. Paleobiology 11, 2–12 (1985)

Halford, G.S., Wilson, W.H., Phillips, S.: Processing capacity defined by relational complexity. Implications for comparative, developmental and cognitive psychology. Behavioural and Brain Sciences 21(6), 803–831 (1998)

Harding, S.: Evolution in Materio. Ph D Thesis University of York (2005)

Haronian, D., Lewis, A.: Elements of a unique bacteriorodopsib neural network architecture. Applied Optics 30, 597–608 (1991)

Hartman, H.: Speculations on the evolution of the genetic code. Origins of Life 6, 423–427 (1975)

Healy, M.J., Caudell, T.P.: Ontologies and worlds in category theory: Implications for neural systems. Axiomathes 16(1), 165–214 (2006)

Hjelmfelt, A., Weinberger, E.D., Ross, J.: Chemical implementation of finite-state machines. PNAS USA 89, 383–387 (1992)

Hummel, J.E., Biederman, I.: Dynamic binding in a neural network for shape recognition. Psychological Review 99, 480–517 (1992)

Hummel, J.E., Choplin, J.M.: Toward an integrated account of reflexive and reflective reasoning. In: Gleitman, L., Joshi, A.K. (eds.) Proceedings of the Twenty Second Annual Conference of the Cognitive Science Society, pp. 232–237. LEA, Mahwah (2000)

Hummel, J.E., Holyoak, K.J.: Distributed representation of structure. A theory of analogical access and mapping. Psychological Review 104, 427–466 (1997)

Inhelder, B., Piaget, J.: The Growth of Logical Thinking from Childhood to Adolescence. Basic Books, New York (1958)

Iordache, O.: Evolvable Designs of Experiments Applications for Circuits. J. Wiley VCH, Weinheim (2009)

Jacob, F., Monod, J.: On the regulation of gene activity. Cold Spring Harbor Symp. Quant. Biol. 26, 193–211 (1961)

Jimenez-Montano, M.A.: Protein evolution drives the evolution of the genetic code and vice versa. Biosystems 54, 47–64 (1999)

Jimenez-Montano, M.A., de la Mora-Basanez, R., Poschel, T.: Poschel T The hypercube structure of the genetic code explains conservative and non-conservative amino acid substitutions in vivo and in vitro. Biosystems 39, 117–125 (1996)

Kargupta, H.: A striking property of genetic-code like transformations. Complex Systems 13, 1–32 (2001)

Katis, P., Sabadini, N., Walters, R.F.C.: On the algebra of systems with feedback & boundary. Rendiconti del Circolo Matematico di Palermo Serie II, Suppl. 63, 123–156 (2000)

Keller, R., Banzhaf, W.: The evolution of genetic code in genetic programming. In: Proc. Genetic and Evolutionary Computation Conf., pp. 077–1082. Morgan Kaufmann Publishers, San Francisco (1999)

Kitano, H.: Designing neural network using genetic algorithm with graph generation system. Complex Systems 4, 461–476 (1990)

Koonin, E.V., Novozhilov, A.S.: Origin and evolution of the genetic code: The universal enigma. IUBMB Life 61(2), 99–111 (2009)

Koza, J.: Genetic Programming: On the Programming of Computers by Means of Natural Selection. MIT Press, Cambridge (1992)

Kuhn, H., Kuhn, C.: Diversified world: drive of life's origin? Angew. Chem. International 42, 262–266 (2003)

Kuhn, H., Waser, J.: Molecular self-organization and the origin of life. Angew. Chem. Int. Ed. Engl. 20, 500–520 (1981)

Kuhn, H., Waser, J.: Hypothesis on the origin of genetic code. FEBS Letters 352, 259–264 (1994)

Kuhnert, L., Algadze, K.I., Krinsky, V.I.: Image processing using light-sensitive chemical waves. Nature 337, 244–247 (1989)

Lehn, J.M.: Supramolecular chemistry- from molecular information towards self-organization and complex matter. Reports on Progress in Physics 67, 249–265 (2004)

Lindenmayer, A.: Mathematical models for cellular interaction in development, I, II. J. Theor. Biol. 18, 280–315 (1968)

Lovell, C.J., Jones, G., Zauner, K.P.: Autonomous Experimentation: Coupling Machine Learning with Computer Controlled Microfluidics. In: ELRIG Drug Discovery, Liverpool, September 7-8 (2009)

Maass, W.: Liquid computing. In: Cooper, S.B., Löwe, B., Sorbi, A. (eds.) CiE 2007. LNCS, vol. 4497, pp. 507–516. Springer, Heidelberg (2007)

Macias, N.J.: The PIG paradigm: The design and use of a massively parallel fine grained self-reconfigurable infinitely scalable architecture. In: Proceedings First NASA/DoD Workshop on Evolvable Hardware (1999)

Mahalik, N.P.: Micromanufacturing and Nanotechnology. Springer, New York (2005)

Mallot, H.A.: Spatial cognition: behavioral competences, neural mechanisms and evolutionary scaling. Kognitionwissenchaft 8, 40–48 (1999)

von der Malsburg, C.: Am I thinking assemblies? In: Palm, G., Aertsen, A. (eds.) Brain theory, pp. 161–176. Spinger, NY (1986)

von der Malsburg, C.: The correlation theory of brain function. In: Domany, E., van Hemmen, J.L., Schulten, K. (eds.) Models of Neural Networks II: Temporal Aspects of Coding and Information Processing in Biological Systems, pp. 95–119. Springer, New York (1994)

von der Malsurg, C.: Vision as an exercise in organic computing. GI Jahrestagung (2), 631–635 (2004)

Mange, D., Sanchez, E., Stauffer, A., Tempesti, G., Marchal, P., Piguet, C.: Embryonics: A new methodology for designing Field-Programmable Gate Arrays with Self-Repair and Self-Replicating Properties. IEEE Transactions on VLSI Systems 6(3), 387–399 (1998)

Mange, D., Stauffer, A., Petraglio, E., Tempesti, E.: Self-replicating loop with universal construction. Physica D 191, 178–192 (2004)

Mann, S. (ed.): Biomimetic Materials Chemistry. VCH, Weinheim (1996)

Matsumaru, N., Dittrich, P.: Organization-oriented chemical programming for the organic design of distributed computing systems. In: Proc. of BIONETICS 2006, Cavalese, December 11-13. IEEE, Los Alamitos (2006)

Miller, J.F.: Evolution in materio. In: International Conference on Evolvable Systems, Prague, Czech Republic (2008)

Miller, J.F., Downing, K.: Evolution in materio: Looking beyond the silicon box. In: Proceedings of the 2002 NASA/DoD Conference on Evolvable Hardware, pp. 167–176. IEEE Computer Society Press, Los Alamitos (2002)

Müller-Schloer, C., von der Malsburg, C., Wurtz, R.P.: Organic Computing. Informatik Spektrum (27), 332–336 (2004)

Nikolajewa, S., Friedel, M., Beyer, A., Wilhelm, T.: The new classification scheme of the genetic code, tRNA usage and early code evolution. J. Bioinf. Comput. Biol. 4, 609–620 (2006)

Nolfi, S., Parisi, D.: Genotypes for Neural Networks. In: Arbib, M.A. (ed.) The Handbook of Brain Theory and Neural Networks. MIT Press, Cambridge (1995a)

Nolfi, S., Parisi, D.: Evolving artificial neural networks that develop in time. In: Moran, F., Moreno, A., Merelo, J.J., Chacon, P. (eds.) Advance in Artificial Life, Proceeding of the third European Conference on Artificial Life. Springer, Heidelberg (1995b)

Nolfi, S., Parisi, D.: Neural networks in an artificial life perspective. In: Gerstner, W., Germond, A., Hasler, M., Nicoud, J.D. (eds.) Proceedings of the 7th International Conference on Artificial Neural Networks (ICANN 1997), pp. 733–738. Springer, Berlin (1997)

Nolfi, S., Tani, J.: Extracting regularities in space and time through a cascade of prediction networks: The case of a mobile robot navigating in a structured environment. Connection Science 2(11), 129–152 (1999)

van Noort, D., Wagler, P., McCaskill, J.S.: The role of microreactors in molecular computing. Smart Mater. Struct. 11, 756–760 (2002)

Patel, A.: The triple genetic code had a doublet predecessor. J.Theor. Biol. 233, 527–532 (2005)

Paton, R.C., Toh, C.-H.: Computational aspects of protein functionality. Comp. Funct. Genom. 5, 85–90 (2004)

Pattee, H.H.: Evolving self-reference: matter, symbols, and semantic closure. Communication and Cognition-Artificial Intelligence 12(1-2), 9–28 (1995)

Pattee, H.H.: Causation, control and the evolution of complexity. In: Anderson, P.B., et al. (eds.) Downward Causation, pp. 63–77. Aarhus University Press, Aarhus (2000)

Perlwitz, M.D., Burks, C., Waterman, M.S.: Pattern Analysis of the Genetic Code. Advances in Applied Mathematics 9, 7–21 (1988)

Pietzowski, A., Satzger, B., Trumler, W., Ungerer, T.: A bioinspired approach for self-protecting in organic middleware with artificial antibodies. In: de Meer, H., Sterbenz, J.P.G. (eds.) IWSOS 2006. LNCS, vol. 4124, pp. 202–215. Springer, Heidelberg (2006)

Plate, T.: Holographic Reduced Representations. IEEE Transactions of Neural Networks 6, 623–641 (1995)
Pollack, J.B.: Recursive Distributed Representations. Artificial Intelligence 46, 77–105 (1990)
Prigogine, I.: From Being into Becoming. W. H. Freeman, San Francisco (1980)
Quick, T., Dautenhahn, K., Nehaniv, C., Roberts, G.: The Essence of Embodiment: A Framework for Understanding and Exploiting Structural Coupling Between System and Environment. In: Proc. Third International Conference on Computing Anticipatory Systems, Liège, Belgium (CASYS 1999) (1999)
Rasmussen, S., Chen, L., Deamer, D., Krakauer, D.C., Packard, N.H., Stadler, P.F., Bedau, M.A.: Transition from nonliving to living matter. Science 303, 963–965 (2004)
Rocha, L.M.: Artificial semantically closed objects, Communication and Cognition. Artificial Intelligence 12(1-2), 63–90 (1995)
Rocha, L.M.: Evidence Sets and Contextual Genetic Algorithms: Exploring Uncertainty, Context and Embodiment in Cognitive and Biological Systems, Ph.D. Dissertation, Binghamton University (1997)
Ronneberg, T., Landweber, L.F., Freeland, S.J.: Testing a biosynthetic theory of the genetic code: fact or artifact? PNAS 97(25), 13690–13695 (2000)
Schmajuk, N.A., Thieme, A.D.: Purposive behavior and cognitive mapping: An adaptive neural network. Biological Cybernetics 67, 165–174 (1992)
Schrauwen, B., Verstraeten, D., van Campenhout, J.: An overview of reservoir computing theory, applications and implementations. In: Proceedings of the 15th European Symposium on Artificial Neural Networks, pp. 471–482 (2007)
Shastri, L., Ajjanagadde, V.: From simple associations to systematic reasoning: A connectionist encoding of rules, variables and dynamic bindings using temporal synchrony. Behavioral and Brain Sciences 16(3), 417–493 (1993)
Siegmund, E., Heine, B., Schulmeyer, P.: Molecular electronics: The first steps towards a new technology. Int. J. of Electronics 69, 145–152 (1990)
Spiro, P.A., Parkinson, J.S., Othmer, H.G.: A model of excitation and adaptation in bacterial chemotaxis. Proc. Natl. Acad. Sci. USA 94, 7263–7268 (1997)
Suzuki, H., Sawai, H.: Chemical genetic algorithms-Coevolution between codes and code translation. In: Artificial Life VIII, pp. 164–172. MIT Press, Cambridge (2002)
Tangen, U., Wagler, P.F., Chemnitz, S., Goranovic, G., Maeke, T., McCaskill, J.S.: An electronically controlled microfluidics approach towards artificial cells. Complexus 3, 48–57 (2006)
Taylor, E.G., Hummel, J.E.: Finding similarity in a model of relational reasoning. Cognitive Systems Research 10, 229–239 (2009)
Taylor, T.: Creativity in evolution: Individuals, Interactions and Environments. In: Bentley, P.J., Corne, D.W. (eds.) Creative Evolutionary Systems. Morgan Kaufmann, San Fransisco (2002)
Thompson, A.: Hardware Evolution: Automatic design of electronic circuits in reconfigurable hardware by artificial evolution. Springer, Heidelberg (1998)
Trifonov, E.N.: Elucidating sequence codes. Three codes for evolution. Annals of the NY. Acad. Sci. 870, 330–338 (1999)
Trifonov, E.N.: Consensus temporal order of amino acids and evolution of the triplet code. Gene 261, 139–151 (2000)
Trullier, O., Meyer, J.A.: Biomimetic Navigation Models and Strategies in Animats. AI Commun. 10(2), 79–92 (1997)

von Uexküll, J.: Theoretische Biologie. Frankfurt a. M.: Suhrkamp Taschenbuch Wissenschaft, xxiv+378 (1973)

Verpoorte, E., de Rooij, N.F.: Microfluidics meets MEMS. Proc. of the IEEE. 91, 930–953 (2003)

Virgo, N., Harvey, I.: Adaptive growth processes: a model inspired by Pask's ear. In: Bullock, S., Noble, J., Watson, R.A., Bedau, M.A. (eds.) Proceedings of the Eleventh International Conference on Artificial Life, pp. 656–661. MIT Press, Cambridge (2008)

Vsevolodov, N.: Biomolecular Electronics, an Introduction via Photosensitive Proteins. Birkhaeuser, Boston (1998)

Zachary, W., Le Mentec, J.C.: Incorporating metacognitive capabilities in synthetic cognition. In: Proceedings of The Ninth Conference on Computer Generated Forces and Behavioral Representation, Orlando, FL, pp. 513–521 (2000)

Weberndorfer, G., Hofacker, I.L., Stadler, P.F.: On the evolution of primitive genetic codes. Orig. Life Evol. Biosph. 33(4-5), 491–514 (2003)

Wilhelm, T., Nikolajewa, S.L.: A new classification scheme of the genetic code. J. Mol. Evol. 59, 598–605 (2004)

Wilson, H.W., Halford, G.S.: Robustness of Tensor Product networks using distributed representations. In: Proceedings of the Fifth Australian Conference on Neural Networks (ACNN 1994), pp. 258–261 (1994)

Winfree, E.: Algorithmic self-assembly of DNA: Theoretical motivations and 2d assembly experiments. Journal of Biomolecular Structure & Dynamics 11, 263–270 (2000)

Wong, J.T.: A co-evolution theory of the genetic code. Proc. Natl. Acad. Sci. USA 72, 1909–1912 (1975)

Wu, H.L., Bagby, S., van der Elsen, J.M.: Evolution of the genetic triplet code via two types of doublet codons. J. Mol. Evol. 61, 54–64 (2005)

Yao, X.: Evolving Artificial Neural Networks. Proceeding of the IEEE 87(9), 1423–1447 (1999)

Zauner, K.P.: Molecular Information Technology. Critical Reviews in Solid State and Material Sciences 30(1), 33–69 (2005)

# Chapter 5
# Systems Sciences and Cognitive Systems

**Abstract.** The evolvable multi-scale engineering design is presented in correlation with general design theory. The role of meta-models for evolvable and creative conceptual design is emphasized.

The potential of active cases base reasoning systems and their interaction with designs of experiments is evaluated.

Evolvable diagnosis strategies for failure analysis and security purposes are proposed.

Manufacturing systems developments from fixed to flexible, reconfigurable and lastly evolvable with reference to assembly operations are presented. Multiple-scale agent architectures based on cognitive science studies allows integrative closure and autonomy.

## 5.1 Evolvability for Engineering Design

### *5.1.1 Modeling Design Processes*

As products become increasingly complicated and technology becomes increasingly advanced, the amount of engineering knowledge and of operations required from engineering designers is extensive (Bar-Yam 2003).

Several major problems in developing the computer aided design, CAD systems are related to the complexity advent in industry. Because designs are subject to a dynamical industrial context they must be able to change with the context.

The evolvable designs, and the evolvable CAD, ECAD, represent the target methodology to confront complexity advent in design. Evolvable designs are those that are more easily modified in accordance to shifting consumer demands, safety constraints and dynamic environment. Those designs that have ability to evolve can change more quickly in concert with the dynamic market and then have a better chance for continued survival on the market.

The design process has been described by various authors and from a variety of points of view. Descriptive models explain how design is done, systematic prescriptive type of models show how design should be done, cognitive models explain the designer thinking process. The correlation with the cognitive and system sciences methods as artificial intelligence, AI or artificial life, AL, methods start to be emphasized in design activities.

Interaction between subjective knowledge and objective world was discussed in significant descriptive theories of design. Yoshikawa (1981) proposed the theory called General Design Theory GDT in which the interaction between a designer and an objective world is formulated as a continuous map between two topological spaces Extended GDT (Tomiyama and Yoshikawa 1987) includes major developments of the GDT. Correlated to such approach are the axiomatic approach of design and other mathematical theories of design (Reich 1995, Braha and Reich 2003).

The systematic design approach describes the engineering activities as a sequence of phases as for instance: clarification of task, conceptual design, embodiment layout and detailed design (Pahl and Beitz 1996). Related to this approach is the universal design theory that view design as a finite number of abstraction levels and a set of structures stages to follow (Grabowski et al. 1998). These methodologies are in the same time descriptive and prescriptive focusing mainly on how design should be done as a procedure.

The cognitive design approach, identify and represents the cognitive or mental activities in design. A class of models refers to logical frameworks for design (Coyne 1988, Takeda et al. 1990). The design process is regarded as an evolutionary process that is the design is improved step by step. The design grammar and the analogies between language and design have been emphasized by Coyne (Coyne 1988). These grammars are interesting for conceptual design work, more concerned with reasoning in terms of engineering concept than physical parts (Pahl and Beitz 1996).

Inspired also by cognitive methods the so-called conceptual cycles of Sowa (Sowa 2000) may be considered as an illustration of Piaget (Piaget 1971) developmental theory that found applications in design modeling.

### *5.1.2 Framework for Engineering Design*

Although engineering design has always been associated with human creativity and skill, the generation of designs can be formalized in a structured manner. Such a formalization of design synthesis enables automatic design synthesis through computation.

The SKUP framework capability for engineering design and relation with GDT will be evaluated in what follows.

For design modeling the states S, in SKUP will be associated to real artifacts or to solutions of the design. The conditions K are associated to symbolic or ideal design, to specifications or knowledge.

For evolvability we need to consider that the designer system K and the designed one S, the operators U and the possibilities P are vectors or tensors. This means multiple conditioning levels or in other terms multiple time scales. The wave equation, WE, may generate and organize the space of conditions K. The multiple scales allow combining sequential and iterative steps in design. Large designs can be decomposed into small designs each having similar structures.

The operator U characterizes the capability to pass from plans or symbols K to reality based on previous reality and on new plans. In one of the simplest case U describes the concatenation of successively realized stage of the plans.

The states S capability to reinitiate some of the plans and to modify the symbolic description is characterized by the possibility P. P describes the selective activation or deactivation of conditions pertaining to K.

Evolvable designs should be based on the complementarity of upward and downward causation. The operator U, from the SKUP is associated to the upward design. Typically upward design is sequential and easier to model. The design may start with the first-order wave equation, WE solutions.

These generate the design schemes.

GDT formulates engineering design as a process initiated by ideal specifications such as functional requirements in the function space. This corresponds to the conditions space K, in the SKUP. The designer is able to match partial structural information in the attributes space corresponding to S.

The operator U is linked to deduction stages in GDT.

The possibilities P from the SKUP are associated to downward design. Complementing the upward typically discrete design, it is the downward design working with continuous parameters states. The appropriate downward dynamical design based on specifications or on dynamical parameter measurements may contribute to the closure and finally the design scheme evolvability. The downward design is local and parallel.

The possibilities P are associated to the abduction step in GDT.

Abduction is the method to integrate knowledge that satisfies the two aspects of the evolutionary design that is creates new product and expands knowledge.

U and P have complementary roles.

The conventional design cycle is focused upon registering and compensating the deviation from established, specified goals. Conventional design is about this single loop design cycle and is adaptive. But in complex situations it happens that this cycle is not achievable in its present form. Developing the conventional design, the evolvable design focuses upon goal formation. The general goal formation is included in the design cycle not outside to the cycle. In other words the evolvable design should be more focused to what happens inside the design cycle.

The transformations within the design cycles may be more important. The multi-scale transformations allow accelerating designs and potentially make the design evolvable. In conventional design, the partial goals formation that is, the design schemes set up is external to the design cycle. The design schemes are established too early and are fixed in most cases. Modifications of design schemes, if performed, are too late. As the design scheme formation is more included in the multi scale design cycle, the design itself may become more evolvable. Dynamic markets require that the design have the capability to adapt itself to an environment within its life span.

Elements of the general SKUP schema may be identified in the study of evolutionary stimulation in conceptual design. The process design starts from pre-inventive structure, in K, and follows a cycle including generating design rules, U, and exploration and understanding of the results, that is P following S. The new state in K includes the evolutionary design. A practical example is the cyclic process of chemical product design (Gani 2004). In this case K denotes the conceptual designs, S the computer aided design, CAD solutions, U-the methods and constraint selection, P-the result analysis and verification. The K-process corresponds to the evolution from pre-designs to post-designs.

### *5.1.3 Multiple Scales Evolvable Designs*

Evolvable designs, ED, should have an adjustable or modifiable architecture since they contain interacting components designs subjected to continuous reorganization after confronting the reality.

It was observed that design steps consist of typical elementary processes. These form the so called design cycle. Design cycle solves a small design problem or divides it into smaller sub-problems. This observation led to a model which is the repetition of design cycles at different scales (Yoshikawa 1981, Takeda et al. 1990).

Yoshikawa (1981) considered that the so-called ontogenetic design can be decomposed into small design cycles. Each cycle has sub-processes focusing on research, development, testing and evaluation aspects.

Following GDT Let us consider that the design cycle has four basic processes or actions: R-research (includes problem identification and suggesting key concepts to solve the problem). D-development (includes developing alternatives from the key concepts by using design knowledge. T-testing (includes evaluation of alternatives, simulations).

E-evaluation and adaptation (includes selection of a candidate for adaptation and modification, conclusions).

K is the symbolic description of the system, in this particular case, the planned choices for elementary design processes such as research-R, development-D, testing-T and evaluations-E for different conditioning levels.

For sub-processes the elements of R are R1, R2, R3 and R4, the elements of D are D1, D2, D3 and D4 and so on. All pertains to K at different conditioning levels.

K elements may be grouped in matrices of DOE, as generated by the wave equation, WE.

Examples of DOE are the matrices associated to Latin-squares or to Walsh-Hadamard designs (Iordache 2009). These should be more efficient than the unstructured designs of experiments as that utilized by Reich (Reich 1995).

Table 5.1 consists of several levels of conditions.

**Table 5.1** Array of conditions for RDTE

| | | | | | | | | | | | | | | |
|---|---|---|---|---|---|---|---|---|---|---|---|---|---|---|
| D22 | | D23 | | D32 | | D33 | | T22 | | T23 | | T32 | | T33 |
| | **D2** | | | | **D3** | | | | **T2** | | | | **T3** | |
| D21 | | D24 | | D31 | | D34 | | T21 | | T24 | | T31 | | T34 |
| | | | **D** | | | | | | | | **T** | | | |
| D12 | | D13 | | D42 | | D43 | | T12 | | T13 | | T42 | | T43 |
| | **D1** | | | | **D4** | | | | **T1** | | | | **T4** | |
| D11 | | D14 | | D41 | | D44 | | T11 | | T14 | | T41 | | T44 |
| | | | | | | | | | | | | | | |
| R22 | | R23 | | R32 | | R33 | | E22 | | E23 | | E32 | | E33 |
| | **R2** | | | | **R3** | | | | **E2** | | | | **E3** | |
| R21 | | R24 | | R31 | | R34 | | E21 | | E24 | | E31 | | E34 |
| | | | **R** | | | | | | | | **E** | | | |
| R12 | | R13 | | R42 | | R43 | | E12 | | E13 | | E42 | | E43 |
| | **R1** | | | | **R4** | | | | **E1** | | | | **E4** | |
| R11 | | R14 | | R41 | | R44 | | E11 | | E14 | | E41 | | E44 |

Here the notations are: R-Research, D-Development, T-Testing, and E-Evaluation.

Then R, D, T and E describe the conditions at the conditioning level m=0.

Then R1, R2, R3, R4 are the conditions of R corresponding to the level m=1. Then R11, R12, R13, R14 are the sub-conditions of R1 and corresponds to the level m=2. They were represented as elements of a cyclic loop.

Table 5.1 contains the semantic network off all the possible conditions K, that is, the selected factors, to be grouped in DOEs.

The design includes the act of redesign which is defined as the act of successive improvements or changes made to a previously implemented design.

This suggests to identify elementary RDTE actions having components as R11-research, r, R12-design, d, R13-test, t and R14-evaluation, e. This reality level is indexed here by n=0 or by m=2 if considered as a conditioning level. It is possible that some components of the elementary actions are unrelated.

The coupling of several RDTE actions corresponds to the reality level n=1 or with different notations to the conditioning level m=1. The resulting actions are R1, R2 and so on.

For the action R2 we have specific components: R21-research, r, R22-design, d, R23-test, t and R24-evaluation, e. Coupling the information resulting from different actions corresponds to the reality level n=2 that is, to the conditioning level m=0. At this level the R, D, T and E are the four summarizing actions. The reality level n=2 is unconditional in this case study.

One outcome of the complexity is that currently the designer may not have time to adequately explore all the design alternatives and select the best alternative.

Consequently, PSM framework will include only some of the conditions K and the corresponding states S, also.

The detailed PSM framework is presented in the following.

## *5.1.4 Schema for Multiple Scales*

Consider only a fragment of the Table 5.1. For three level evolution, m=0, m=1, m=2 the SKUP consists of the vectors S = ($s^0$, $s^1$, $s^2$); K = ($k^0$, $k^1$, $k^2$); U = ($u^0$, $u^1$, $u^2$); P = ($p^0$, $p^1$, $p^2$).

Table 5.2 illustrates the PSM framework, with conditions and states for two-levels only, m=0 and m=1. The conditions at the level m=0 are R, D, T and E. Let R=$k_0^0$, D=$k_1^0$, T=$k_2^0$ and E=$k_3^0$. The upper index refers to level while the lower index refers to the time step. It should be emphasized that the time steps at different levels may be different and this is a key feature for different levels of evolvability. The states and conditions at the level m=0 are represented by a loop with high thickness border cells.

The system initial state is $s_0^0$. With possibility $p^0(k_0^0|s_0^0)$ the condition $k_0^0$ is selected. This condition is a digit symbolizing a specific research R. This may be a matrix corresponding to R-DOE. Based on this, the operator $s_1^0=u^0(k_0^0, s_0^0)$ allows the transition to the new state $s_1^0$. This state is the realization of the research. Then with possibility $p^0(k_1^0| s_1^0)$ the new condition, $k_1^0$ arises. This condition symbolized by a digit corresponds to the selection of developments D. In the new condition, the operator $u^0(k_1^0, s_1^0) = s_2^0$ allows the system to reach the state $s_2^0$. This corresponds to the completion of design and materials.

Observe that: $s_1^0=u^0(k_0^0, s_0^0)$ implies $s_2^0=u^0(k_1^0, u^0(k_0^0, s_0^0))$.

With possibility $p^0(k_2^0|s_2^0)$ the testing, $k_2^0$ that is T, is selected and finally the new state results $s_3^0=u^0(k_2^0, s_2^0)$ results. This may corresponds to the processed product. It represents the succession of realized design, materials and processes.

Observe that: $s_3^0 = u^0(k_2^0, u^0(k_1^0, u^0(k_0^0, s_0^0)))$.

This result will be modified at the level m=0 in the condition E denoted by $k_3^0$. After this the state is $s_4^0=u^0(k_3^0, s_3^0)=u^0(k_3^0, u^0(k_2^0, u^0(k_1^0, u^0(k_0^0, s_0^0))))$.

The states are resulting not necessarily in a recursive way since, in practical cases the operators may varies with the step.

The states at the level m=0 are: $s_0^0$, $s_1^0$, $s_2^0$, $s_3^0$, $s_4^0$. The interpretation of the high thickness border cells trajectory is the process description as follows: from the state $s_0^0$ through condition $k_0^0$ towards the state $s_1^0$, then through condition $k_1^0$ towards the state $s_2^0$, and so on. Any trajectory is a succession of states and conditions. The role of initial state specification $s_0^0$ is outlined by such presentations.

If the experiment analysis shows that the factor E is the more significant factor, the analysis may be continued at the level m=1 for different test conditions E1=$k_0^1$, E2=$k_1^1$, E3=$k_2^1$, E4=$k_3^1$. This means to perform four different evaluations.

The states and the conditions at the level m=1 are in medium thickness border cells. The system initial state at the level m=1 is $s_0^1$. With possibility $p^1(k_0^1| s_0^1)$ the condition $k_0^1$ arises. The condition is a digit symbolizing a specific test. Based on this, the operator $u^1(k_0^1, s_0^1)=s_1^1$ describes the transition to the new state $s_1^1$. Then with possibility $p^1(k_1^1| s_1^1)$ the new condition, $k_1^1$ arises. In the new condition, the operator $u^1$ $(k_1^1, s_1^1) = s_2^1$ allows the system to reach the state $s_2^1$.

Observe that: $s_2^1=u^1(k_1^1, u^1(k_0^1, s_0^1))$ and $s_3^1=u^1(k_2^1, u^1(k_1^1, u^1(k_0^1, s_0^1)))$.

The states at the level m=1 are: $s_0^1$, $s_1^1$, $s_2^1$, $s_3^1$. The conditioning at the level m=1 is represented by the loop E1E2E3E4 tat is: $k_0^1$, $k_1^1$, $k_2^1$, $k_3^1$.

The interpretation of the medium thickness border cell trajectory is as follows: from the initial state $s_0^1$ through condition $k_0^1$ to the state $s_1^1$, then through condition $k_1^1$ to the state $s_2^1$, and so on.

Due to representation restrictions Table 5.2 illustrates only two successive levels, in this case m=0 and m=1. Suppose that a more detailed study is necessary. The same framework may be used to outlines levels m=1 and m=2.

Observe that generally $K=((k_0^0, k_1^0,\ldots), ((k_0^1, k_1^1,\ldots)), S=((s_0^0, s_1^0,\ldots), (s_0^1 s_1^1,\ldots))$;

$U = (u^0, u^1, u^2,\ldots,), P = (p^0, p^1, p^2,\ldots,)$

The design activity is confronted with knowledge data-base acquisition and development. Modern products and technologies should surpass the complexity emerging by non-linear interactions of large number of rules, restrictions or objectives. In the same time it is a need to construct larger-scale knowledge bases to put together data bases and models from different domains. This represents just a part of the designer activity that includes also, searching data base, analysis, product improving and so forth. An evolvable CAD, ECAD is the targeted strategy to confront complexity and the objective of this research. The entire design system should be evolvable. This involves problem solving, support of designer and interface. Local databases store all knowledge about the behavior of the agent and the community to whom the agents belongs. The information stored in these databases involves constraints, objectives, procedures, rules and experience, and organizational structures and techniques. It may be organized by logical rules. The CYC project, the KIF-knowledge integrated format and KIEF-knowledge intensive engineering framework represent major attempts in this direction (Takeda et al. 1995). These projects collect knowledge and provide mechanisms for sharing and reusing knowledge.

**Table 5.2** Two levels schema for engineering design

| | | | | | | | | |
|---|---|---|---|---|---|---|---|---|
| | | | | $s_2^0$ | | | | |
| | | | | | | | | |
| | | **D** | | | | **T** | | |
| | | | | | | | | |
| $s_1^0$ | | | | | | $s_2^1$ | | $s_3^0$ |
| | | | | | E2 | | E3 | |
| | | **R** | | $s_1^1$ | | **E** | | $s_3^1$ |
| | | | | | E1 | | E4 | |
| | | | | $s_0^0$ | | $s_0^1$ | | |

## *5.1.5 Perspectives*

### 5.1.5.1 Three Levels Evolutionary Designs

Most of the current computer aided design, CAD systems employ the hierarchical decomposition strategy that is a form of analysis thought process, corresponding to categorical product interpretation of tensor product for the K model. Such a strategy can lead to refinements of existing design but do not leads always to new, evolutionary designs. Moreover it happens that the design becomes too large.

The switch from categorical product to coproduct controls the size of the search space and allows the emergence of new designs. Similar ideas have been emerged in the study of divergence/convergence strategies in engineering design. The divergence step, correlated to categorical product, implies understanding the problem and creating solutions. The convergence step, correlated to coproduct, selects among solutions for further development.

The process is illustrated in Fig. 5.1.

A general categorical presentation of the architecture was presented in Fig. 4.9.

In this case the notations are: K1-cognitive design, conventional (convergence), K2-evolutionary design (divergence), and S-design entities.

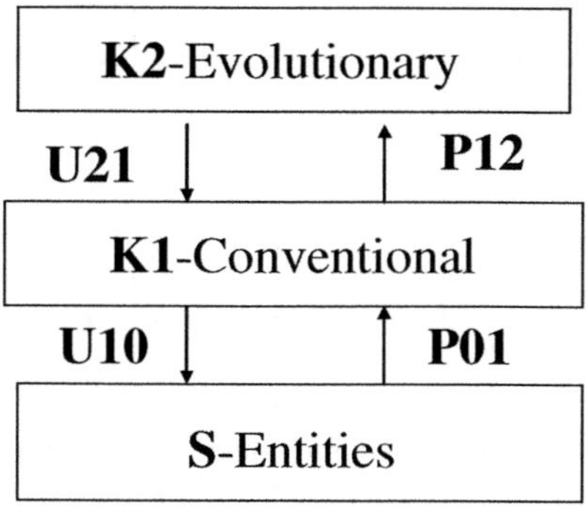


**Fig. 5.1** Three levels hierarchical framework for engineering design

#### 5.1.5.2 Four Realms: Meta-meta-models for Engineering Design

Difficulties arising in designing complex industrial products and operations require evolvable engineering design schemes. Evolvability in turn, is based on closure, at several complexity levels, for instance the closure between the dynamical laws of the material aspects and the complementary symbolic aspects of a physical organization. Full evolvability and autonomy requires the integrative closure that is a link between top and bottom levels.

Studies of evolvable and autonomous systems in complexity conditions challenged the traditional engineering fixed design methodology. Traditional engineering is based on a clear distinction between the design and the production phase and requires a system's performance and construction to be specified. For complexity conditions this is neither possible nor recommendable. Implementation of cyber-physical systems engineering concepts (Lee 2007, Carreras et al. 2009) outlined the need for a design for complexity that is a design for systems that fundamentally and continually adapt and evolve in a changing environment.

One problem that may benefits from this new type of model is the easy formulation of schedules in cases when the number of manufacturing cells and interconnected operations increases. The wave model offered the minimal number of cells and offers feasible schedules (Black 1991).

The challenge is to effectively build a fully evolvable design scheme based on integrative closure. This implies complementary and continuous back and forth between the wave equation, WE, results, that is, a specified design scheme, and the physical data of design process.

For this, the evolvable computer aided design, ECAD, should be able to face hard restrictions with respect to measurement analysis. The uses of ECAD lie primarily in exploring beyond the scope of conventional designs tools not competing with them. The main concerns are related to robustness, results analysis and scalability.

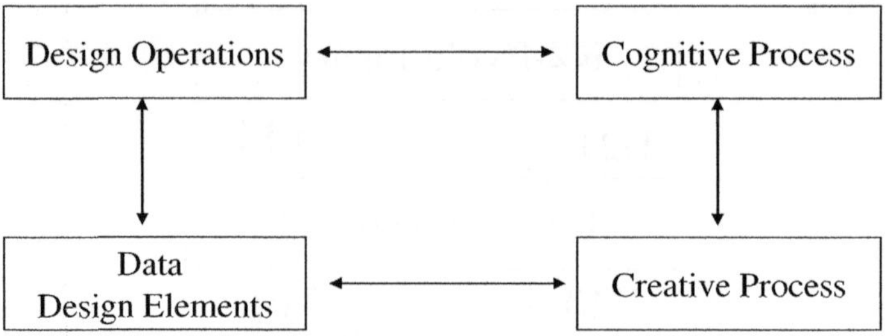


**Fig. 5.2** Creative conceptual designs

The elements of the four realms schema are easy identified in the study of creative stimulation in conceptual design (Benami and Jin 2002). In this case the Data are operated at Design Operation level. To ensure evolvability, this level should be interconnected to Cognitive and then Creative levels (Fig. 5.2).

Another four level approach to ED may be detected in the use of meta-meta-model and meta-model concepts for integrated modeling in design (Tomiyama et al. 1989, Kiriyama et al. 1992).

In this case the levels to be considered are: first level quantitative features, second level qualitative relationships, third level translation and interaction of concepts and fourth level the meta-meta-model (Fig. 5.3).

The meta-meta-model represents relation between the aspect models.

Engineering of complex systems focuses on meta-designing of genotype associated to concepts instead of directly designing the phenotype associated to quantitative aspects.

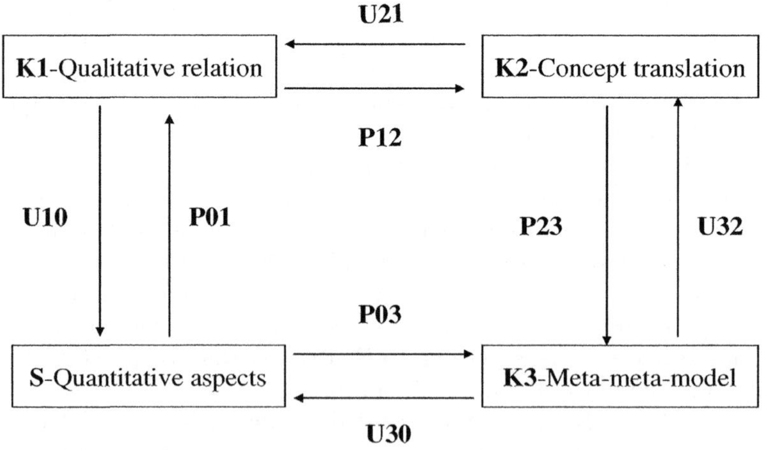


**Fig. 5.3** Meta-meta-model for evolvable engineering design

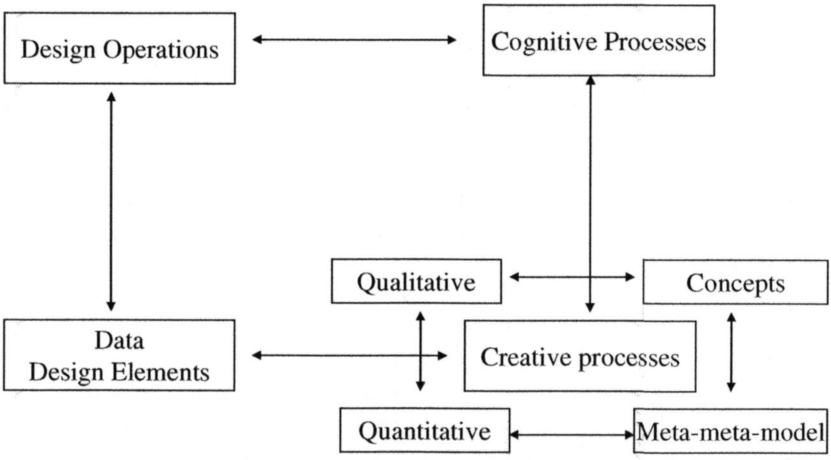


**Fig. 5.4** Four realms and sub-realms for creativity

The integrative closure including a link between K3-Meta-meta-model and S-Quantitative aspects is critical for evolvability and autonomy of the design system.

The entire structure shown in Fig. 5.2 may be just one realization of the creative process of four realm diagram shown in Fig. 5.3. This aspect is clarified by the Fig. 5.4 representing a kind of superposition of Fig. 5.2 and Fig. 5.3.

The achievement of a particular creative design parallels and recapitulates the history of general design systems from data to creative stage. It is an ontogenetic versus phylogenetic relationship.

The relation between ontogenetic design and phylogenetic design has been discussed also by Braha and Maimon (1998). Purposeful adaptation of artificial thing, that is, the ontogenetic design was seen as an interface between the inner environment of artifacts and the outer environment, the surroundings in which it operates.

Ontogenetic design evolution refers to the design process that share the characteristic of observed evolutionary phenomenon which occurs between the time when a problem is assigned to the designer and the time the design is passed to the manufacturer.

During this period the design evolves and changes from the initial form to an acceptable form, towards a fit between the design and the requirements.

#### 5.1.5.3 n-Graphs for Multiple Scale Engineering Design

Fig. 5.5 shows a representation of multiple scales frames presented in Table 5.1 using n-graphs. The n-graphs are computing tools able to describe asynchronous systems with multiple entrances and exits (Appendix A5). Asynchrony allows faster processing.

Different scales are associated to different levels in n-graph.

The reality level n=0 corresponds to the 0-graphs or sets. They represent r, d, t or e individual, undefined and uncorrelated actions, elementary process or objects. The level n=1 correspond the 1-graphs. These are directed graphs including the morphisms that is, the relations between r, d, t and e.

The morphisms are 1-cells. Their coupling allows the initiation of engineering design.

A $1^{st}$ order evolutionary step is represented by the transition to the level n=1.

This level is associated to 1-categories.

The level n=2 corresponds to the 2-graphs. These are graphs plus the 2-cells between paths of same source and target. These 2-cells express relation between relations, in particular the natural association of the quadruple r, d, t, e elements in just one macro actions denoted here by D-design, R-research, E-evaluation, or T-test.

There exist two different compositions of the 2-cells. The vertical corresponds to sequential 2-cells, while the horizontal corresponds to parallel 2-cells.

A $2^{nd}$ order evolutionary step is represented by the transition to the level n=2.

This level is associated to 2-categories.

The level n=3 corresponds to the 3-graphs. These are 2-graphs that include 3-cells that is, the relations between 2-cells. Fig. 5.5 shows the cells association as an evaluation action **E**. The 3-graphs represent graphs modification and should be subjected to conditions of natural transformations too.

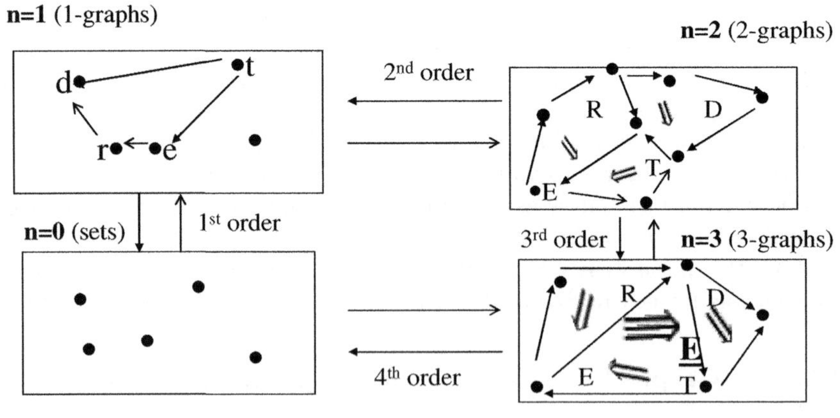


**Fig. 5.5** n-graphs for multiple scale engineering design

A $3^{rd}$ order evolutionary step is represented by the transition to the level n=3.

This level is associated to 3-categories. The $4^{th}$ order step represents a challenge for engineering design.

#### 5.1.5.4 Nested Frameworks for RDTE

Fig. 5.6 shows a categorical framework for RDTE as presented in Table 5.1

The four realms are identified as K0-Research, K1-Design, K2-Tests and K3-Evaluations. Fig. 5.6 outlines the possibility of integrative closure since it includes the link between K0 and K3. This allows evolvability and autonomy.

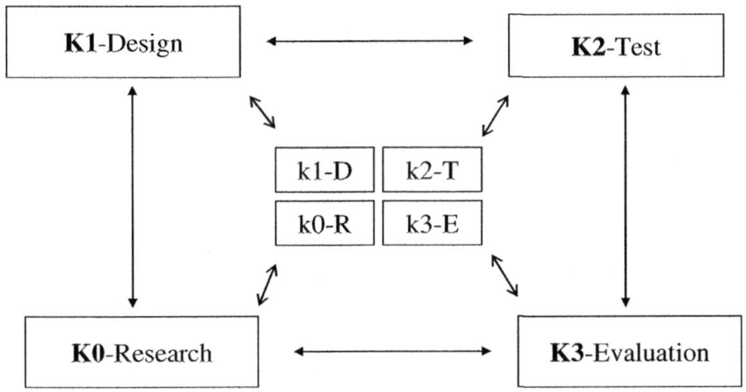


**Fig. 5.6** Nested frameworks for RDTE

Theoretically the architecture is not confined to four realms.

Fig. 5.6 shows nested and self-similar architectures.

A similar four realm cyclic structure is repeated starting from the whole central system that may be built by four sub-realms denoted here by k0, k1, k2 and k3.

Fig. 5.6 emphasizes the integrative closure as a process that can develop self-similar patterns in design.

## 5.2 Case Based Reasoning

### *5.2.1 Case Based Reasoning Method*

Conventional CBR is a problem solving paradigm that solves a new problem by remembering a previous similar situation and by reusing information and knowledge of that situation (Aamodt and Plaza 1994, Aha et al. 2001). More specifically, CBR uses a database of problems to resolve new problems. The database can be built through the knowledge engineering (KE) process or it can be collected from previous cases.

CBR traces its roots to the studies of learning and memory. Schank (1982) developed a theory of learning and reminding based on retaining of experience in a dynamic evolving memory structure. This model of dynamic memory was the basis for some of the earliest CBR systems.

In a problem-solving system, each case would describe a problem and a solution to that problem. The reasoning engine solves new problems by adapting relevant cases from the library. Moreover, CBR can learn from previous experiences. When a problem is solved, the case-based reasoning engine can add the problem description and the solution to the case library. The new case that in general represented as a pair <problem, solution> is immediately available and can be considered as a new piece of knowledge.

The CBR process can be represented by a schematic cycle, as shown in Fig. 5.7 Aamodt and Plaza (1994) have described CBR typically as cyclical process comprising the four steps:

1. Recall the most similar cases
   During this process, the CBR engine searches the database to find the most approximate case to the current situation.
2. Reuse the cases to attempt to solve the problem
   This process includes using the retrieved case and adapting it to the new situation. At the end of this process, the user might propose a solution.
3. Revise the proposed solution if necessary
   Since the proposed solution could be inadequate, this process can correct the first proposed solution.
4. Retain the new solution as a part of a new case

This process enables CBR to learn and create a new solution and a new case that should be added to the case base.

It should be noted that the Recall process in CBR is different from the process in a database. If we want to query data, the database only retrieves some data using an exact matching while a CBR can retrieve data using an approximate matching.

As shown in Fig. 5.7, the CBR cycle starts with the description of a new problem, which can be solved by recalling previous cases and reusing solved cases, if possible, revising the solution and giving a suggested solution, retaining the restored case and incorporating it into the case base.

However, this cycle rarely occurs without human intervention that is usually involved in the retain step. Many application systems and tools act as a case retrieval system, such as help desk systems and customer support systems.

The CBR provides support for applications if the input data tend to repeat similar patterns from time to time. When the factors recur, the studied system

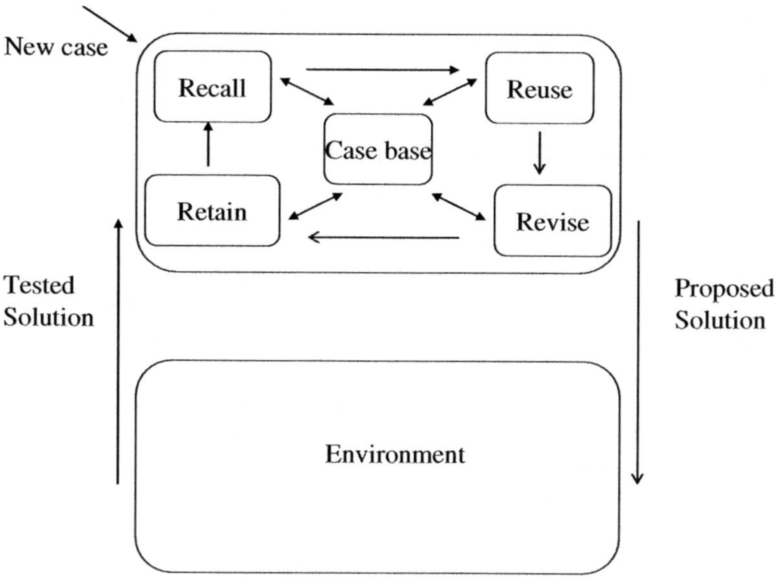


**Fig. 5.7** CBR basic framework

is likely to display regularly repetitive patterns. This repetitiveness explains why it is reasonably to apply CBR in complex problem solving situations.

Traditional CBR have limited potential. For example in common versions, CBR involves just one user and don't answer in real-time to explosive amount of user data, to the unexpected cases, or to non-linear interacting cases and questions.

It is a need to implement CBR frameworks in which answers to multiple questions are gathered from multiple information sources, in real time.

For continuously addressing multiple-goals, multiple arrays of CBR cells systems are needed. For such arrays it is difficult to arrange the architecture or scheme of problem-solving, to schedule, to elaborate and to run rules, to adjust them to continuous changing environment.

Problem solving methodologies as case-based reasoning CBR are confronted with high complexity situation due to chaotic or random character of data, and to severe time restrictions. The method to confront the high complexity is that of evolvability. This implies improving the conventional passive CBR, to an evolvable one, ECBR. ECBR should be active and autonomous, able to take control of its environment, able of responding to random unexpected problems and to large amounts of knowledge in real-time.

ECBR schemes may be generated by the developed here partial differential model, WE, and compared to existing schemes. Schemes with variable number of cells and multi-scale schemes are resulting. Connections between the problem solving schemes, and the designs of experiments are of interest.

Applicability in domains including: process diagnosis and failure analysis, financial analysis, data mining, sensor operation and interpretation is expected.

### 5.2.2 Case Based Reasoning Frameworks

Concepts from theoretical biology, developed to characterize evolvability in artificial and natural systems represent the inspiration source for ECBR building.

The wave equation, WE, generates schemes with variable number of stages and multi-scale schemes. Connections between these problem solving schemes, and the well-known designs of experiments will be outlined. Cyclic problem solving arrays with evolvability based on multi-scale schemes organized by self-similar replication at different conditioning levels will be presented more in detail.

As biology suggests, the evolvable knowledge schemes should be embodied or situated.

The PSM framework outlines the active interaction between conditions and real states.

The classic scheme of problem solving in CBR is a cycle with four steps (Aamodt and Plaza 1994, Melendez et al. 2001). The CBR cycle steps are: Recall, Reuse, Revise, and Retain. The steps will be denoted by "c", "u", "v", and "t".

The four steps represent a CBR cell.

A platform or a scheme in which in any time step any of the four operations is activated is of interest. The advantage is continuous data input and output for the scheme. The scheme should contain four CBR cells indexed here by #0, #1, #2 and #3.

The functioning of the cells at successive time steps is represented in Table 5.3.

This table is in fact a solution of the wave equation were the following identifications are of use: 0-c, 1-u, 2-v, and 3-t. Here c, u, v and t represent the cell condition.

T is the time step, 0, 1, 2 or 3 and Z denotes the space of operations, 0, 1, 2 or 3.

Z describes also the travel along the classification scheme.

Observe that the positions of the c-recall, u-reuse, v-revise, t-retain are changed in the direction of circular information flow at a regularly point in time. At any given time for any cell, only one of the inputs corresponding to c-recall, u-reuse, v-revise, or t-retain is in the active mode. The 4 cells allow continuous input and output.

### 5.2.3 Schema Modification

Let us restrict as a first example, to the C(4) solution shown in Table 5.3.

Observe that the above examined cycling operations schedules are in fact designs of experiments, DOE. Table 5.3 contains a Latin square. Running DOE based scheme allows fast identification of significant data for classification regime acceleration.

The DOE factors are time steps, cell and operations. The time is multiple of the same time-step.

Standard DOE table may be developed by indicating the conditions associated to any element of the 4x4 Latin square (Table 5.4).

Experimental results of DOE application may be the interesting object selection, the efficiency, the resolution, and so forth. Typical control tasks in classifications are to obtain the highest throughput of the cell, highest efficiency or to reduce time consumption.

The DOE selects the significant results and also the significant factors by standard ANOVA calculations. This is in fact Fourier analysis over the real field, for the functioning parameters. The evaluation of performances may be based on real data or on real-field dynamical model of the process.

Next step will be to reorganize the scheme or to reproduce the experiment in the direction of beneficial results. The new experiment means a new DOE based on modulo-m algebra calculation and the WE model. Physically this means to generate a new, modified classification scheme. This scheme may be one with a different number of cells or a device with the same number of cells but with modified parameters.

Suppose for instance that the experiment underlined in Table 5.4 gives the worst result (cell #2, at the third step 3, lumped operation "u-reuse"). Suppose that u-reuse is the operation offering the expected product or result and that cell factor is the only significant factor. In that case the cell may be changed with a modified one, possibly from the same array of cells. The classification scheme is supposed to be redundant. If all other factors are significant, an improvement strategy may consists in the modification of the cell (#2 by #2'), of the time step, 3 by 3', of the operation u-reuse by u'-reuse followed by the introduction of a new operation for instance s-restore in a new cell #4 (Table 5.5). This kind of situation appears in case-base maintenance situations. For maintenance, restore step which selects and applies maintenance operations is necessary.

Schemes as presented Table 5.3 or Table 5.5 are solutions of the wave equation, WE. The classification scheme evolution appears in fact as a continuous

**Table 5.3** Scheduling for CBR

| Z\T | 0 | 1 | 2 | 3 |
|---|---|---|---|---|
| #0 | c | u | v | t |
| #1 | u | v | t | c |
| #2 | v | t | c | u |
| #3 | t | c | u | v |

**Table 5.4** DOE associated to CBR

| Exp | Cell | Time Step | Operation |
|---|---|---|---|
| 1 | #0 | 0 | c |
| 2 | #0 | 1 | u |
| 3 | #0 | 2 | v |
| 4 | #0 | 3 | t |
| 5 | #1 | 0 | u |
| 6 | #1 | 1 | v |
| 7 | #1 | 2 | t |
| 8 | #1 | 3 | c |
| 9 | #2 | 0 | v |
| 10 | #2 | 1 | t |
| 11 | #2 | 2 | c |
| 12 | #2 | 3 | u |
| 13 | #3 | 0 | t |
| 14 | #3 | 1 | c |
| 15 | #3 | 2 | u |
| 16 | #3 | 3 | v |

**Table 5.5** Modified CBR

| Z\T | 0 | 1 | 2 | 3' | 4 |
|---|---|---|---|---|---|
| #0 | c | u' | v | t | s |
| #1 | u' | v | t | s | c |
| #2' | v | t | s | c | u' |
| #3 | t | s | c | u' | v |
| #4 | s | c | u' | v | t |

oscillation between the WE generated DOE schemes and the real field evaluations of the resulting data. The complementarity or disjoint-ness between the finite-field scheme and, the real field data represents the key mechanism for evolvable classification.

Data mining schemes in which the duration of some steps is higher than that of other steps are frequently encountered.

## *5.2.4 Multiple Scales*

A multi-scale scheme is considered in what follows. We may limit the Table 5.3 to vectors containing distinct elements only.

They represent solutions at specified time or stage in the problem solving development.

**Table 5.6** Singlets $y_0$

| Z\T | 0 | 1 |
|---|---|---|
| 0 | c | u |
| 1 | t | v |

They results as solution of the WE too.

Denote by Y(T) = $y_0$ = (c, u, v, t).

Since there are only four elements it is possible to represent $y_0$as a 2x2 matrix as that shown in Table 5.6. It results as a kind of cyclic folding of the vector $y_0$. Obviously other type of folding may be of interest.

Let us consider more scales in the scheme of conditions. The method used in Sect. 2.2.3 is used in the following. The doublets are resulting by direct product and concatenation $y_0 \times y_1$. Table 5.7 represents a direct product of 2x2-matrices, Y (T) = $y_0 \times y_1$ with $y_0 = y_1$ The new letters have been put adjacent to the first, to the right side. The initial problem was solved by splitting in 4 steps (c, u, v, t). However any of these steps is a new problem that may be solved by the same algorithm.

In this way the sub-problems cc, cu, cv, ct and so on are resulting.

They have significance as described by their notation and as suggested by the task-method decomposition a possible interpretation is as follows (Aamodt and Plaza 1994): cc-identify features, cu-initially match, cv-search, ct-select, tc-extract, tu-index, tv-adjust indexes, tt-update knowledge and so on.

The transition from a level to another may be triggered by the presence of data of interest in the expected product, by specific shapes of the recorded signals, and so forth The higher-level problem solving operations should take place with a timing that ensures and support the cyclic functioning at the previous level. The problem solving scheme should have an adjustable or unsettled construction since it contains interacting modules subjected to continuous reorganization after confronting the reality. It is not only a spatial scheme but a temporal one as well.

**Table 5.7** Doublets $y_0 \times y_1$

| Z\T | 00 | | 01 | | 10 | | 11 |
|---|---|---|---|---|---|---|---|
| 00 | cc | | cu | | uc | | uu |
| | | c | | | | u | |
| 01 | ct | | cv | | ut | | uv |
| | | | | | | | |
| 10 | tc | | tu | | vc | | vu |
| | | t | | | | v | |
| 11 | tt | | tv | | vt | | vv |

Due to the size of search space the multi-scale scheme is confronted to the apparent improbability of chance to produce any successful solution of the problem solving. But in fact the problem solving trajectory in the multi-scale scheme is not a blind search. The multi-scale scheme allows modifying the searchable domain and the search velocity by adding more levels and scales to the search process. Any new problem solving level appears as adding more sensors and effectors to the system. Moreover interaction with real data accelerates scheme construction and discovering new solutions.

### 5.2.5 *Schema for Multiple Scales*

The elements of the SKUP for multi-scale scheme will be presented in what follows. This is of help for classification schemes design and processes visualization.

A section of the general Table 5.7, illustrating PSM structure at two levels only, m=0 and m=1 will be considered. The SKUP elements are:

$$\mathrm{S} = (\mathrm{s}^0, \mathrm{s}^1);\ \mathrm{K} = (\mathrm{k}^0, \mathrm{k}^1);\ \mathrm{U} = (\mathrm{u}^0, \mathrm{u}^1);\ \mathrm{P} = (\mathrm{p}^0, \mathrm{p}^1)$$

The scheme includes at the first level m=0 the operations c, u, v, t and at the second level, m=1, the operations vc, vu, vv and vt. In this particular case, the second level is resulting by a separate re-cycle processing after operation v-revise.

The states and conditions at the level m=0 are in the high thickness border cells.

The states and the conditions at the level m=1 are in medium thickness border cells.

Table 5.8 includes the conditions K and the real valued states S. The conditions at the level m=0 are t=$k^0_0$, c=$k^0_1$, u=$k^0_2$ and v=$k^0_3$. The upper index refers to level while the lower index refers to the time step. Time steps at different levels are different. The system initial state is $s^0_0$. With possibility $\mathrm{p}^0(k^0_0|s^0_0)$ the condition $k^0_0$ is selected. This is a digit symbolizing a specific operation t-retain. Based on this, the operator $\mathrm{u}^0(k^0_0, s^0_0) = s^0_1$ allows the transition to the new state $s^0_1$. Then with possibility $\mathrm{p}^0(k^0_1|\, s^0_1)$ the new condition, $k^0_1$ arises. This condition symbolized by a digit corresponds to the selection of c-recall. In the new condition the operator $\mathrm{u}^0(k^0_1, s^0_1) = s^0_2$ allows the system reach the state $s^0_2$. With possibility $\mathrm{p}^0(k^0_2|s^0_2)$ the operation, $k^0_2$ that is u-reuse, is selected and finally the product $\mathrm{u}^0(k^0_2, s^0_2) = s^0_3$ results. It will be operated at the level m=0 in the condition v-revise denoted by $k^0_3$. Then the state is $s^0_4$. The states at the level m=0 are represented by the square: $s^0_0$, $s^0_1$, $s^0_2$, $s^0_3$. The conditions at the level m=0 are represented by the square t-retain, c-recall, u-reuse, v-revise, tat is: $k^0_0$, $k^0_1$, $k^0_2$, $k^0_3$.

If experiments shows that v-revise is the critical operation, the classification may be limited at the level m=1 for the operations vt=$k^1_0$, vc=$k^1_1$, vu=$k^1_2$ and vv=$k^1_3$.

**Table 5.8** Two-level schema for CBR

| | | | | | | | | |
|---|---|---|---|---|---|---|---|---|
| | | | | $s_2^0$ | | | | |
| | | | | | | | | |
| | | **c** | | | | **u** | | |
| | | | | | | | | |
| $s_1^0$ | | | | | | $s_2^1$ | | $s_3^0$ |
| | | | | | vc | | vu | |
| | | **t** | | $s_1^1$ | | **v** | | $s_3^1$ |
| | | | | | vt | | vv | |
| | | | | $s_0^0$ | | $s_0^1$ | | |

The system initial state at the level m=1 is $s_0^1$. With possibility $p^1(k_0^1|s_0^1)$ the condition $k_0^1$ arises. This is a digit symbolizing a specific operation. Based on this the operator $u^1(k_0^1, s_0^1) = s_1^1$ describes the transition to the new state $s_1^1$ and so on. Each condition supposes the selection of other condition for operations.

The states at the level m=1 are represented by the square: $s_0^1$, $s_1^1$, $s_2^1$, $s_3^1$. The conditions at the level m=0 are represented by the square vt, vc, vu, vv, tat is: $k_0^1$, $k_1^1$, $k_2^1$, $k_3^1$.

The potentialities are defined by vectors as $P = (p^0, p^1)$. The component $p(k^m)$ is an evaluation of the condition $k^m$. An example of evaluation is to take $p(k^m)$ equal to 0 or 1. The value zero corresponds to situation in which that condition is ineffective while the value 1, corresponds to active conditions.

Transfer between different levels may be controlled by external criteria.

K elements, representing the symbolic conditions indicating the types of operations at two levels are in fact cyclic classification schemes. S appears as sequences of more or less classified problems. Operators U characterize the capability to pass from intended conditions of classification to the reality of classification steps. The possibility P describes the capability of states S to reactivate the classification scheme and to modify the symbolic K description that is the classification scheme elements. P shows the activation of some areas of the operations shown in Table 5.8 and the inactivation of others.

## *5.2.6 Perspectives*

### 5.2.6.1 Three Levels Evolvable CBR

Elements of adaptability and evolvability may be detected for some knowledge systems presented in literature. For instance, conversational CBR, received substantial attention (Aha et al. 2001). This essentially interactive CBR, involves the refinement of diagnoses through interaction with the user or other CBRs.

These systems attempt to find the quickest ways to increase the accuracy of diagnosis through estimating information gain.

Another research is that of active knowledge systems with conceptual graphs (Li and Yang 1999, Delugach 2003). In that case the concept may play the role of factors.

Difficulties arising in cyclic operations in complex situations require adaptable and evolvable classification schemes. These are in turn, based on closure concepts implies the disjoint or complementary description and closure between the dynamical laws of the material aspects tat is real data and the symbolic aspects tat is condition data of the physical organization. This is the concept of semantic closure, restricted to two levels architectures.

The challenge is to effectively build entirely evolvable classification schemes based on complementarity and continuous back and forth between the wave model results that is a specified classification scheme, and the physical or real data of classification process itself.

A number of useful areas of applications have been identified for evolvable classification technologies.

Main examples are process and quality control, diagnosis and failure analysis, engineering design, financial analysis, emergency situations, Data mining with augmented semantics, evolvable agent operations, temporal reasoning, sensor operation and interpretation.

Evolvable knowledge schemes offer the prospect of devices to suit a particular individual. Evolvable technologies have the potential to adapt autonomously to changes in their environment. This could be useful for situations where real-time control over systems is not possible such as for space applications. Evolvable control systems are required in such cases. Evolvable knowledge devices may be of help in the study of central concepts as self-repair and development.

An open question is if this kind of evolvable knowledge systems can be implemented in reality. A generic architecture and algorithms should be proposed so that the particular system builder do not starts from beginning and may become evolvable.

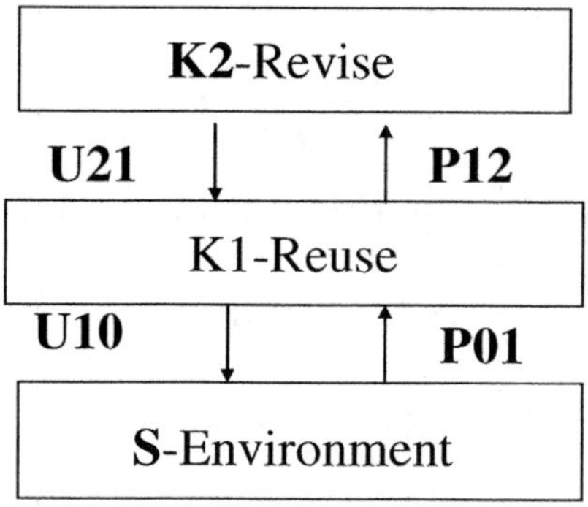


**Fig. 5.8** Three levels framework for evolvable CBR

Fig. 5.8 and Fig. 5.9 show the categorical framework for a CBR system with three hierarchical levels or three realms.

S represents the real system.

The elements of the SKUP categorical framework are as follows:

S-environment representing the processes

K1 and K2-corresponds to the CBR cycle

U10: K1→S action

P01: S→K1 sense of data and monitor

K is structured to provide an approximation of what it is considered as a dynamic memory model that basically consists of retaining experiences as cases for further reuse.

Cases are registers containing a description of a problem and its solution.

The elements of the categorical framework may be: K1-Reuse, K2-Revise, U21-Reuse action P12-Retain procedure.

It is possible to run on different time scales for different SKUPs. Several K1, K2 cycles may be performed before the coupling in the larger SKUP loop. K1 and K2 cycles negotiate among themselves as to which should be active. This allows anticipative control of the process.

The system interacts with its environment, through its data base that acquires new cases in response to changes in the environment and through the actions that it performs.

The framework shown in Fig. 5.9 allows cognitive evolvability and autonomy.

#### 5.2.6.2 Nested Frameworks for Evolvable CBR

Applications of CBR methodology for autonomous service failure diagnosis have been proposed (Montani and Anglano, 2006). This kind of CBR approach allowed self-healing in software systems.

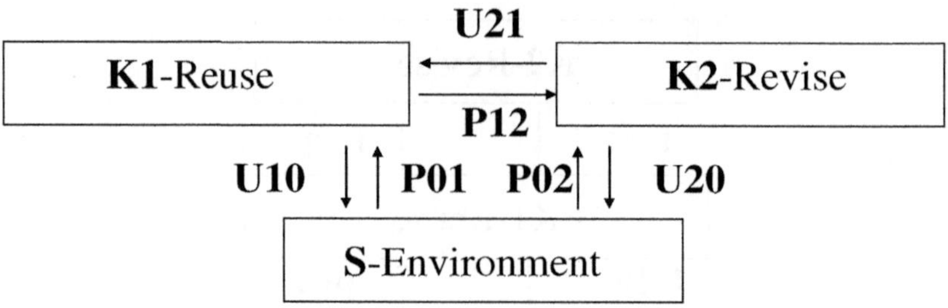


**Fig. 5.9** Three realms framework for evolvable CBR

Fig. 5.10 shows a four realms categorical framework for CBR as presented in Table 5.8. The four realms are K0-Retain-T, K1-Recall-C, K2-Reuse-U and K3-Revise-V.

K0 reflects the environment response.

The architecture shown in Fig. 5.10 outlines the possibility of integrative closure including the link between K0 and K3 and allowing evolvability and autonomy.

This link may be established by implementing autonomic computing paradigm (Kephart and Chess 2003). This studies methods for increasing environment-awareness and automatic responsiveness. Autonomic computing methods promise to facilitate CBR tasks and facilitate information capture (Montani and Anglano 2006).

Theoretically the cyclic architecture is not confined to four realms.

Fig. 5.10 shows nested and self-similar architectures.

A similar structure is repeated starting from the whole central system that may be built by four sub-realms denoted here by k0, k1, k2 and k3.

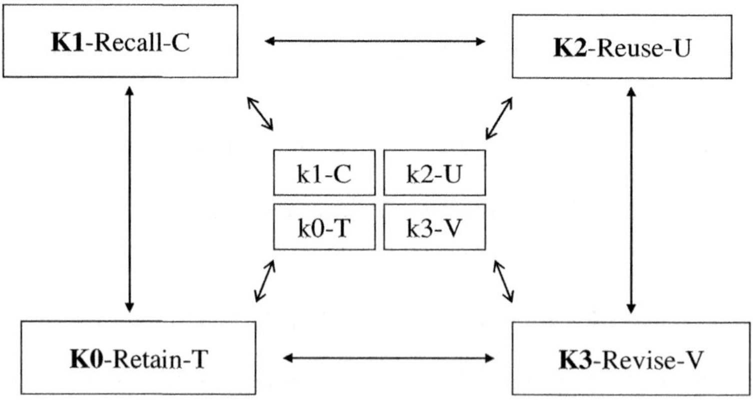


**Fig. 5.10** Nested frameworks for evolvable CBR

## 5.3 Failure Analysis

### *5.3.1 Complexity Challenges*

Failure analysis, FA, failure mode and effect analysis, FMEA, root cause analysis, RCA are useful quality and reliability tools in different industries.

FA and RCA are structured analytic methodology used primarily to examine the underlying contributors to an adverse event or condition. FMEA focuses on prevention and proactive risk management as RCA is concentrated on the occurrence of adverse events.

FMEA differs from FA in that it is a structured methodology used to evaluate a process prior to its implementation. Its purpose is to identify on an a priori basis the ways in which that process might potentially fail, with the goal in mind being to eliminate or reduce the likelihood and outcome severity of such a failure.

The complexity advent imposes significant modification of basic concepts and methods such as FA, FMEA, RCA, reliability and quality systems, problem solving methodologies, testing strategy, time concepts and frames.

Some of the difficulties of conventional FA methods are as follows:

- They focus on short-term customer satisfaction not on process improvement.
- FA is fixed not reviewed during the life of the product.
- In several companies, FA are developed too late and don't improve the processes.
- FA are not conceived as dynamic tool that will be developed.
- FA is not able to identify complex failure modes involving multiple failures or subsystems, or to discover expected failure intervals of particular failure modes.
- FA don't take into account the timing and scheduling in failure analysis and process improvement.

The constructivist strategy to confront complexity frontier is based on evolvability. Evolvable failure analysis, EFA are presented as the approach to meet the requirements imposed to the industry in high complexity domains.

An EFA system is an FA system that has the characteristics:

- Addresses multiple problems, tasks, failures that can be correlated to the real system
- Can change autonomously both the FA scheme as the real process dynamic behavior
- Is capable to control and to take advantage of the unexpected events of their environment in increasingly complex ways
- Have emergent, not entirely pre-programmed, behavior
- Shows multiple scale, parallel, evolution potentialities
- Can incorporate and accommodate new information

### 5.3.2 Basic Framework for Diagnosis

The two level basic categorical frameworks are able to gather some of the elements of adaptability for failure analysis (Fig. 5.11).

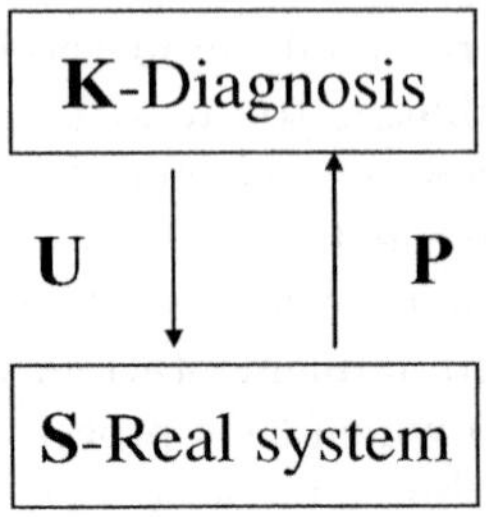


**Fig. 5.11** Two levels diagnosis

The starting step is a failure analysis scheme in K. It is based on the expertise for several case studies. This summarizes the experiment and the possible factors.

To this summary we may associate a tree-like diagram being in fact similar to the standard fishbone root-causes. Obviously we may start from the existing root-cause diagrams. Then a comparison with reality S is proposed.

In this step couples of factors are selected and tested.

It is a process described by the operators U showing how the real state for a given scheme in K.

In the next step the significant factors may be grouped.

It is a process allowing establishing affinity diagrams based on similarities.

It results a structure similar to the root-cause diagram but having reversed direction arrows. It looks more like a decision diagram.

This step is associated to possibilities P.

Finally a structure in K is resulting.

Iordache (2009) presented in detail examples of failure analysis based on SKUP frame and DOE resulting as solution of the wave equation WE.

The basic SKUP shown in Fig. 5.11 may be perceived in other failure analysis method.

The SKUP steps corresponds to the Boyd's OODA loop (Dettmer 2003).

In this case the observe part is linked to S, the decide part is associated to K, the orient part is linked to P and the act part is liked to U.

The same elements may be identified in the Goldratt, constraint management model (Dettmer 2003).

In this case the mismatches analyze part is linked to S, the creation of transformation design and plan part is associated to K, the review the strategy and the definition the new paradigm is linked to possibilities P and the strategy deployment part is liked to the operators U.

## *5.3.3 Perspectives*

### 5.3.3.1 Three Levels Frameworks

Recent work has pointed out that diagnosis strategies represent a necessary tool for complex systems diagnosis. Nejdl et al. (1995) introduced a formal meta-language to express strategic knowledge in an explicit way.

Fig.5.12 shows the categorical framework for a diagnosis system with three levels.

S represents the real system.

K1 and K2 are the two cognitive levels. K1 represents the diagnosis level.

K2 represents the meta-level of strategies. The strategies are defined at the meta-level.

U10:K1→S describes the actions towards the real level while P01: S→K1, summarizes the observations from S evaluations.

The information change between the basic level and the meta-level of diagnosis is characterized by the operator U21 and the possibilities P12.

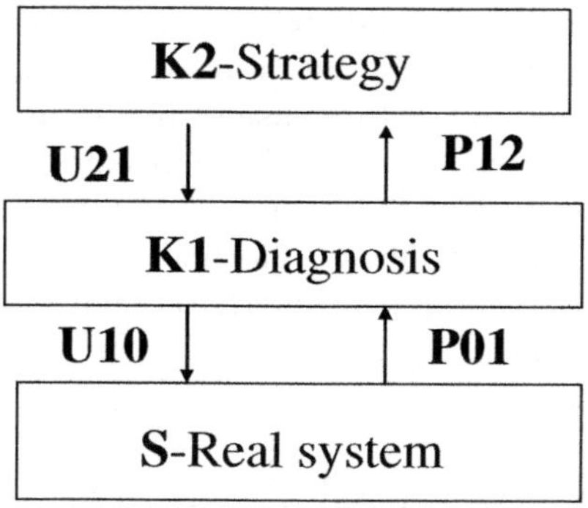


**Fig. 5.12** Three levels hierarchical diagnosis

A similar failure-driven driven modeling approach that incorporates ideas from developmental learning is due to Sakhanenko et al. (2007). It is based on the architecture with several levels of control and of learning for adapting and evolving models to represent changing and evolving situations (Piaget 1970).

The loop S, K1, P01, U10 is linked to assimilation mechanism while the loop K1, K2, P12, U21 is linked to accommodation mechanism.

Assimilation supposes integrating new information into pre-existing structures. Accommodation supposes changing and building new structures to understand information.

### 5.3.3.2 Four Realms Frameworks

A developed four-level categorical approach for security of distribution information systems was presented by Sisiaridis et al. (2008). The four levels correspond to Data, Schema, Construct and Concept (Fig. 5.13). The improvement is representing by the integrative closure allowing the emergence and autonomous testing of new concepts.

Restricting the levels interactions to the operators U10, U21, U32 leave the choice of control to the users and are appropriate for low-level security risks. The bottom-up approach, emphasizing the possibilities P01, P12 and P23 allows risk analysis and are more suited to high level security risks.

The signification of the functors U and possibilities P is explicit. U10, U21, U32 and U30 corresponds to implementation operations.

Observe that: U10: K1-Schema→S-Data, U21:K2-Constructs→K1-Schema, U32: K3-Concepts→K2-Constructs, and U30: K3-Concepts→S-Data.

P01, P12, P23 and P03 are synthesis steps.

P01: S-Data→K1-Schema, P12: K1-Schema→K2-Constructs, P23: K2-Constructs→K3-Concepts, and P03: S-Data→K3-Concepts.

Fig. 5.13 emphasizes the role of integrative closure via U30 and P03. This interconnection may make the system quite evolvable and autonomous.

The link via U30 and P03 may be established by implementing pervasive computing (Estrin et al. 2002). In a case of failure analysis and self-healing, as sensors are installed on a device, the information can be automatically captured during preventive maintenance. It may be possible to broaden the range of environmental and device information captured and transmitted automatically. Pervasive computing methods facilitate information capture and failure analysis tasks.

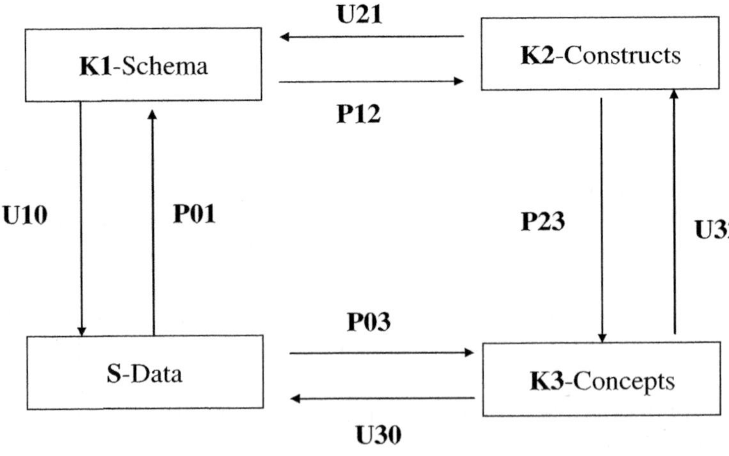


**Fig. 5.13** Four realms network for security of distribution information systems

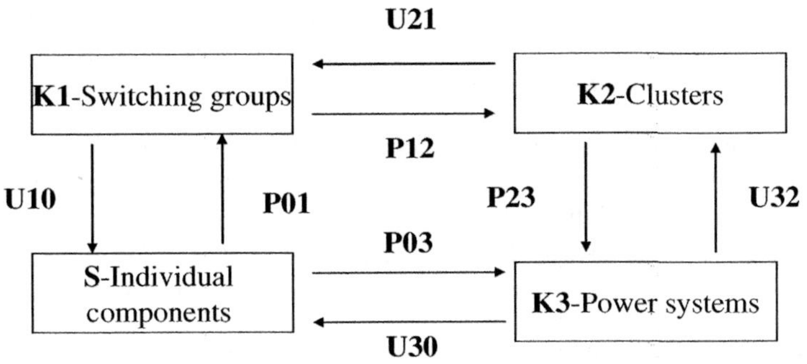


**Fig. 5.14** Four realms network for failure diagnosis

Another example of evolved failure analysis making use of the four-level architectures is shown in Fig. 5.14 (Rayudu et al. 2000).

The first reality level represents behavior of individual components and their present status. The second level, characterizes the switching groups and this refers for instance to isolators, protective relays, circuits breakers, and so forth.

The representation of entities bounded by a set of switching groups called clusters make the third level. The cluster level incorporates behavior knowledge concerning connected switching groups and the operational equipment between them.

The fourth level represents the whole network in terms of clusters. This level encompasses the strategic problem solving knowledge related to the complete power network. It is an integrative closure for failure diagnosis, allowing system evolvability, self-repairing and autonomy. The operators U and P describe the testing procedures and the action in case of failure. The possibilities P describe the testing procedures and the information transfer between levels.

## 5.4 Multi Agent Manufacturing Systems

### *5.4.1 Multi Agent Systems*

Agents are participants as individual elements within a complex system. Each agent may have its own set of internal states, skills, rules, and strategies that determine its behavior. Agents generally exist in a hierarchy. For example, an employee in a corporation interacts with other agents at a higher hierarchical level in the organizational environment. The agents receive information from within and outside their environment. Agents may develop their own schema through interaction and find regularities in the data and compress these perceived regularities into internal models that are used as the basis for action. An agent may be defined as a device or a self-directed program object which has its own value system and the means to solve certain tasks independently and then communicate its solution to a larger problem solving organization.

The main categories of agents are:

- Autonomous agents, capable of effective independent actions
- Objective directed agents, when autonomous actions are directed towards the achievement of defined tasks
- Intelligent agents, with ability to learn and adapt
- Cooperative agents, assisting other agents to perform a task

Examples of agents are neurons in brain, antibodies in case of immune systems, ants in colonies, wolfs in packs, investors in the case of stock market, people in social networks, and so forth. In each case agents have relatively limited set of rules, and the complexity of the collective behavior emerges from the large number of interactions among each other and their environment. There is constant action and reaction to what other agents are doing, thus nothing in the system is essentially fixed.

The distributed manufacturing environments and the flexibility and reaction to disturbances requirements are crucial reasons for moving to new organization paradigms.

Next generation of manufacturing control systems comprises the high adaptation and reaction to the occurrence of disturbances and to environment changes. On the other hand these control systems should optimize the global performance of the system which requires a global view of the entire system. These requirements imply the development of new manufacturing control systems with more autonomy, robustness against disturbances, able to handle to the changes and disturbances much better than the actual systems. New paradigms should focus on the ability to respond promptly and correctly to external changes, without external interventions. Distributed manufacturing architectures, multi-agent-based manufacturing systems represent a potential answer to complexity challenges (Parunak and Brueckner 2001).

Table 5.9 compares the multi agent systems MAS, with conventional hierarchical approaches. The autonomous multi agent systems may have some disadvantages. Theoretical optima cannot be guaranteed. Predictions can be made only at the aggregate level. Systems of autonomous agents can become computationally unstable. On the other hand, an autonomous approach appears to offer significantly advantages over conventional systems. Because each agent is close to the point of contact with the real world, the system computational state tracks the state of the world closely, without need for a centralized database.

Multi agent systems offer a way to relax the constraints of centralized planned, sequential control. They offer production systems that are decentralized rather than centralized, emergent rather than planned and concurrent rather than sequential.

**Table 5.9** Multi-agent versus conventional systems

| Characteristics | Conventional | Multi agent systems |
|---|---|---|
| Model source | Military | Biology, economy |
| Optimum | Yes | No |
| Prediction level | Individual | Aggregate |
| Computational stability | High | Low |
| Match to reality | Low | High |
| Requires central data | Yes | No |
| Response to change | Fragile | Relatively robust |
| System reconfiguration | Hard | Easy |
| Calculus | Complicated, long | Simple, short |
| Time required to schedule | Slow | Real time |
| Processing | Sequential | Concurrent, parallel |

## 5.4.2 Frameworks for Manufacturing Systems

PSM and EDOE methodology offers suggestions for agent based architectures for manufacturing applications (Iordache 2009).

The knowledge processor is a knowledge base system that stores and processes the necessary knowledge for an agent to play the role the agent society has designed for it.

The typical conceptual model of an agent comprises four components surrounding the knowledge processor. The four elements are:

- Perception, a channel for an agent to receive information from the external world
- Actuator, an interface for an agent to modify or influence the states of an agent community
- Communication, a mechanism for an agent to exchange views with other members in the agent society
- Objectives, a list of roles for an agent to play

In terms of EDOE, the knowledge processor plays the role of the center, K.

These local databases store all knowledge about the behavior of the agent and the community were the agents belongs. The information stored in these databases involves constraints, objectives, procedures, rules and experience, and organizational structures and techniques. It may be organized by logical rules.

The four factors of the center K are: Communication C, Perception P, Objectives O and Actuator A.

Modules have to be associated to component DOE.

The communication module deals with the need to regulate the interaction between distributed agents and defines a communication language.

The communication module may have sub-modules such as: contents C1, message C2, physical information C3 and so on. These are analogous to the factors of component designs.

It results that the generic architecture for an agent may be represented as an EDOE.

The multi-agent system is an open and distributed system that is formed by a group of agents combined with each other through a network of cooperatively solving a common problem. Often, several agents do same simple thing. It is possible that other agents don't use all modules defined in the generic architecture.

The architecture is in fact a complex EDOE framework and can be operated according to the EDOE methodology.

It is easy to identify the SKUP elements.

The product of implementing multi-agent architecture is described by S.

The elements of K are denoted by C, P, O, A at the level m=0 and C1, C2, C3 and so on at the level m=1. Operators U describe the scheduling, while possibilities P take into account the execution step. The overall system performance is not globally planned but emerges through the dynamic interaction of the agents in real time.

Operators U and possibilities P express that the system does not alternate between cycles of central scheduling and final execution. Rather than this mechanism, the schedule emerges from the concurrent independent decisions of the local agents.

Elements of the SKUP may be detected in other multi-agent systems.

For instance, in the study of coordination, the agents with their specific roles represents the elements of K, the so-called rational actions are associated to the elements of U, the perceptions are elements of P while the states S are represented in this case by pheromones (Parunak and Brueckner 2001).

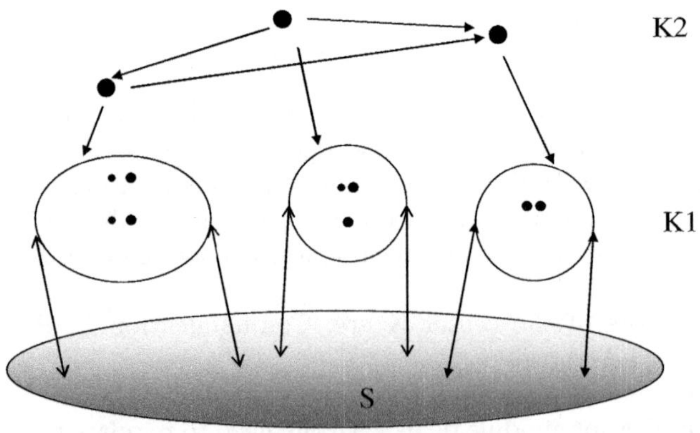


**Fig. 5.15** Three levels hierarchical framework for multi-agent-system

Fig. 5.15 shows a three levels hierarchical framework for multi-agent-system.

A general presentation of this architecture was shown in Fig. 4.9.

In this case the notations are: S-environment, K1-agents and K2-meta agents.

#### 5.4.2.1 Holonic Manufacturing Systems

The holonic manufacturing system is the paradigm that translates the holon concept developed for living systems and social organizations into a set of appropriate concepts for manufacturing industries (Tharumarajah et al. 1996, Valckenaers et al. 1997, Ulieru 2002).

The term holon describes the basic unit of organization in living organisms and social organizations. The holon can represent a physical or logical activity such as a machine on order or a human operator. The holon can be broken into several other holons, a procedure which allows the reduction of problem complexity. A manufacturing holon comprises a control part and an optional physical processing part. Multiple holons may dynamically aggregate into a single-higher level holon.

The holarchy is a system of holons that can co-operate to achieve an objective. The holonic manufacturing system is a holarchy that integrates the entire range of manufacturing activities.

The holonic manufacturing systems paradigm is part of the next generation of distributed control and introduces the hierarchical control within a heterarchical structure. This innovation makes available the combination of robustness against disturbances, presented in heterarchical control with the stability and global performances optimization presented in hierarchical control. The implementation of this concept requires that decision power must be distributed between the central and local holons that is there exists a switch between hierarchical and heterarchical control. In categorical terms this corresponds to a switch from product to coproduct constructions. A categorical presentation of the architecture was shown in Fig. 4.9. Usually the categorical product is associated to cognitive or interactive steps while the categorical coproduct is associated to reactive steps.

The function of central holon is to advise the local holons. When disturbances occur the autonomy of holons increase, while during normal functioning, the autonomy of local holons decreases and they follows the central holons input as for example the scheduling plans.

The holonic manufacturing system design starts with a forward control step in which the definition of all appropriate holons and their responsibility is established. In comparison with traditional methodologies a rather vague responsibility than a precise function for each holon is established. This facilitates the backward control. Complementary controls ensure systems evolvability.

Ulieru et al. (2002) studied the four layer holonic control architecture.

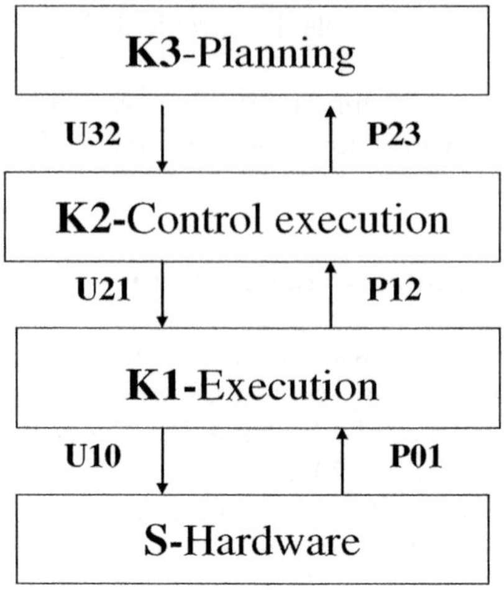


**Fig. 5.16** Four levels hierarchical framework for holonic control

The level S corresponds to resources. It represents the physical platform.

The K1 layer is concerned with the execution of the application.

In this case the K2 layer achieves the arranging for the distribution of applications across multiple resources.

The K3 layer is concerned with planning issues such as reconfiguration or execution control.

This architecture reflects the multi-resolution structure of the holonic enterprise. As we move down the layers shown in Fig. 5.16, time scales become shorter.

Higher layers are more sophisticated but slower, while lower layers are fast and light-weight.

The multi layer holonic assembly system was studied also by Sugi et al. (2003). The holarchy of the system consists of execution holons layer identified as K1, assembly operation holons layer identified as K2 and top management operation holon identified here as K3 layer. If the management holon is ordered to assembly a specified product, this assembly task is decomposed into subtasks for lower management holons. An operation holon secures appropriate execution holons, which corresponds to real manufacturing devices, using the contract net protocol. Then the operation holon makes the execution holons execute a job such as assembling parts. The decentralized nature of the system enabled to realize plug and produce, a system function that supports easy addition and removal of manufacturing devices. Sugi et al. (2003) developed techniques for plug and produce such as distributed resource allocation method for installation of new robots and an automated calibration for mutual positional relationship between an existing robot and a newly added one.

Plug and produce function is related to interoperability at several levels and n-categorical modeling.

What is missing for hierarchical holonic architecture as that shown in Fig. 5.16 is the interaction of the top layer K3 and real execution layer S, that is, the integrative closure.

The four-level approach to holonic systems presented in Fig. 5.17 challenge this difficulty. It represents the mandatory development of the current hierarchical approach (Naumenko and Wegmann 2003, Baina and Morel 2006).

In this case the lowest level denoted here by K0 presents different subjects for modeling, each of them called a universe of discourse. The next level K1 contains viewpoints, for instance mechanisms and models.

The next level K2 focuses on meta-models and the highest level K3 focuses on meta-meta-models.

The meta-meta-model should be designed to allow unification under a common framework.

Each application can be considered as a specific use of a viewpoint defined in the K1 level which is based on meta-models described in K2.

The universe of discourses K0 concerns the manufacturing enterprise.

To describe this universe it is possible to use holonic views K1 that are instantiations of the meta-model K2 defining specific type of holons and their relationship within the context. K2 is an instantiation of K3. K3 level corresponds to the meta-meta-model.

Application interoperability can be resolved by achieving the K3→K0 connection and the integrative closure. This integrative closure makes the system evolvable.

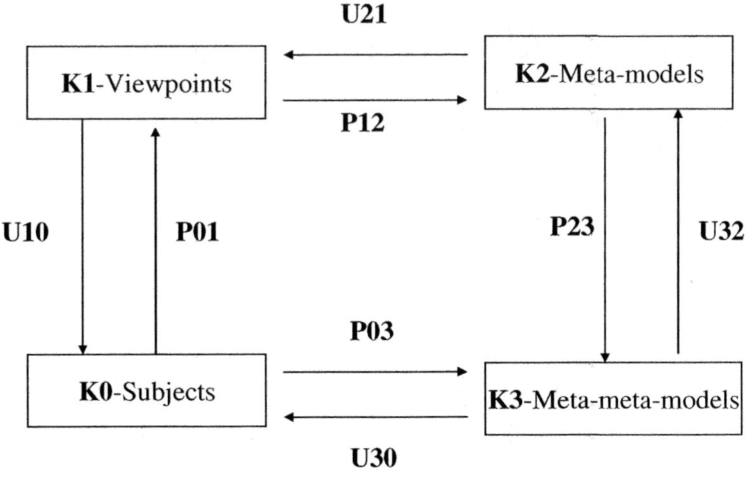


**Fig. 5.17** Four realms multi-agent-system modeling

### *5.4.3 Evolvable Manufacturing Systems*

The globalization of markets, shortening of product life cycles, decrease of dimensions for products and outsourcing were identified as major threats for industry. Answers to such threats were paradigms as evolvable assembly system, EAS (Onori 2002), evolvable production systems EPS (Onori et al. 2006, Frei et al. 2007), and evolvable manufacturing systems EMS (Iordache 2009).

The design process of assembly systems, EAS has been modeled by a hierarchy of four levels: S-Environment, K1-Domain knowledge, K2-Inference knowledge and K3-Task knowledge (Lohse et al., 2005). The domain knowledge level defines all the specific concepts needed by the inferences. The inference knowledge level defines what inferences are needed to fulfill the reasoning task. The task knowledge level defines the reasoning tasks required to achieve a specific goal. All three levels of knowledge, K1, K2 and K3 have been modeled in a Protégé interface to allow the dynamic definition and adaptation of the assembly system design process.

As discussed the four levels hierarchy doesn't allows complete evolvability and autonomy.

EPS represents a concrete solution to the requirements from the market such as stated within the agile, reconfigurable and distributed approaches. They include high responsiveness, low down-times, ability to handle small series with many variants, and on the fly changeability. Together with ontology-based process specific modules, a distributed control system using the multiple agent paradigm allows to quickly and cost effectively adapt to ever changing production requirements.

EPS have similarities with the bionic, fractal, holonic, biological and reconfigurable manufacturing systems, but there exists major differences too.

Besides considering system morphology, EPS strongly links product, processes, system and environment by the means of detailed ontologies.

EPS focuses on self-organization and implies the ability of complex systems to co-evolve with continuously changing requirements. EPS are expected to allow the user to build any required system and to modify this at wish.

Some features of the production systems necessary to achieve evolvability are:

- Modularity since many small, dedicated units that can easily be integrated into different systems/cells
- Process-orientation for units
- Variable granularity and fluidity process related. This implies multiple scales
- Control system, distributive
- Interoperability
- Multi-agent technology to capture emergent behavior

Evolvable systems may be considered as a natural development of flexible and reconfigurable manufacturing systems (Table 5.10).

**Table 5.10** Comparison of different management systems

| Criterion\System | **Specialized** | **Flexible** | **Reconfigurable** | **Evolvable** |
|---|---|---|---|---|
| Skills | One | Set of fixed skills | More skills adapted | No particular product focus |
| Flexibility | Low | Discrete | Continuous | Emergent |
| Capability | High efficiency for one situation | Cope with different situations | Cope with differences. Can be adapted | Very agile |
| Concerns | Rigid | Cannot cope with new | Unexpected are not coped | Difficult to define a generic mechanism |

Table 5.10 suggests considering the different stage in the historical development of manufacturing systems as the necessary stages in categorification.

The first stage corresponds to specialized manufacturing, to single installation and in the same time to sets or 0-categories (Appendix A4).

A $1^{st}$ order evolutionary step is represented by the transition to flexible manufacturing systems.

Flexibility approach allows doing diverse tasks with the same installation. This is associated to 1-categories.

A $2^{nd}$ order evolutionary step is represented by the transition to reconfigurable manufacturing systems.

Reconfiguration is supposed to make use of several installations. It is linked to the 2-categories. Reconfigurable manufacturing systems incorporate principles of modularity, integrability, flexibility, scalability, convertibility and diagnosability. Some flexible and reconfigurable systems failed because they don't take into account that if any system is to be flexible then its constituents need to be far more flexible.

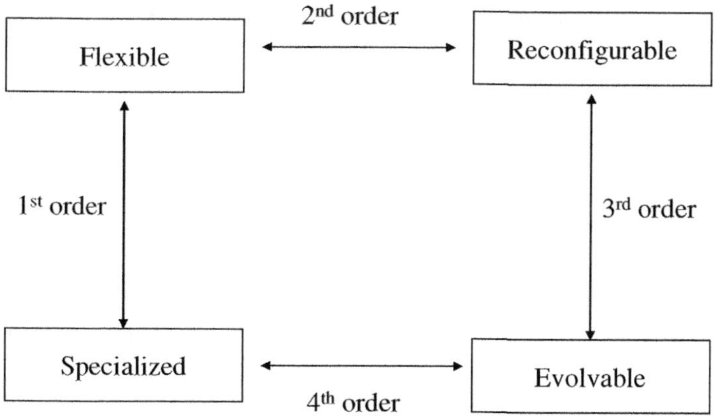


**Fig. 5.18** Four stages for evolvable manufacturing systems

A $3^{rd}$ order evolutionary step is represented by the transition to evolvable manufacturing systems.

Evolvability achieves the full flexibility and is related to the 3-categories concept implementation.

Observe that EAS, EPS, EMS considers the production unit as an artificially living entity and emphasizes on evolution rather than adaptation.

Usually the adaptability implies an adjustment on the time scale of the life cycle of the organism. It characterizes 1-category frames. But this is not enough to challenge the high complexity. Evolvability should imply the capacity for genetic-like change to invade new life-like cycles on several time-scales, by higher categorification steps.

In a dynamic environment, the lineage that adapts first wins. Fewer mutations steps mean faster evolution. The request is for some production or management systems built to minimize the number of mutations required to find improvements.

Fig. 5.18 reinforces the idea of categorification process by imposing to different realms to be categories. By successive categorification the legacy equipment and associated software will still be utilizable. An n-graph model may be naturally associated to the framework shown in Fig. 5.18. Categorical issue implies that EMS achieves specific fluidity properties. It should have fluidity at different levels of complexity. Consider that the production line is composed from several components that can be plugged in or out. These are 1-cells and the corresponding fluidity is the so-called fine fluidity or 1-fluidity corresponding to flexible manufacturing and to 1-categories.

When a manufacturing line is composed of several cells and these cells are modules or 2-cells that can be plugged in or out this is the thin fluidity or 2-fluidity. It corresponds to reconfigurable manufacturing and to 2-categories.

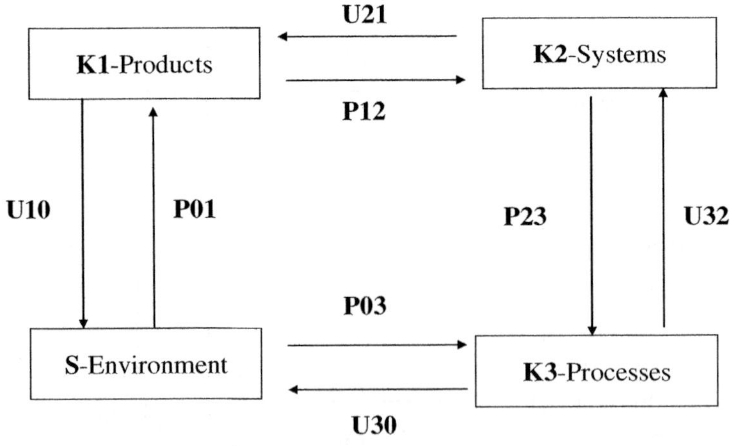


**Fig. 5.19** Four sub-realms network for evolvable manufacturing systems

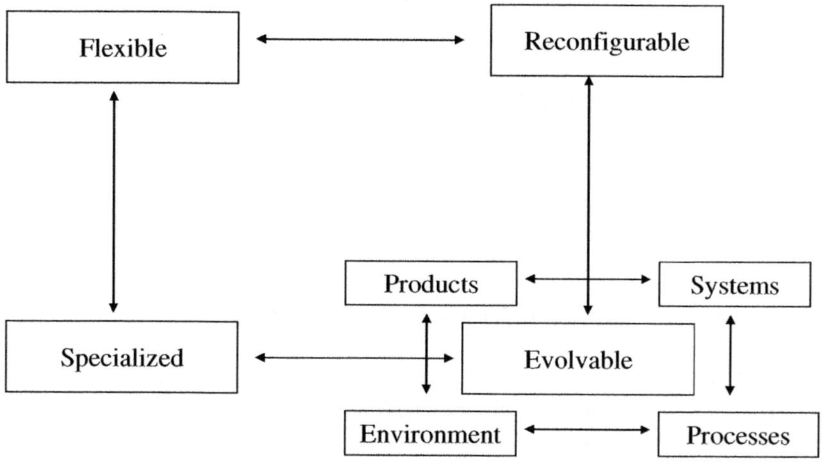


**Fig. 5.20** Four realms and sub-realms for evolvability

The thick fluidity or 3-fluidity will refer to the whole system that is 3-cells to be plugged in or out. This corresponds to evolvable manufacturing and to 3-categories.

The autonomic and organic computing (Kephart and Chess 2003, Bauer and Kasinger 2006) were identified as fundamental concepts for achieving evolvable manufacturing systems. Although autonomic computing was designed for software systems, the related ideas can be projected into a modular production system. Automatic computing in this context refers to computing elements disseminated throughout the production system which beyond the normal mechanical, electrical and sensorial units includes computational power.

Organic computing focuses on completing the closure by studying the $4^{th}$ order evolutionary step.

Fig. 5.19 outlines a four sub-realms network for evolvable manufacturing systems.

The environment refers to real and artificial aspects, including the available materials.

Products sub-realm denotes the products and product related activities. Production sub-realm denotes the production system skills, modules. Processes sub-realm refers to all processes, for example assembly.

The signification of the functors U and possibilities P is explicit.

U10, U21, U32 and U30 corresponds to top-bottom implementation operations

In this case U30=U10oU21oU32.

P01, P12, P23 and P03 are bottom-top synthesis steps. Observe that: P03=P01oP12oP23.

Fig. 5.19 emphasizes the role of integrative closure via U30 and P03.

Onori (2002) highlighted the interaction between products and systems illustrated by a generic product life cycle view.

The entire structure shown in Fig. 5.19 may be just one realization of the four stage diagram shown in Fig. 5.18. This aspect is clarified by the Fig. 5.20.

It is a superposition of Fig. 5.18 and Fig. 5.19.

Observe that the construction of a specific evolvable manufacture parallels and recapitulates the general history of manufacturing systems from specialized to evolvable.

## 5.4.4 Belief Desire Intention Agents

The belief desire intentions, BDI, agent introduced a formal meta-language to express agent rationality in an explicit way. BDI architecture is one of numerous architectures that enact deliberative agents. The BDI agent architecture is an attempt to encapsulate the hidden complexity of the inner functioning of an individual agent.

The agent shown in Fig. 5.21 is structured in four elements: beliefs, goals, plans and intentions (Rao and Georgeff 1991). We will refer to a BGPI structure for BDI agent.

Fig. 5.22 outlines a possible categorical framework for a procedural reasoning system with three levels.

S represents the environment.

K1 and K2 are the two cognitive levels. In this case K1-includes goals while K2-includes plans. The strategies are defined at the K2-level.

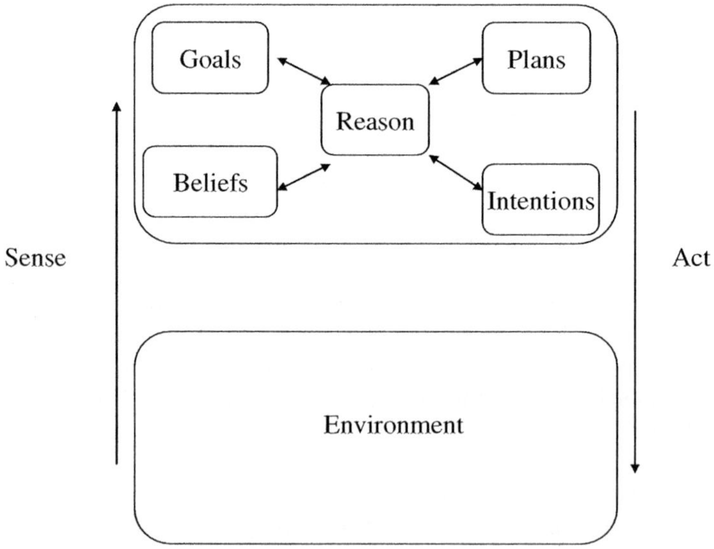


**Fig. 5.21** Structure of BDI agents

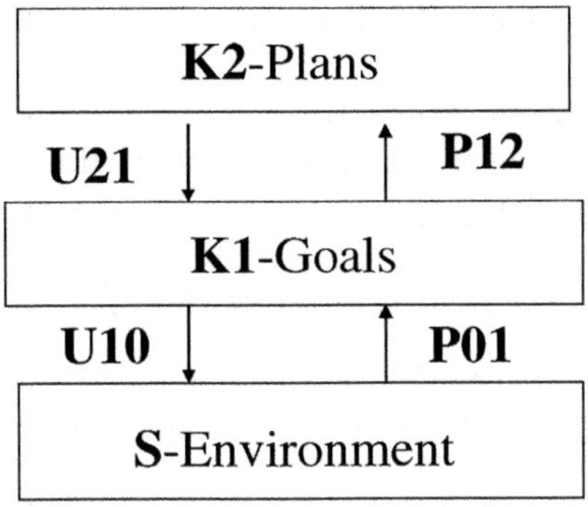


**Fig. 5.22** Three levels hierarchical framework for cognitive agents

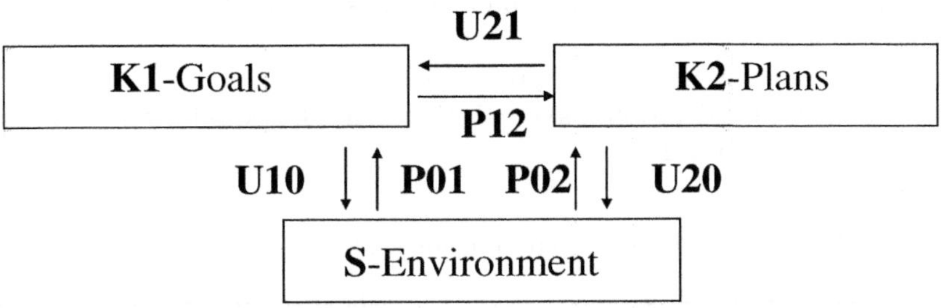


**Fig. 5.23** Three realms framework for cognitive agents

U10: K1→S describes the actions from goals towards the environment level while P01: S→K1 summarizes the sensed info about S and forwards this toward goals.

The information change between the basic level and the K2 of strategies is characterized by the operator U21 and the possibilities P12.

The BDI architecture with three realms, and links between K2 and S is shown in Fig 5.23.

The framework shown in Fig. 5.23 allows cognitive evolvability and autonomy.

## *5.4.5 Multiple Levels Cognitive Architecture*

### 5.4.5.1 Multiple Levels Agents

Innovative multiple-scale agent architectures have been proposed by Goschnick (Goshnick 2003, Goschnick and Sterling 2002).

Based on Jung analytical psychology (Jung 1997) Goschnick developed a cognitive architecture named Shadow Board. This implies:

- Decomposition of a user's multiplicity of roles into a hierarchy of sub-agency
- Relaxing of the autonomy of the sub-agents under control of an autonomous agent-the so called Ego/Self agent which is autonomous
- Wrapping of the external services and agencies including the web services and utilizing them as if they were internal sub-agents
- Ability to apply ontology at the localized level

The Shadow architecture may be considered as resulting by WE equation.

Table 5.11 shows the agent based structure of the cognitive architecture at different levels.

The notations are: B-beliefs, G-goals, P-plans, and I-intentions.

The central agent $\begin{matrix} G & P \\ B & I \end{matrix}$ is the so-called Ego/Self agent (Goshnick 2003).

It is autonomous in the sense that its parts B, G, P, I don't depend on others agents.

This central agent should be considered as a whole.

It is an executive decision maker. Decisions are based on the knowledge of sub-selves B, G, P and I.

**Table 5.11** Array of conditions for BGPI multi-agent system

| | | | | | | | | | | | | | | |
|---|---|---|---|---|---|---|---|---|---|---|---|---|---|---|
| G22 | | G23 | | G32 | | G33 | | P22 | | P23 | | P32 | | P33 |
| | **G2** | | | | **G3** | | | | **P2** | | | | **P3** | |
| G21 | | G24 | | G31 | | G34 | | P21 | | P24 | | P31 | | P34 |
| | | | **G** | | | | | | | | **P** | | | |
| G12 | | G13 | | G42 | | G43 | | P12 | | P13 | | P42 | | P43 |
| | **G1** | | | | **G4** | | | | **P1** | | | | **P4** | |
| G11 | | G14 | | G41 | | G44 | | P11 | | P14 | | P41 | | P44 |
| | | | | | | | **G P**<br>**B I** | | | | | | | |
| B22 | | B23 | | B32 | | B33 | | I22 | | I23 | | I32 | | I33 |
| | **B2** | | | | **B3** | | | | **I2** | | | | **I3** | |
| B21 | | B24 | | B31 | | B34 | | I21 | | I24 | | I31 | | I34 |
| | | | **B** | | | | | | | | **I** | | | |
| B12 | | B13 | | B42 | | B43 | | I12 | | I13 | | I42 | | I43 |
| | **B1** | | | | **B4** | | | | **I1** | | | | **I4** | |
| B11 | | B14 | | B41 | | B44 | | I11 | | I14 | | I41 | | I44 |

The Ego/Self agent is able to call the four main agents, B, G, P and I individually or in team. The sub-selves are again divided in sub-sub-selves as for instance B is divided in four agents B1, B2, B3 and B4. Then the splitting in four is continued one level more. From B1 it results B11, B12, B13 and B14 and so on. They may be identified with an elementary BDI agents B11-belief, b, B12-goal, g, B13-plans, p and B14-intentions, i. Consider that this elementary level is indexed by n=0. It is possible that the elementary agents are unrelated.

The coupling in agents corresponds to the level n=1. The resulting agents are B1, B2 and so on.

For the agent B2 we have specific activities: B21-belief, b, B22-goal, g, B23-plans, p and B24-intentions, i.

The coupling of information corresponds to the level n=2. At this level the B, G, P and I are the four agents. The reality level n=3 corresponds to the Ego/Self agent.

All levels have been illustrated by Table 5.11.

### 5.4.5.2 n-Graphs for Multiple Levels BGPI

Fig. 5.24 shows a representation of multiple scales frames using n-graphs.

The reality level n=0 corresponds to the 0-graphs or sets. They are represented by b, g, p or i individual uncorrelated objects. The level n=1 correspond the 1-graphs. These are directed graphs including the morphisms that is, the connections between b, g, p and i.

The morphisms are 1-cells. Their coupling allows the functioning of agents.

The level n=2 corresponds to the 2-graphs. These are graphs plus the 2-cells between paths of same source and target. These 2-cells express the

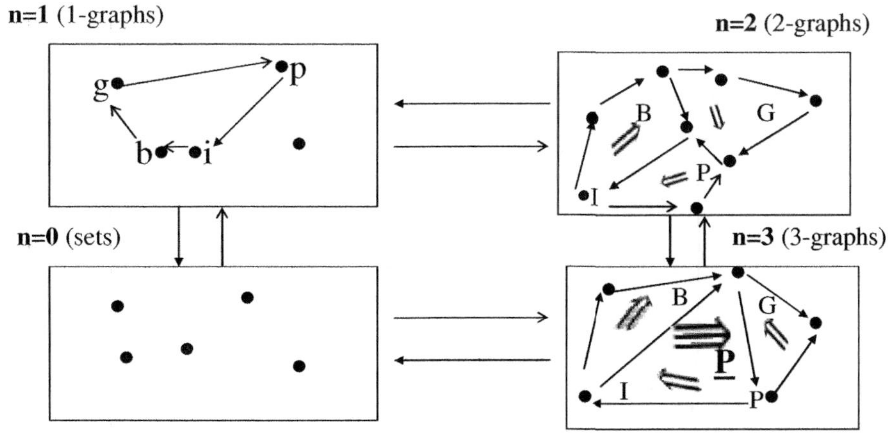


**Fig. 5.24** n-graphs for multi-scale framework

natural association of the quadruple b, g, p, i elements in just one macro agent denoted here by B-belief, G-goal, P-plans or I-intention. The level n=3 corresponds to the 3-graphs. These are 2-graphs that include the 3-cells that is, the cells between 2-cells. Fig. 5.24 shows a complete association as a plan **P**. The 3-graphs represent graphs modification and are subjected to conditions of natural transformations too.

#### 5.4.5.3 Nested Frameworks for BGPI

Fig. 5.25 shows a categorical presentation for BGPI architecture as presented in a different form in Table 5.11.

In this presentation K0 includes environment and the Beliefs, K1-Goals, K2-Plans, K3-Intentions.

The architecture shown in Fig. 5.25 outlines the possibility of integrative closure allowed by the link between K0 and K3 opening the road for evolvability and autonomy.

The architecture is not confined to four realms.

Fig. 5.25 shows also nested and self-similar architectures.

A similar structure is repeated starting from the whole central system built by four sub-realms denoted here by k0, k1, k2 and k3.

It should be noted that similar architectures are of interest for autonomic and organic computing (Trumler et al. 2004, IBM 2005, Bauer and Kasinger 2006).

The logical structure of an autonomic element is similar to that of BDI or BGPI agents.

For autonomic computing, the BGPI structure is replaced by the so-called MAPE loop whose elements are M-Monitor, A-Analyze, P-Plans, E-Execute.

Autonomic computing systems are composed of four levels identified as S-Managed resources, K1-Autonomic managers, K2-Orchestred autonomic

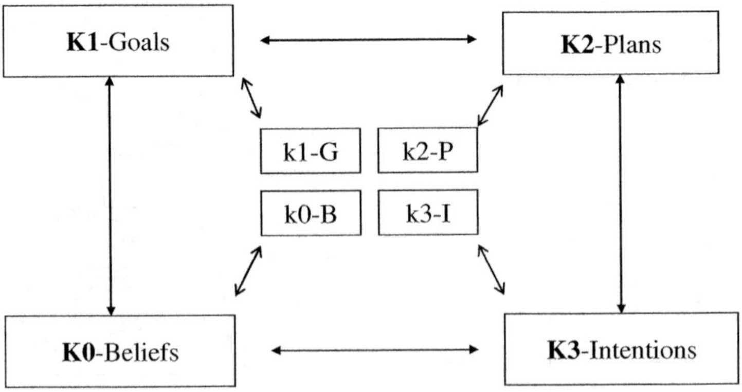


**Fig. 5.25** Nested frameworks for BGPI

managers, K3-Manual manager. The closed loop in which K3 is replaced by an automatic device was presented by IBM (2005).

For the organic computing middleware architecture (Trumler et al. 2004), the four levels may be identified as: S-Transport interface, K1-Event dispatcher, K2-Service interface and proxy, K3-Organic manager. In the middleware architecture the organic manager is, linked to the levels below it.

## References


Aamodt, A., Plaza, E.: Case-based reasoning: Foundational issues, methodological variations, and system approaches. AI Communications 7(1), 39–59 (1994)

Aha, D.W., Breslow, L.A., Munoz-Avilla, H.: Conversational case-based reasoning. Applied Intelligence 14(1), 9–32 (2001)

Baina, S., Morel, G.: Product centric holons for synchronization and interoperability in manufacturing environments. In: 12th IFAC Symposium on Information Control Problems in Manufacturing, St-Etienne, France (2006)

Bar-Yam, Y.: When Systems Engineering Fails-Toward Complex Systems Engineering. In: International Conference on Systems, Man & cybernetics, vol. 2, pp. 2021–2028. IEEE Press, Piscataway (2003)

Bauer, B., Kasinger, H.: AOSE and organic computing-how can they benefits from each other. Position paper. AOSE III. Springer, Heidelberg (2006)

Benami, O., Jin, Y.: Creative stimulation in conceptual design. In: Proceedings of DETC 2002 ASME 2002 Design Engineering Technical Conference, Montreal, Canada, vol. 20, pp. 1–13 (2002)

Black, J.: The Design of the Factory with a Future. McGraw-Hill, New York (1991)

Braha, D., Maimon, O.: A mathematical theory of design: foundations, algorithms and applications. Kluwer, Boston (1998)

Braha, D., Reich, Y.: Topological structures for modeling engineering design processes. Res. Eng. Design 14, 185–199 (2003)

Carreras, I., Miorandi, D., Saint-Paul, R., Chlamtac, I.: Bottom-up design patterns and the energy web. IEEE Transactions on Systems, Man Cybernetics, Part A, Special issue on Engineering Cyber-Physical Systems (2009)

Coyne, R.: Logic Models of Design. Pitman, London (1988)

Delugach, H.S.: Towards Building Active Knowledge Systems With Conceptual Graphs. In: Ganter, B., de Moor, A., Lex, W. (eds.) ICCS 2003. LNCS (LNAI), vol. 2746, pp. 296–308. Springer, Heidelberg (2003)

Dettmer, H.W.: Strategic Navigation: A Systems Approach to Business Strategy. ASQ Quality Press (2003)

Estrin, D., Culler, D., Pister, K., Sukhatme, G.: Connecting the physical world with pervasive networks. IEEE Pervasive Computing, 59–69 (2002)

Frei, R., Barata, J., Di Marzo Serugendo, G.: A Complexity Theory Approach to Evolvable Production Systems. In: Sapaty, P., Filipe, J. (eds.) Proceedings of the International Workshop on Multi-Agent Robotic Systems (MARS 2007), pp. 44–53. INSTICC Press, Portugal (2007)

Gani, R.: Chemical product design: challenges and opportunities. Comp. & Chem. Engng. 28, 2441–2457 (2004)

Goschnick, S.B.: Enacting an Agent-based Digital Self in a 24x7 Web Services World. In: Zhong, N., Raś, Z.W., Tsumoto, S., Suzuki, E. (eds.) ISMIS 2003. LNCS (LNAI), vol. 2871, pp. 187–196. Springer, Heidelberg (2003)

Goschnick, S.B., Sterling, L.: Psychology-based Agent Architecture for Whole-of-user Interface to the Web. In: Proceedings of HF2002 Human Factors Conference: Design for the Whole Person - Integrating Physical, Cognitive and Social Aspects, Melbourne (2002)

Grabowski, H., Rude, S., Klein, G. (eds.): Universal Design Theory. Shaker Verlag, Aachen (1998)

IBM, An architectural blueprint for automatic computing (2005)

Iordache, O.: Evolvable Designs of Experiments. Applications for Circuits. J. Wiley VCH, Weinheim (2009)

Jung, C.G.: Man and his symbols. Dell Publishing Company, NewYork (1997)

Kephart, J.O., Chess, D.M.: The vision of autonomic computing. IEEE Computer 36(1), 41–50 (2003)

Kiriyama, T., Tomiyama, T., Yoshikawa, H.: Qualitative Reasoning in Conceptual Design with Physical Features. In: Faltings, B., Struss, P. (eds.) Recent Advances in Qualitative Physics, pp. 375–386. MIT Press, Cambridge (1992)

Lee, E.A.: Computing foundations and practice for cyber-physical systems: A preliminary report. Tech Report Univ of California Berkeley/EECS-2007-72 (2007)

Li, S., Yang, Q.: Active CBR, Integrating case-based reasoning and active database, TR-1999-03, School of Computing Science, Simon Fraser University, Burnaby BC, Canada (1999)

Lohse, N., Valtchanov, G., Ratchev, S., Onori, M., Barata, J.: Towards a Unified Assembly System Design Ontology using Protégé. In: Proceedings of the 8th Intl. Protégé Conference, Madrid, Spain (2005)

Melendez, J., Colomer, J., Macaya, D.: Case based reasoning methodology for supervision. In: Procceedings of the European Control Conference, ECC 2001, Oporto, Portugal, pp. 1600–1605 (2001)

Montani, S., Anglano, C.: Case-Based Reasoning for autonomous service failure diagnosis and remediation in software systems. In: Roth-Berghofer, T.R., Göker, M.H., Güvenir, H.A. (eds.) ECCBR 2006. LNCS (LNAI), vol. 4106, pp. 489–503. Springer, Heidelberg (2006)

Naumenko, A., Wegmann, A.: Two approach in system modeling and their illustration with MDA and RM-ODP. In: ICEIS 2003, the 5th International Conference on Enterprise Information System (2003)

Nejdl, W., Froehlich, P., Schroeder, M.: A formal framework for representing diagnosis strategies in model-based diagnosis systems. In: Int. Joint Conf. on Artif. Int., IJCAI, vol. 95, pp. 1721–1727. Morgan Kaufmann Publishers, Inc., San Francisco (1995)

Onori, M.: Evolvable Assembly Systems-A New Paradigm. In: IST 2002 33rd International Symposium on Robotics, Stockholm, pp. 617–621 (2002)

Onori, M., Barata, J., Frei, R.: Evolvable Assembly System Basic Principles. BASYS Niagara Falls, Canada (2006)

Pahl, P.G., Beitz, W.: Engineering design, a systematic approach. Springer, London (1996)

Parunak, H.V.D., Brueckner, S.: Entropy and Self-Organization. In: Multi-agent Systems, Proceedings of the Fifth International Conference on Autonomous Agents, pp. 124–130. ACM Press, New York (2001)

Pattee, H.H.: Causation, control and the evolution of complexity. In: Anderson, P.B., et al. (eds.) Downward Causation, pp. 63–77. Aarhus University Press, Aarhus (2000)

Piaget, J.: Genetic Epistemology. Columbia University Press, New York (1970)

Piaget, J.: The construction of Reality in the Child. Ballantine Books, New York (1971)

Rao, A., Georgeff, M.: Modelling rational agents with a BDI architecture. In: Allen, J., Fikes, R., Sandewall, E. (eds.) Proceedings of Knowledge Representation and Reasoning. Morgan Kaufman Publishers, San Mateo (1991)

Rayudu, R.K., Samarasinghe, S., Maharaj, A.: A co-operative hybrid algorithm for fault diagnosis in power transmission. IEEE Journal of power Systems Engineering, 1939–1944 (2000)

Reich, Y.: A critical review of General Design Theory. Res. Eng. Des. 7, 1–18 (1995)

Sakhanenko, N.A., Luger, G.F., Stern, C.R.: Managing Dynamic Contexts Using Failure-Driven Stochastic Models. In: Wilson, D., Sutcliffe, G. (eds.) Proceedings of the Florida Artificial Intelligence Research Society of AAAI, FLAIRS-2, pp. 466–472. AAAOI Press (2007)

Schank, R.: Dynamic memory: a theory of reminding and learning in computers and people. Cambridge University Press, Cambridge (1982)

Sisiaridis, D., Rossiter, N., Heather, M.A.: Holistic Security Architecture for Distributed Information Systems - A Categorical Approach. In: European Meeting on Cybernetics and Systems Research, Symposium Mathematical Methods in Cybernetics and Systems Theory, EMCSR-2008, University Vienna, pp. 52–57 (2008)

Sowa, J.F.: Knowledge Representation: Logical, Philosophical, and Computational Foundations. Brooks Cole Publishing Co., Pacific Grove (2000)

Sugi, M., Maeda, Y., Aiyama, Y., Harada, T., Arai, T.: A Holonic architecture for easy reconfiguration of robotic assembly systems. IEEE Trans. on Robotics and Automation 19(3), 457–564 (2003)

Takeda, H., Tomiyama, T., Yoshikawa, H., Veerkamp, P.J.: Modeling design processes. Technical Report CS-R9059, Centre for Mathematics and Computer Science (CWI), Amsterdam, Netherlands (1990)

Takeda, H., Iino, K., Nishida, T.: Agent organization and communication with multiple ontologies. International Journal of Cooperative Information Systems 4, 321–337 (1995)

Tharumarajah, A., Wells, A.J., Nemes, L.: Comparison of the bionic, fractal and holonic manufacturing systems concepts. International Journal of Computer Integrated Manufacturing (9), 217–226 (1996)

Tomiyama, T., Yoshikawa, H.: Extended General Design Theory. In: Design Theory for CAD, Proceedings from IFIP WG 5.2, Amsterdam (1987)

Tomiyama, T., Kiriyama, T., Takeda, H., Xue, D.: Metamodel: A key to intelligent CAD systems. Research in Engineering Design 1(1), 19–34 (1989)

Trumler, W., Bagci, F., Petzold, J., Ungerer, T.: Towards an organic middleware for the smart dooplate project. GI Jahrestagung 2004(2), 626–630 (2004)

Ulieru, M.: Emergence of holonic enterprises from multi-agents systems: A fuzzy evolutionary approach. In: Loia, V. (ed.) Soft Computing Agents, pp. 187–215. IOP Press, Amsterdam (2002)

Ulieru, M., Brennan, R.W., Walker, S.S.: The holonic enterprise: a model for Internet-enabled global manufacturing supply chain and workflow management. Integrated Manufacturing Systems 13(8), 538–550 (2002)

Valckenaers, P., Van Brussel, H., Bongaerts, L., Wyns, J.: Holonic Manufacturing Systems. Integr. Comput-Aided Eng. 4(3), 191–201 (1997)

Yoshikawa, H.: General Design Theory and a CAD system. In: Man-Machine Communications in CAD/CAM, Proceedings, IFIP W.G5.2, Tokyo, pp. 35–38. North-Holland, Amsterdam (1981)

# Chapter 6
# Perspectives

**Abstract.** Integrative points of view are presented with reference to the classification of sciences, cybernetics and its extensions, and to the categorification as the new wave of transdisciplinarity, coming after complexity.

The cyclic view of sciences is correlated to different orders of cybernetics approach.

Intra, inter, and trans disciplinarity are presented as natural steps in problem solving.

The polystochastic modeling place and the correlation with other methodologies and research directions is evaluated.

A table offers a synthetic view of the cognitive frameworks and of methodologies, proposed in the book as conceptual tools capable to manage different degrees of complexity for different domains of reality and of science.

This may be of help to see where future studies might be going and to promote new domains of applications.

## 6.1 Cybernetics and Classification of Sciences

Creating common methodologies and languages for discussing and solving problems for a wide range of systems making use of different disciplines is critical for complexity studies. In practice, one of the main questions is how the languages of different sciences and methodologies result in scientific interventions and actions into the real world.

Answering such questions one can start by referring to cybernetics.

Cybernetics studies the assumptions and procedures of different particular disciplines as for instance modelling, controlling and predicting (Wiener 1945).

The interdisciplinary character of cybernetics was explicit from its founding.

The cybernetic approach is correlated to the study of levels of reality and also to the classification of sciences problem (Hartmann 1952, Piaget 1972, 1977, Poli 2001, Nicolescu 2002).

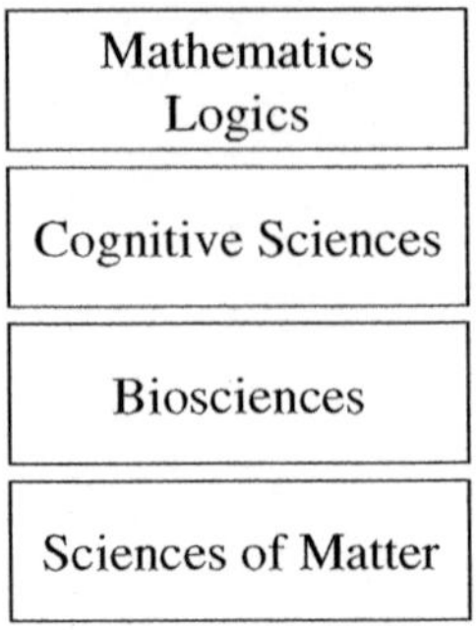


**Fig. 6.1** Hierarchy of sciences

According to Hartmann there are four main levels of reality: physical, biologic, cognitive or psychological, and intelligent or spiritual. These levels correspond to the four main domains of sciences: sciences of matter, biosciences, cognitive sciences, and lastly logics and mathematics.

The usual view on systems and related sciences classification is a linear and hierarchical one. Fig. 6.1 shows a hierarchical view of sciences.

Numerous researchers disagree with this linear model of sciences. According to Piaget (1967) the sciences are cyclically related since there is nowhere to look for an explanation of logical and mathematical phenomena other than in psychological or cognitive activity. The Piaget circular model of sciences goes as follows: psychological that is cognitive phenomena are dependent on biological phenomena, that in turn depend on physical and chemical phenomena that, in their turn are stated in mathematical and logical laws and with these laws we are back at the beginning namely at the cognitive phenomena (Fig. 6.2, Table 6.1). The concordance between the mathematical and the experimental science would not be the product of an accident but rather it would be because mathematical knowledge and empirical knowledge are both weaved from the same psychological or cognitive cloth.

The field of cybernetics is usually described as developed in two stages namely the $1^{st}$ order and the $2^{nd}$ order cybernetics.

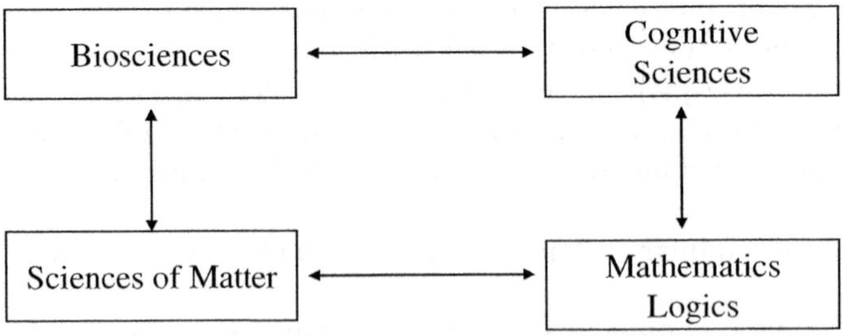


**Fig. 6.2** Cycle of sciences

**Table 6.1** Cycle of sciences and cybernetics

| | | |
|---|---|---|
| **Biosciences** | **↔2nd order** | **Cognitive sciences** |
| Biology | Reflexivity. Self-organization | Psychology. Sociology |
| Anatomy | | Engineering Design |
| **↕ 1st order** | | **↕ 3rd order** |
| Homeostasis Feed-back | | Virtual. Anticipative |
| **Sciences of mater** | **↔ 4th order** | **Mathematics** |
| Physics | Embodiment. Evolvable | **Logics** |
| Chemistry | | |

The $1^{st}$ order cybernetics developed as the science of control and communication for machines and animals outlines the feedback concept. It focused on the idea of homeostasis, the ability of systems to maintain steady states despite disturbances in the environment. The concepts and the applications of $1^{st}$ order cybernetics are interdisciplinary between the sciences of mater and biosciences (Table 6.1).

The second phase of developing cybernetics, focused on the attempt to incorporate reflexivity into the systems, that is, to acknowledge that the observer is part of the observed system. This led, to the deep study of reflexivity and self-organization concepts. The concepts and applications of the $2^{nd}$ order cybernetics are interdisciplinary between biosciences and cognitive sciences (Table 6.1).

Possible $3^{rd}$ order and $4^{th}$ order cybernetics will confront higher levels of complexity that arrived in science and technology.

The idea of emergence is fundamental here-the thought that complex systems, when recursively structured, can spontaneously evolve in directions their designers did not anticipate. It is the case of some proactive or evolutionary methods and devices. The virtual is a key characteristic of $3^{rd}$ order cybernetics. According to $3^{rd}$ order cybernetics some systems can change goals without pre-programming. This means that the observer is considered as a proactive or anticipative component that not only observes but also decides and acts. Noticeably, the observer is not necessary a human one.

The focus of $3^{rd}$ order cybernetics is beyond cognitive sciences level and includes virtual, conceptual, proactive, anticipative technologies, and cyberspace. The $3^{rd}$ order cybernetics concentrates on virtual systems, on building information systems. The concepts and applications of the $3^{rd}$ order cybernetics are interdisciplinary between cognitive sciences and the logical and mathematical sciences.

The $4^{th}$ order cybernetics should confront and surpasses the challenge of high complexity in technology and sciences. The $4^{th}$ order cybernetics may be one of the embodied, fully evolvable, creative and autonomous systems. It implies that a system will immerge into its environment, of which it is part. A $4^{th}$ order cybernetic system is embedded, integrated into the context and context aware. As outlined by Table 6.1 the $4^{th}$ order cybernetics can be understood and described in terms of the complement of the first, second and $3^{rd}$ order cybernetics considered as a whole.

The $4^{th}$ order cybernetics may be linked to emerging new scientific domains as synthetic biology (Endy 2005), artificial life (Bedau et al. 2000), and organic computing (Würtz 2008).

Synthetic biology, studies the design and construction of new biochemical systems, such as the genetic circuitry. Just as the engineers design electrical circuits based on known physical properties of materials and then fabricate functioning circuits and processors, the synthetic biologists design and build biological circuits.

Artificial life studies the life as it could be, while organic computing studies the life-like computing structures and processes. Programmable artificial cell evolution is a project illustrating such innovative research directions (Chemnitz et al. 2008).

Organic computing starts from the principle that the problems of organization in different domains as molecular biology, neurology, computer science, manufacturing, ecology and sociology can be studied scientifically in a unified way (Würtz 2008). Technical usage and embedding of general principles observed in natural systems is the long term objective of organic computing.

Elements of higher order cybernetics have been outlined in the study of the regulations due to Piaget (Piaget 1977), third-wave cybernetics (Hayles 1999), social systems (Luhmann 1997), conceptual systems and cyber semiotics (Brier 1998) and of viable systems (Schwarz 1997, Yolles 2006).

It should be observed that any new type of cybernetics embeds elements of the previous ones. The higher order should be inclusive and self-aware on previous levels.

After the integrative closure, the material embodiment of logics, mathematics and computing capacity will allow operating the material realm at multiple levels simultaneously. This may support the emergence of another type of sciences of mater, of biosciences and so on. Consequently a spiral of sciences instead of cycle of sciences and associated systems may be taken into account as a more appropriate image of knowledge development (Iordache 2009).

## 6.2 Transdisciplinarity

It has been argued in many ways that the problem solving for complexity domain is an activity which cannot succeed on the basis of one point of view, or the knowledge of one discipline, but that it needs cooperation of a number of disciplines to develop valid knowledge.

Researchers still respect the idea that each discipline has its own level of explanation. It is considered that assembling parts from a system does not give the whole since the whole of each system needs its own point of view. Each level of explanation has it own long-established background and from that there seems to be a natural hierarchy between the disciplines. Confronted with an explosion of new disciplinary knowledge, it is difficult for any specialist to understand more that the tiniest fraction of his specialized domain.

The management of the cooperation of different disciplines for complex problem solving is the main concern. Consequently, it is necessary to find ways to radically simplify and unify knowledge about problems and problem solving.

Piaget and Garcia (1989) methodology starts from the bold hypothesis that there exists a parallelism between the particular problem solving and the historical development of the involved sciences. The short history of an individual problem solving that is the ontogeny, is considered as parallel to the evolutionary long history of a lineage that is, the phylogeny. Piaget explained the isomorphism between psychogenesis and the historical development in sciences by the general equilibration based on assimilation accommodation mechanism and instantiated as the so-called intra-inter-trans process.

The intra-inter-trans process is the functional mechanism that proceeds from simple object analysis, the so-called "intra" step, to the analysis of relations between objects via transformations, that is the "inter" step, and to the building of cognitive structures, that is the "trans" step.

This general mechanism is relevant to both particular problem solving and to scientific activity itself. Piaget considered that the general intellectual development involves the same sequence of steps. In particular he reconstructs intellectual development from pre-operational thinking, the "intra" stage, via concrete-operational thinking, the "inter" stage, towards formal-operational thinking, that is the "trans" stage.

In a broader Piagetian view, the claim is that this kind of stages can be traced in different domains and at all levels of development.

The intradisciplinarity step corresponds to single disciplinarity or to multidisciplinarity realm. It represents the first step of the problem solving.

Disciplinary research is not able to fully cope with complex problems because these problems do not fit into the system of scientific disciplines. Energy, health, ecology, security and financial problems can't be solved by disciplinary approaches. A scientific understanding of complex problems is mandatory but the increasing specialization and fragmentation of scientific disciplines prevents disciplinary research from working.

Multidisciplinarity makes use of different disciplines and suppose that studying complex problem is not just in one discipline only, but in several, at the same time. Any issue in question will be enriched by incorporating the perspectives of several disciplines.

Multidisciplinary approach brings a benefit to the disciplinary study, but this benefit is still in the restricted service of the source disciplines. The multidisciplinary approach overflows disciplinary boundaries while its goal remains limited to the frameworks of disciplinary research.

It should be noted that multi-scale models are often multidisciplinary. There exists a growing number of tools and methods for engineering systems but little fundamental conceptual analysis leading to general frameworks that help guide modeling of multi-scale systems.

The next step to be considered in problem solving is that of interdisciplinarity. This involves cooperating disciplines and has a different goal than multidisciplinarity. It concerns the transfer of methods from one discipline to another. Like multidisciplinarity, the interdisciplinarity run over the disciplines. Confronted with problems between two disciplines the interdisciplinarity has even the potentiality of generating new disciplines.

The next step in problem solving is that of transdisciplinarity. The definition of problems to solve is, for this step, relatively independent of disciplinary perspectives.

Transdisciplinarity concerns that which is at once between the disciplines, across the different disciplines and beyond disciplines (Nicolescu 2006). Transdisciplinarity was considered not as a discipline but rather as a process of problem solving able to increase knowledge by integrating and transforming different perspectives (Klein et al. 2001).

Highly complex problems do not belong to only one of the three main types or disciplinarity sketched above but contain elements of each type. The focus is on providing technical solution to a given problem rather than on gaining scientific knowledge. There is no opposition between intradisciplinarity (including disciplinarity and multidisciplinarity), interdisciplinarity and transdisciplinarity. In fact, there is no transdisciplinarity without interdisciplinarity and this in turn without multiple disciplinarity. Disciplinary competence remains the essential precondition for transdisciplinarity tasks but it alone does not suffice to deal with complexity.

Fig. 6.3 shows an illustrative problem solving cycle (Murase 2008). It makes use of the analogy with n-graphs (Appendix A5). The disciplines are represented here by signs as "•", for primarily theoretical part and "o" for primarily experimental part.

Initially the parts are separated but start to form well defined disciplines in the $1^{st}$ order stage, "intra". They may be coupled in the $2^{nd}$ order stage, "inter" to form interacting disciplines. The $3^{rd}$ order stage, "trans" corresponds to the coupling of two or more sciences in wide-ranging frameworks going beyond disciplines and solving disciplinary isolation.

The $4^{th}$ order, last stage, shown in Fig. 6.3 may represent an integrative or self viewpoint. After a complete cycle intra-inter-transdisciplinarity the self viewpoint is open towards a new disciplinary approach and a new cycle. This

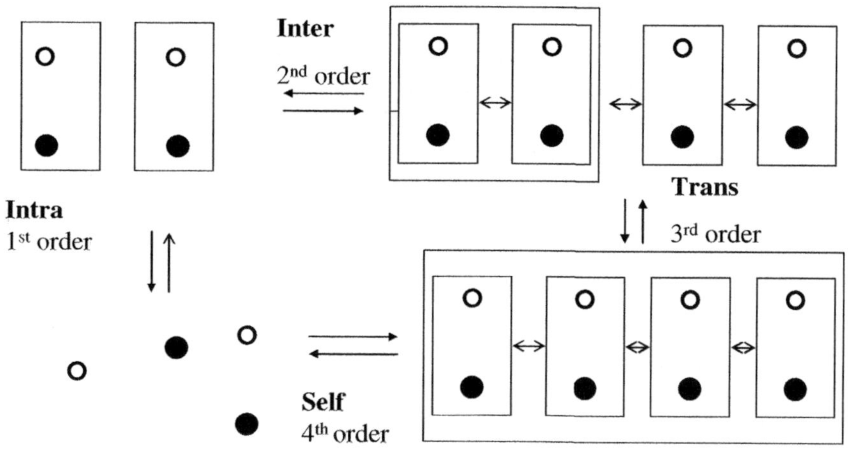


**Fig. 6.3** Intra, inter, trans disciplinary

$4^{th}$ order stage and arrow completes the knowledge cycle and the problem solving. It corresponds to the creative stage in intellectual development and supposes the ability to formulate post-disciplinary notions as for instance new goals.

How the integrative or self disciplinary viewpoint turns back into a new disciplinary life is an open problem. A suggestion is that evolvable problem solving could restart and follow the same steps on a higher methodological plane that is at a higher dimension in modeling. This is the categorification way (Appendix A4). New open problems concern the timing of travel back and forth across the levels of abstraction, alternating categorification and decategorification in a specific dynamics for specific problems.

It was observed that the transdisciplinary research tends to be reinvented and reformulated about every one or two decades. In the second half of the last century there have been waves of transdisciplinarity in cybernetics in the 1945s, control systems in the 1960s, chaos theory in the 1975s and complexity theory in the 1990s (Strogatz 2003). It would be of interest to explain why this approach in waves is followed and proves to be efficient, instead of a more constant effort, why each wave of interdisciplinary lost its initial impetus as a unifying force, and what comes after the complexity pulsation.

The transdisciplinary waves are imposed by the social and economic context. The recent history of transdisciplinary problem-focused researches dates from the 1940s, initially in defense research. The 1960s represent the start of increased funding for transdisciplinary research in areas of economic competition as engineering, manufacturing and medicine.

Study of control systems became mandatory. It was evident during the 1975s context of unpredictability and chaos in market and the 1990 context of environmental research that new discourses of transdisciplinary problem solving are necessary and emerging (Klein 2004).

The case studies presented in this book sketched how categorification was imposed by the complexity advent, how it can serve for complex problem solving, and how category theory could serve as a "lingua franca" that lets us translate between certain aspects of different subjects in different domains and eventually, build a general science of systems and processes (Baez and Stay 2008).

Such observations allow us to assume that the categorification may represent the new wave of transdisciplinarity, a new C-theory, coming after cybernetics, control, chaos and complexity (Strogatz 2003).

## 6.3 Synopsis

At the end of this incursion in complexity domain it is appropriate to evaluate the PSM place and the correlation with other methodologies and research directions presented in the book.

Fig. 6.4 and the table 6.2 offers a synthetic view of the cognitive frameworks and of methodologies, revealed in the book as conceptual tools to manage different degrees of complexity for different domains of reality and of science. The rectangles in Fig. 6.4 represent reality levels or sub-levels. Fig. 6.4 shows that after an increasing number of hierarchical levels, integrative cyclic closure structures limited to three or four realms and sub-realms have to be considered. The integrative closure shown in Fig. 6.4c or Fig. 6.4e is not seen as a final stage. As illustrated by Fig. 6.4f, we may consider a process that can develop self-similar patterns for cognitive architecture.

The rows in Table 6.2 correspond to the main domains of sciences as shown in Fig. 6.1.

A supplementary level of transdisciplinarity was included on top of this hierarchical perspective. Periodicity refers to the fact that for the same column we may detect pattern similarities despite the fact that the issues pertain to different domain of sciences.

Column a, in Table 6.2 (Fig. 6.4) refers to low-dimension methodologies and devices. Column a, includes adaptive devices as material systems, genetic code versus amino acids relation, evolutionary algorithms and theoretical concepts as semantic closure for biosystems domain. The multi-agent systems MAS, and theoretical concepts as the modeling relation, illustrates the cognitive domain.

As mathematics, the column a, includes automata, Turing machines and learning systems. All these are low dimensional models of computation that correspond to 1-categories.

The $1^{st}$ order cybernetics appears to be the transdisciplinary approach associated to the a-frameworks.

It is possible to conceive highest dimension developments for the mathematical concepts shown in column a. Examples are the higher n-dimensional automata, nDA (Pratt 1991) and the n-categories approach for rewriting systems (Johnson 1991, Burroni 1993).

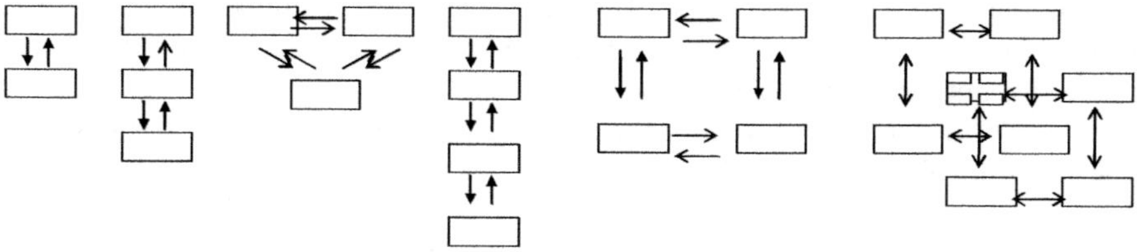


**Fig. 6.4** Synthesis of cognitive frameworks

There exists several attempts to define n-categories models but the knowledge about n-categories is still in progress and sometimes controversial (Baez and Stay 2008).

Column b, in Table 6.2 corresponds to the three levels hierarchical, b-frameworks.

Evolutionary devices as Pask's device (Cariani1993) may represent the material systems. For biosystems one may considers the central dogma of biology, structured genetic algorithms-GA (Dasgupta and Gregor 1992), contextual GA (Rocha 1997) and chemical GA (Suzuki and Sawai 2002).

As cognitive systems or models, the von Uexküll functional circle (1973), the mesoscopic cognition frame (Doursat 2007), and the universal modeling language-UML studies may be mentioned.

**Table 6.2** Periodic table of methods and cognitive frameworks

| | a | b | c | d | e | f |
|---|---|---|---|---|---|---|
| **Trans-Disciplinary** | ·1st order Cybernetics | ·2nd order Cybernetics | ·Three realms Ontology | ·3rd order Cybernetics ·Four levels Ontology | ·4th order Cybernetics | |
| **Mathematics Logics** | · 1-category ·Automata ·Turing Machine ·Learning systems | ·2-category ·2DA ·Petri nets ·Strings rewriting | ·2-category | ·3-category ·3DA ·Terms rewriting | ·3-category | |
| **Cognitive Systems Sciences** | ·MAS ·Modeling Relation | ·von Uexküll Functional Circle ·Mesoscopic Cognition ·UML | ·CBR ·BDI ·SASO | ·Symbolic-Connectionism ·K-sets ·Holonic Systems · Autonomic computing ·MOF | ·GDT ·EMAS ·IRDS · Organic Computing ·Piaget Schema | ·Digital self ·EDOE, PSM ·Nested MML ·Piaget Garcia Schema |
| **Biosciences Bioinspired Sciences** | ·Genetic code-Amino-acids ·Evolutionary Algorithms ·Semantic Closure | ·Central Dogma of Biology ·Structured GA ·Contextual GA ·Chemical GA | ·Operons | | · Synthetic Biology · Artificial Life | |
| **Mater Sciences** | ·Adaptive Devices | ·Evolutionary Devices | ·Evolution in *materio* | ·Embryonics Devices | | |

As mathematical methods, the 2-categorical formulation of the 2-dimensional automata, 2DA, string rewriting systems (Johnson 1991), and the Petri nets may be regarded.

The $2^{nd}$ order cybernetics is the transdisciplinary approach associated to the three-level, b-frameworks.

Column c, refers to a cyclic, which is closed version of the previous framework that is to three realms c-frameworks. This corresponds in part to the evolution in *materio* project as biomimetic devices (Miller 2008) and to operon models in genetics (Jacob and Monod 1961). The conventional case based reasoning-CBR systems (Aamodt and Plaza 1994) and the self adapting self organizing-SASO framework (Di Marzo Serugendo et al. 2007) represents examples of cognitive systems. The three realms ontology allows a transdisciplinary study of this three realms cyclical framework linked to 2-categories (Poli 2001).

Column d, refers to four-level hierarchical d-frameworks. This includes embryonics project as biomimetic devices (Mange et al. 1998), some symbolic-connectionist models (Hummel and Holyak 1997), the K-sets models for neurodynamics (Freeman 1995), holonic systems (Valckenaers et al. 1997), autonomic computing (IBM 2005) and meta-object facility-MOF studies, as examples of cognitive systems and models.

For mathematical frames, the 3-categorical formulation of the dimensional automata, 3DA, and the term rewriting systems (Johnson 1991), may be considered. The transdisciplinary studies are associated to $3^{rd}$ order cybernetics and to Hartmann four levels ontology.

Column e, includes synthetic biology (Endy 2005) and artificial life as biosystems.

The engineering general design theory-GDT (Tomiyama and Yoshikawa 1987), some evolvable MAS, EMAS, information resource dictionary systems-IRDS (Rossiter and Heather 2003), the organic computing studies focusing on embodiment (Würtz 2008) and the four realms schema of Piaget (1980) may represent the cognitive sciences domain. The e-frameworks were proposed in different sections of this book as n-graphs mathematical model inspired by study of computads (Street 1987) or polygraphs (Burroni 1993) and as the coming up $4^{th}$ order cybernetics corresponding to transdisciplinarity.

The f-frameworks from column f, represent refinements and developments of the cyclic variant of the four level e-frameworks. Potential developments correspond to the splitting of a realm in four sub-realms and to the inclusion of a central realm that may in turn split in four sub-realms and so on. This parallelism allows making use for the new f-framework of the same software support as for the e-frameworks. The column f may include cognitive systems as the digital self MAS developed by Goschnick (2003) on the basis of Jung analytical psychology, some evolvable designs of experiment-EDOE and PSM frameworks (Iordache 2009), nested meta-modeling language-MML (Alvarez et al. 2001), and systems based on the psychogenetic schema due to Piaget and Garcia (1980).

Several f-frameworks have been just sketched in this book as for instance: the evolvable SMB, the creative engineering design, the BGPI and the evolvable manufacturing systems.

Moving from left to right in the periodic table 6.2, means to perform categorification steps, and to increase the system dimensionality and capability to confront higher levels of complexity.

The periodic table 6.2 highlights the interconnection of different theoretical concepts and research directions. This may be of help to see where future studies might be going and to promote applications.

Many of the presented frameworks and methods are based on less than four levels and these levels are incompletely connected.

For instance some evolutionary devices, and also some cognitive tools as the conventional CBR, BDI lack the fourth level. This lacking may explains the difficulties reported with Pask's device, in evolutionary hardware or the deceptive applications of genetic algorithms. Pask's devices and evolutionary hardware are confronted with reproducibility problems. Genetic algorithms work well in some calculations and not in others, but it was not clear why and when this happens. It may be supposed that some artificially constructed evolutionary systems are less evolvable than other.

Embryonics project frameworks, holonic enterprise management systems, autonomic computing systems and MOF, show four levels hierarchies but do not focus the interconnection between top and lowest levels.

The integrative closure of these hierarchies is a necessary next step toward fully evolvable and autonomous technological systems, management systems, enterprises and organizations.

## References


Aamodt, A., Plaza, E.: Case-based reasoning: Foundational issues, methodological variations, and system approaches. AI Communications 7(1), 39–59 (1994)

Alvarez, J., Evans, A., Sammut, P.: MML and the meta-model framework. In: Workshop on Transformations in UML (WTUML 2001), Genoa (2001)

Baez, J., Stay, M.: Physics, Topology, Logic and Computation: A Rosetta Stone. In: Coecke, B. (ed.) New structure for physics. Lecture Notes in Physics. Springer, Heidelberg (2008)

Bedau, M.A., McCaskill, J.S., Packard, N.H., Rasmussen, S., Adami, C., Green, D.G., Ikegami, T., Kaneko, K., Ray, T.S.: Open problems in artificial life. Artificial Life 6, 363–376 (2000)

Brier, S.: Cybersemiotics: A transdisciplinary framework for information studies. BioSystems 46, 185–191 (1998)

Burroni, A.: Higher-dimensional word problems with applications to equational logic. Theoretical Computer Science 115, 43–62 (1993)

Cariani, P.: On the Design of Devices with Emergent Semantic Functions. Ph.D. Dissertation, Binghamton University (1989)

Cariani, P.: To evolve an ear: epistemological implications of Gordon Pask's electrochemical devices. Systems Research 10, 19–33 (1993)

Chemnitz, S., Tangen, U., Wagler, P.F., Maeke, T., McCaskill, J.S.: Electronically programmable membranes for improved biomolecule handling in micro-compartments on-chip. Chemical Engineering Journal 135S, 276–279 (2008)

Dasgupta, D., McGregor, D.R.: Designing Neural Networks using the Structured Genetic Algorithm. In: Proceedings of the International Conference on Artificial Neural Networks (ICANN), Brighton (1992)

Di Marzo Serugendo, G., Fitzgerald, J., Romanovsky, A., Guelfi, N.: A Generic Framework for the Engineering of Self-Adaptive and Self-Organising Systems. Technical Report CS-TR-1018, School of Computing Science, University of Newcastle (2007)

Doursat, R.: Of tapestries, ponds and RAIN. Toward fine-grain mesoscopic neurodynamics in excitable media. In: International Workshop on nonlinear brain dynamics for computational intelligence, Salt Lake City (2007)

Endy, D.: Foundations for engineering biology. Nature 438(7067), 449–453 (2005)

Freeman, W.J.: Tutorial on neurobiology: from single neurons to brain chaos. Inter. J. of Bifurcation and Chaos 5(3), 849–858 (1995)

Goschnick, S.B.: Enacting an Agent-based Digital Self in a 24x7 Web Services World. In: Zhong, N., Raś, Z.W., Tsumoto, S., Suzuki, E. (eds.) ISMIS 2003. LNCS (LNAI), vol. 2871, pp. 187–196. Springer, Heidelberg (2003)

Hartmann, N.: The new ways of ontology. Greenwood Press, Westport (1952)

Hayles, K.: How we became posthumans. Virtual bodies in cybernetics. literature and informatics, Chicago (1999)

Hummel, J.E., Holyoak, K.J.: Distributed representation of structure. A theory of analogical access and mapping. Psychological Review 104, 427–466 (1997)

IBM, An architectural blueprint for automatic computing (2005)

Iordache, O.: Evolvable Designs of Experiments Applications for Circuits. J. Wiley VCH, Weinheim (2009)

Jacob, F., Monod, J.: On the regulation of gene activity. Cold Spring Harbor Symp. Quant. Biol. 26, 193–211 (1961)

Johnson, M.: Linear term rewriting systems are higher dimensional string rewriting systems. In: Rattray, C.M.I., Clark, R.G. (eds.) The Unified Computation Laboratory, pp. 101–110. Oxford University Press, Oxford (1991)

Klein, J.T.: Interdisciplinarity and complexity: An evolving relationship. E:CO 6(1-2), 2–10 (2004)

Klein, J.T., Grossenbacher-Mansuy, W., Haeberli, R., Bill, A., Scholz, R.W., Welti, M. (eds.): Trandisciplinarity: Joint problem solving among science, technology, and society. An effective way for managing complexity. Birkhaeuser, Basel (2001)

Luhmann, N.: Die Gesellschaft der Gesellschaft. Frankfurt am Main, Suhrkamp (1997)

Mange, D., Sanchez, E., Stauffer, A., Tempesti, G., Marchal, P., Piguet, C.: Embryonics: A new methodology for designing Field-Programmable Gate Arrays with Self-Repair and Self-Replicating Properties. IEEE Transactions on VLSI Systems 6(3), 387–399 (1998)

Miller, J.F.: Evolution in materio. In: International Conference on Evolvable Systems, Prague, Czech Republic (2008)

Murase, M.: Endo-exo circulation as a paradigm of life: towards a new synthesis of eastern philosophy and western science. In: Murase, M., Tsuda, I. (eds.) What is life? The next 100 years of Yukawa's dream, Progrees of Theoretical Physics, suppl. 173, pp. 1–20 (2008)

Nicolescu, B.: Manifesto of Transdisciplinarity. SUNY Press, New York (2002)

Nicolescu, B.: Transdisciplinarity-Past, Present and Future. In: Haverkort, B., Reijntjes, C. (eds.) Moving Worldviews - Reshaping Sciences, Policies and Practices for Endogenous Sustainable Development, pp. 142–166. COMPAS Editions, Holland (2006)

Piaget, J.: Classification des sciences et principaux courants épistémologiques contemporains. In: Piaget, J. (ed.) Logique et connaissance scientifique, Gallimard, Paris, pp. 1151–1224 (1967)

Piaget, J.: The epistemology of interdisciplinary relationships. In: Interdisciplinarity: Problems of teaching and research in universities, Paris, OECD, pp. 127–139 (1972)

Piaget, J.: L'épistémologie des régulations: introduction. In: Lichnerrowicz, A., Perroux, F., Gadoffre, G. (eds.) L'idée de régulation dans les sciences: 2e vol. des Séminaires interdisciplinaires du Collège de France: A. Paris: Maloine: Doin: I-XIII (1977)

Piaget, J., Garcia, R.: Psychogenesis and the History of Science. Columbia University Press, New York (1989)

Poli, R.: The basic problem of the theory of levels of reality. Axiomathes 12(3-4), 261–283 (2001)

Pratt, V.R.: Modeling concurrency with geometry. In: Proceedings 18th Ann. ACM Symposium on Principles of Programming Languages, pp. 311–322 (1991)

Rocha, L.M.: Evidence Sets and Contextual Genetic Algorithms: Exploring Uncertainty, Context and Embodiment in Cognitive and Biological Systems, PhD Dissertation, Binghamton University (1997)

Rossiter, N., Heather, M.: Four-level Architecture for Closure in Interoperability. In: EFIS 2003, 5th International Workshop on Engineering Federated Information Systems, Coventry, UK, pp. 83–88 (2003)

Schwarz, E.: Toward a holistic cybernetics. From science through epistemology to being. Cybernetics and Human Knowing 4(1), 19–23 (1997)

Street, R.: The algebra of oriented simplexes. J. Pure Appl. Algebra 49, 283–335 (1987)

Strogatz, S.: Sync.: The Emerging Science of Spontaneous Order. Hyperion (2003)

Suzuki, H., Sawai, H.: Chemical Genetic Algorithms-Coevolution between Codes and Code Translation. In: Artificial Life VIII, pp. 164–172. MIT Press, Cambridge (2002)

Tomiyama, T., Yoshikawa, H.: Extended General Design Theory. In: Design Theory for CAD, Proceedings from IFIP WG 5.2, Amsterdam (1987)

von Uexküll, J.: Theoretische Biologie. Frankfurt a. M.: Suhrkamp Taschenbuch Wissenschaft (1973)

Valckenaers, P., Van Brussel, H., Bongaerts, L., Wyns, J.: Holonic Manufacturing Systems. Integr. Comput. -Aided Eng. 4(3), 191–201 (1997)

Yolles, M.I.: Organisations as Complex Systems: An Introduction to Knowledge Cybernetics. Information Age Publishing, Inc., Greenwich (2006)

Wiener, N.: Cybernetics or Control and Communication in the Animal and the Machine. MIT Press, Cambridge (1945)

Würtz, R.P.: Organic Computing: Series: Understanding Complex Systems. Springer, Heidelberg (2008)

# Appendix A
# Appendices

## Categories

**Abstract.** Basic category theory notions as objects, morphisms, functors, diagrams, natural transformations, limits, colimits, product and coproduct, pullback and pushout, adjointness, monads are introduced at the informal level.

The high complexity imposes a higher categorical approach.

The higher categories, that is the n-categories, orientals, the periodic table hypothesis, monoidal, braided, sylleptic and symmetric categories, the categorification and the coherence, the computads or polygraphs, the operads and multicategories are informally introduced.

Applicability domains for rewriting systems, are outlined.

## A.1 Categorical Framework

In the tradition of Felix Klein's Erlangen Program, category theory was developed as a way of studying different mathematical structures in terms of their admissible transformations (MacLane 1971, Marquis 2009).

The categorical approach presented here is informal and includes some adaptations to the applications specific needs.

Basically a category contains objects and morphisms associated to arrows, such that there exists an arrow from each object to itself, the arrows may be composed and the arrow composition is associative. Morphisms or arrows are structure preserving mappings between objects. Examples of objects are sets, processes, structures, partial orders, concepts, and so forth. The category of sets, denoted by Set, has sets as objects and functions between sets as morphisms. Category theory put emphasizes on morphisms that is on processes. Category theory emphasizes the relational point of view considering that everything can be defined as an arrow between objects and actually objects can be defined using only arrows. This is one of the main differences between the set theory and category theory. Whereas the first focuses on describing objects with inner structure that is separating them into parts and elements, the latter characterizes an object by its connections, focusing on the role of the object within the net of relationships.

It is possible to imagine a category in which the objects are categories and the morphisms are mappings between categories. The mappings between categories preserving the categorical structures, namely identities and composition, are called functors. A functor between two categories maps objects and morphisms of one category to objects and morphisms of the other in such a way that morphism between two objects is mapped to morphism between the mapped objects. Thus a functor appears as the transformation which maintains the framework of the involved categories.

A category can be seen as a diagram that is a graph, where objects are the vertices of the graph and morphisms are the paths in the graphs. A diagram commutes, if for all paths with equal domain and codomain the value of the diagram functors is equal. This expresses the fact that the results of compositions are equal.

Commutative diagrams represent the categorical equivalent of a system of equations, but are more general in nature. Diagrammatic presentations provide a convenient tool to study the passage between designs and their implementations.

It is possible to define the category in which the objects are functors. A natural transformation is a morphism between two functors. It provides a way to switch from one mapping of a structure to another in a manner that is interchangeable with the two images of any morphism. The naturality allows holding functorial implementation together and the knowledge coherence.

Observe that the focused relationship is that between objects for categories, between categories for functors and between functors for natural transformations.

A change of structure can be modeled as a functor between the two categories modeling the structure. Deeper structure transformations can be performed by defining natural transformations between functors, which allows a reengineering of the model of a system.

The efficiency of category theory lies in the possibility of universal constructions as for instance limits, and colimits. The colimit is a formalization of assembly of objects and morphisms. A colimit for a diagram can be thought of as a structure that completes the diagram to a minimal commutative diagram containing it. The colimit puts everything together. It describes gluing or fusion. The tool for describing putting them together is called a cocone. In the category Set the colimit corresponds to the least set. Limits are the dual notion to colimits, which is the one notion obtained from the other by reversing the arrows and interchanging initial and terminal for objects. Intuitively a limit extracts the abstraction. Given a diagram, an element is called a limit if there are morphisms from that element to all vertices of the diagram, and if for any other element satisfying the same property there is a unique morphism from it to the limit. In the category Set the limit corresponds to the biggest set.

Limit can be seen as an emergent concept summing up in itself the properties of its constituents. This allows considering a hierarchy where at any level the objects are the limits of objects of the lower level. This is consistent with the opinion that complexity is a relative notion depending on the level of observation. The tool to obtain limits is called a cone.

The coproduct and the product represent the categorical notions corresponding to disjoint union and to Cartesian product in the category Set. The coproduct is a special type of colimit and the product is a special type of limit. The pushout gives composition of objects having the same domain under two morphisms. The pushout is a universal property of two morphisms. The coproduct is a universal property of any set of objects.

The pullback gives decomposition of objects having the same image or codomain under two morphisms.

A Cartesian closed category is one which is closed under all kinds of universal constructions for example limits, and colimits.

To any canonical construction from one type of structures to another, an adjunction between the associated categories, will corresponds. Adjoint functors are pairs of functors which stand in a particular relationship with one another. A functor can be left or right adjoint to another functor that maps in the opposite direction. A pair of adjoint functors typically arises from a construction defined by a universal property, and it can be seen as a more abstract and powerful view on universal properties.

If F and G represent a pair of adjoint functors, with F left adjoint to G right adjoint, then the composition GoF will be a monad. A monad is a functor from a category to itself, in other words an endofunctor. The categorical dual of monads, FoG will be a comonad. Every adjunction gives rise to a monad. Generally, the adjunctions relate categories of different natures. The monad theory try to capture what is that adjunction preserves. The monads generalize closure operators on partially ordered sets to arbitrary categories.

The monad investigated by Giry (Giry 1981) as part of the categorical foundation of probability theory is of interest for PSMs.

The traced monoidal category studied by Joyal et al. (1996) provides a category for abstractly capturing feedback, recursion and cyclic structures. Such type of categories may be useful for PSM studies since the monoidal category elements are clearly linked to the elements of the quadruple, SKUP. The elements K, U and P are simply associated to the monad elements while S is associated to the so-called trace or feed-back.

## A.2 Higher Categories

One category frame is not enough to describe the complexity of cognitive or evolvable systems. For this reason, n-categories, multi-categories, operads and other higher dimensional categorical concepts should be developed (Leinster 2004).

The n-categories are high-order generalizations of the notion of category.

Roughly, an n-category is the algebraic structure consisting of a collection of objects, a collection of morphisms between objects, a collection of 2-morphisms between morphisms and so on up to n, with various rational and practical ways of composing theses j-morphisms, j<n (Baez 1997).

The 0-category is a set, while 1-category is a category. An n-category consists of 0-cells (objects, types), 1-cells (morphisms), 2-cells (morphisms between morphisms) and so on, all the way up to n-cells together with composition operations.

There are numerous studies dedicated to n-categories and even to $\infty$-categories, called also $\omega$-categories. These studies start to have an impact on sciences including that of matter, biosciences, cognitive sciences and mathematics.

Applications would imply morphisms between morphisms, that is, processes between processes, or in other words meta-processes and so on. As n increases, the construction of n-categories step by step may be difficult to conceive and need analysis on how higher categories are effectively working.

Consider for example the case of 2-categories of which the category of categories denoted by Cat, is the standard example (MacLane 1971). In Cat, the 0-cells are categories, the 1-cells are functors, and the 2-cells should be natural transformations.

Any 2-category C makes use of three items $C_0$, $C_1$, and $C_2$. Elements of $C_i$ are called i-cells i=0, 1 or 2. The 2-category is the three categories structure that consists of the so-called base category having $C_0$ as objects and $C_1$ as arrows, the horizontal category having $C_0$ as objects and $C_2$ as arrows, and the vertical category having $C_1$ as objects and $C_2$ as arrows.

The 2-cells are arrows in both the horizontal and the vertical category, thus they composes with two different composition operators, horizontal or vertical.

Cat is a strict 2-category, that is, all laws hold exactly, not just up to isomorphism.

Vertical composition corresponds to a sequential operation, while horizontal composition corresponds to a parallel operation. The 2-category is a category with morphisms between morphisms, that is, 2-morphisms.

There are also many weak categories. For example a bicategory is a notion used to extend the notion of 2-category to handle the cases where the composition of morphisms is not strictly associative, but only associative up to an isomorphism.

Bicategories may be considered as result of the weakening of the definition of 2-categories. A similar process for 3-categories leads to tricategories, and more generally to weak n-categories for n-categories.

Informally a tricategory C is done by:

- A class $C_0$ of objects
- For any pair A, B $\in C_0$ a bicategory C(A, B)
- For any triplet A, B, D $\in C_0$ a bifunctor of composition $c_{ABD}$: C(A, B) x C(B,D) $\rightarrow$ C(A,D)
- For any object a bifunctor $u_A$: 1 $\rightarrow$ C(A,A)

These elements verify several axioms (Gordon et al. 1995).

Higher-dimensional categories may be defined inductively in terms of the hom enriched categories (Street 1987, Street 2004). For instance a 2-category C is defined as a Cat-enriched category which means that if x and y are objects of C then the hom C(x, y) is a category its objects being the arrows from x to y and its arrows the 2-cells.

For any symmetric monoidal category V there is a symmetric monoidal category V-Cat whose objects are categories with homs enriched in V. Starting with the category Set of sets and using Cartesian product for the monoidal structure we can iterate the process V$\rightarrow$V-Cat yielding the following sequence of definitions:

Set, Cat: = Set-Cat, 2-Cat: = Cat-Cat, 3-Cat: = (2-Cat)-Cat, ...

All terms have Cartesian product as monoidal structure.

Sets are called 0-categories, categories are called 1-categories, (Set-Cat)-Cat are called 2-categories and so on. There are inclusions: Set $\subset$ Cat $\subset$ 2-Cat $\subset$ 3-Cat $\subset$ ...

The union of this chain is the category $\omega$-Cat of strict $\omega$-categories. Therefore the $\omega-$categories are understood as the directed limit of a sequence of iterated enrichments.

When V is closed, it is enriched in itself. Each n-Cat is Cartesian closed and hence n-Cat is itself naturally an (n+1)-category.

The n-cells in an $\omega$-category can be defined recursively. The 0-cells of a set are its elements, the (n+1)-cells of C are the n-cells of some hom n-category C(x, y) for x, y objects of C. The theory of $\omega$-categories, or $\propto$-categories, seeks to formalize the ideas of thing (object, device), process, meta-process (process of processes), meta-meta-process (process of processes of processes) and so on.

The orientals represents a significant example of strict $\omega$-categories constructed by Street (Street 1987). Orientals are oriented simplices: the n-th oriental is the simplicial n-simplex equipped with source and target relations, assigning to each k-face a set of (k-1)-faces called its source and a set of (k-1)-faces called its target, subject to some natural axioms. Each oriental freely generates a structure of a strict $\omega$-category O($\Delta$n), such that k-morphisms in O($\Delta$n) are pasting diagrams of k-faces in $\Delta$n. $\Delta$ denotes the simplex category.

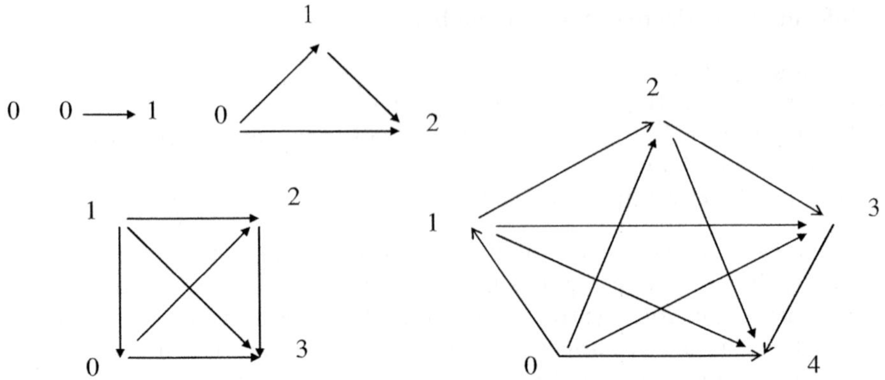


**Fig. A.1** Orientals

One of the axioms is the globularity axiom, which says that the source of a source, that is, the union of sources of all (k-1)-faces in the source of a k-face, equals the source of the target, and similarly that the target of a source equals the target of the target. Thus, the orientals mediate between the simplicial and the globular frames of $\omega$-categories.

The first orientals are presented in Fig. A1.

The construction of the orientals is designed to be compatible with face and degeneracy maps. Therefore the orientals arrange themselves into the so-called cosimplicial $\omega$−category, that is, a functor O: $\Delta \rightarrow \omega$ Cat from the simplex category $\Delta$, to the category of strict $\omega$-categories, $\omega$ Cat.

## A.3 Periodic Table

PSM developments impose to understand and to run computations concerning processes between processes between processes, and so on.

The stabilisation hypothesis may be of help for this difficult task (Baez and Dolan 1995, Leinster 2004). This hypothesis refers to k-tuply monoidal n-categories.

A k-tuply monoidal n-category is an n-category in which objects can be multiplied in k ways, all of which interchange with each other up to isomorphism. This implies that these k ways all end up being equivalent, but that the single resulting operation is more and more commutative as k increases. The stabilization hypothesis states that by the time we reach k=n+2, the multiplication has become maximally commutative.

The stabilization hypothesis says that each column in the periodic table of n-categories stabilizes at a certain precise point. The Baez and Dolan periodic table for classifying n-categories is presented in Table A.1.

It contains the conjectured description of (n+k)-categories with only one j-morphism for j<k. The idea of the periodic table linked to stabilisation hypothesis is to study degenerate forms of n-category that is, n-categories

**Table A.1** Periodic table of categories

| | n=0 | n=1 | n=2 | n=3 |
|---|---|---|---|---|
| k=0 | sets | categories | 2-categories | 3-categories |
| k=1 | monoidal | monoidal categories | monoidal 2-categories | monoidal 3-categories |
| k=2 | commutative monoids | braided monoidal categories | braided monoidal 2-categories | braided monoidal 3-categories |
| k=3 | " " | symmetric monoidal categories | sylleptic | sylleptic |
| k=4 | " " | " " | symmetric monoidal 2-categories | sylleptic |
| k=5 | " " | " " | " " | symmetric monoidal 3-categories |
| k=6 | " " | " " | " " | " " |

that are trivial below a certain dimension k. Such an n-category only has non-trivial cells in the top (n-k) dimensions, so we can perform a dimension shift and regard this as an (n-k) category. The previous k-cells become the new 0-cells, the previous (k+1)-cells become the new 1-cells, and the previous n-cells become the new (n-k) cells. This is called a k-fold degenerate n-category.

Basically the Table A.1 shows that (n+k) category with only one j-morphism for j<k can be reinterpreted as an n-category. But, it will be an n-category with k ways to multiply that is a k-tuply monoidal n-category. For example if n=1, k=1, a 2-category with one object is a monoidal category.

The Table A.1 outlines properties as: monoidal, braided, sylleptic, and symmetric.

In the first row (k=0), a 0-monoidal n-category is simply an n-category.

In the next row (k=1), a 1-monoidal n-category is a monoidal n-category.

For instance, a 1-monoidal 0-category is a one-object category (a monoid), and a 1-monoidal 1-category is a one-object 2-category (a monoidal category). A monoidal 2-category can be defined as a one-object 3-category, or can be defined directly as a 2-category with tensor.

The third row (k=2) allows observing that a degenerate monoidal category is a commutative monoid and a doubly-degenerate 3-category is a braided monoidal category.

Concerning the first column (n=0) it was observed that one-object braided monoidal category is a commutative monoid together with extra data, for the braiding, satisfying some axioms.

This gives the entry for k=3, n=0, and the same applies all the way down the rest of the column. Similar results may be established for the second column (n=1). Observe that for k≥3, the k-monoidal 1-category is just a symmetric monoidal category. Again the column stabilizes, and again the point of stabilization is the most symmetric object possible.

The same is valid for subsequent columns. The sylleptic characterization could be completed by more terms—for instance, the first would be called sylleptic monoidal 2-category. It was observed that a braided category is a monoidal category with additional structure a sylleptic category is a braided category with additional structure and so on (Crans 2000).

## A.4 Categorification and Coherence

Categorification is the process of finding category-theoretic analogs of set-theoretic concepts by replacing elements with objects, sets with categories, functions with functors and equations between functions by natural isomorphisms between functors, which in turn should satisfy certain equations of their own, called coherence laws (Mac Lane 1971, Baez and Dolan 1998).

The correspondence between set theory and category theory is presented in Table A.2.

Decategorification is the reverse process of categorification. Decategorification is a systematic process by which isomorphic objects in a category are identified as equal. Categorification is more complicated than decategorification, and requires insight into individual situations.

**Table A.2** Analogy between sets and categories

| **Set theory** | **Category theory** |
|---|---|
| Set elements | Objects |
| Sets | Categories |
| Functions | Functors |
| Equalities between morphisms | Natural isomorphisms of functors |

The term vertical categorification refers roughly to a process in which ordinary categories are replaced by higher categories. Categorification implies moving from left to right in the periodic table while decategorification implies moving in the reverse direction.

In category theory, the objects or identity arrows are elements within category, whereas the category compares objects, the functors compares categories and the natural transformation compares functors. For example, a monoid is a set with a product satisfying the associative law and a unit element satisfying the left and right unit laws. The categorified version of a monoid is a monoidal category. This is a category C with a product: $\otimes$:C x C$\to$C and a unit object 1$\in$C. For categorization we need to impose associativity and the left and right unit laws as equational laws only up to isomorphism. As part of the structure of a weak monoidal category we specify a natural isomorphism a $_{x,y,z}$: (x$\otimes$y)$\otimes$z$\to$x$\otimes$(y$\otimes$z) called the associator together with the natural isomorphisms: $l_x$: 1$\otimes$x$\to$x and $r_x$: 1$\otimes$x$\to$x.

Associativity means that, within a sequence containing two or more of the same sequencing operations in a row, the order that the operations are

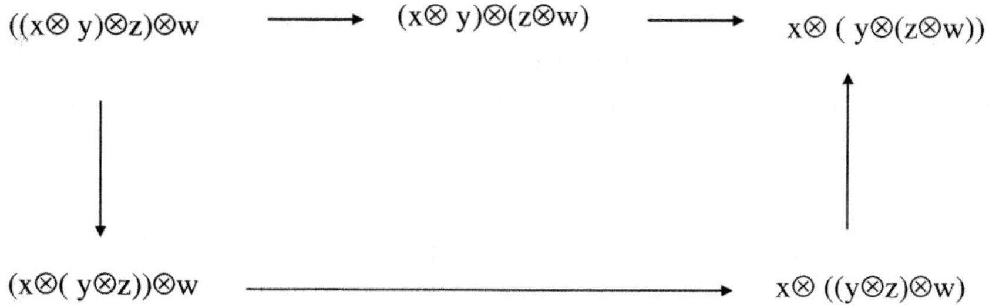


**Fig. A.2** Pentagon relation

performed does not matter as long as the sequence to be operated is not changed.

Using the associator one can construct isomorphisms between any two parenthesized versions of the tensor product of several objects. For example there are five ways to parenthesize the tensor product of four objects, which are related by the associator as shown in Fig. A2. The coherence law called the pentagon identity, say that the diagram shown in Fig. A2 commutes. Pentagon relation concerns monoidal categories and associativity.

Suppose that we are looking to commutativity, that is, we want to categorify the notion of commutative monoid.

Consider a weak monoidal category equipped with a natural isomorphism: $B_{x,y}$: x⊗y→y⊗x called the braiding and then impose coherence laws called hexagon identities (Fig. A3).

In physics there are processes allowing switching two systems by moving them around each other. The monoidal category in which we can do this sort of switch is called braided.

The first hexagon equation says the switching the object x past y⊗z all at once is the same as switching it past y and then past z.

The second hexagon is similar. It says switching x⊗y past z all at once is the same as doing it in two steps.

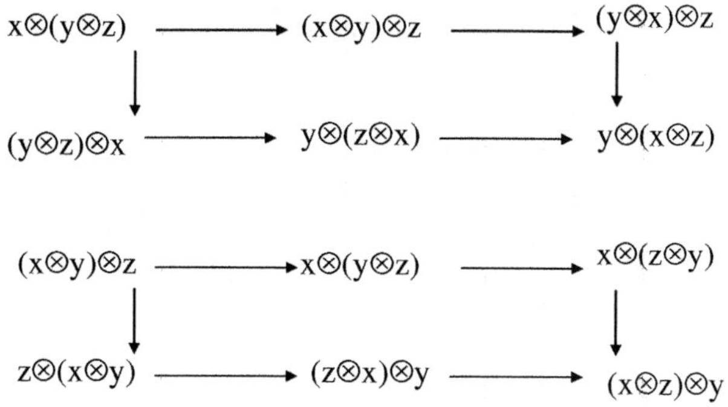


**Fig. A.3** Hexagon relations

Hexagon relation concerns braided monoidal categories and braiding.

Consider as an example from the periodic table the case n=1, k=2 of a doubly monoidal 1-category, a braided monoidal category. The braiding is: $B_{x,y}$: x⊗y→y⊗x.

The process of proving an equation becomes an isomorphism. This happens when we move one step right in the periodic table.

For codimension k=3 we should consider braiding versus inverse braiding. This introduces the notion of syllepsis.

Observe that we a faced with a hierarchy of higher braiding one for each codimension k≥2, each satisfying a hierarchy of laws.

A different proof of commutativity becomes a different isomorphism. $B_{y,x}^{-1}$: x⊗y→y⊗x This explains the existence of knots. A triply monoidal 1-category is a symmetric monoidal category. In this case we need three dimensions of space instead of just two. This makes the two ways of moving x past y equal. So the situation is more commutative. This happens when we move one step down in the periodic table.

It is interesting to lift the monoidal structure up a dimension into tricategories.

A tricategory may be defined on the basis of bicategories and these on the basis of categories. To characterize coherence the edges of the MacLane pentagon shown in Fig. A2 becomes five sides of a cube as shown in Fig. A4. This is the so-called parity 3-cube (Street 2004).

In this setting the state composition "⊗" is the Gray tensor product (Crans 1999, 2000).

Gray defined for 2-categories a product analogous to the conventional product for 1-categories (Kelly and Street 1974).

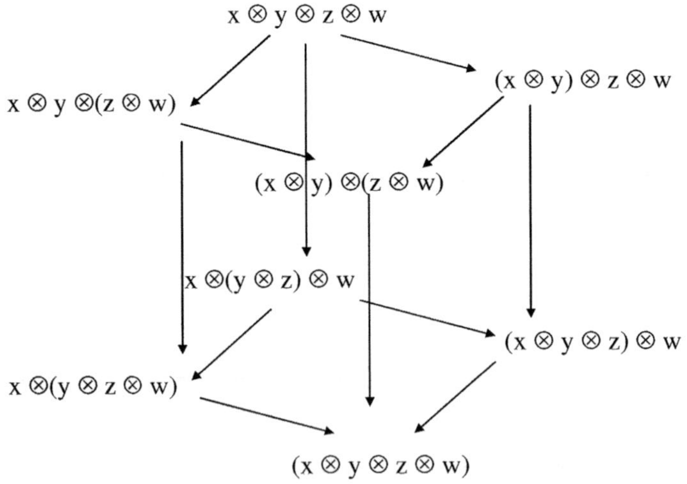


**Fig. A.4** Parity cube relations

Given two 2-categories C, D the Gray tensor C⊗D is informally defined to be the 2-category:

- With 0-cells given by products A⊗A' for all pairs (A, A') ∈ $C_0$ x $D_0$
- With 1-cells given by products A⊗f' and f⊗A' for all pairs (A, f') ∈ $C_0$ x $D_1$ and (f, A') ∈ $C_1$ x $D_0$
- With 2-cells generated by products A⊗$\phi$', f⊗f' and $\phi$⊗A' for all pairs (A, $\phi$') ∈ $C_0$ x $D_2$, (f, f') ∈ $C_1$ x $D_1$ and ($\phi$, A') ∈ $C_2$ x $D_0$ were f⊗f' denotes a 2-cell with specific properties (Kelly and Street 1974).

The horizontal composition of two 2-arrows results in a three dimensional arrow.

It should be emphasized the dimension raising aspect related to the Gray product.

## A.5 Computads or Polygraphs

Computads or polygraphs represent higher categorical diagrams of interest in the study of re-writing systems. They have been introduced by Street (1987) for the purposes of the presentation of strict n-categories. The informal presentation from Fig. A5 is in line with Burroni's polygrahs (Burroni 1993).

Consider a set $G_0$ whose elements are called atomic types. Denote by $G_0{}^*$ the free monoid on the set $G_0{}^*$ and $i_0$: $G_0 \rightarrow G_0{}^*$ the corresponding injection. The elements of $G_0{}^*$ are called types. Let us consider another set $G_1$, whose elements are called generators, together with two functions $s_0$, $t_0$: $G_1 \rightarrow G_0{}^*$ which to every generator associates a type called respectively its source and target. Denote also the sources and targets $s_1$, $t_1$: $G_2 \rightarrow G_1{}^*$, $s_2$, $t_2$: $G_3 \rightarrow G_2{}^*$, $s_0{}^*$, $t_0{}^*$: $G_1{}^* \rightarrow G_0{}^*$, $s_1{}^*$, $t_1{}^*$: $G_2{}^* \rightarrow G_1{}^*$, and the injections $i_1$: $G_1 \rightarrow G_1{}^*$, $i_2$: $G_2 \rightarrow G_2{}^*$.

Fig. A5 illustrates the inductive construction of the polygraphs.

The diagram restricted to: $G_0$, $G_0{}^*$ and $G_1$ corresponds to a 0-polygraph, the diagram restricted to: $G_0$, $G_0{}^*$, $G_1$, $G_1{}^*$ and $G_2$ corresponds to a 1-polygraph, the diagram restricted to: $G_0$, $G_0{}^*$, $G_1$, $G_1{}^*$, $G_2$, $G_2{}^*$ and $G_3$ corresponds to a 2-polygraph, and so on.

From commutativity condition it results: $s_0=s_0{}^*i_1$ and $s_0=t_0{}^*i_1$.

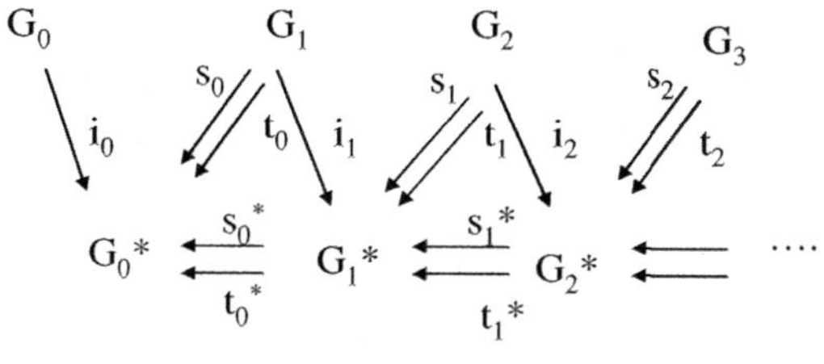


**Fig. A.5** Polygraphs

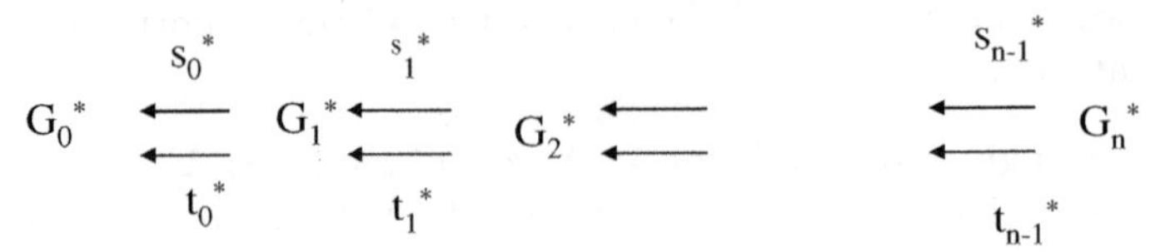


**Fig. A.6** Polygraphs and n-categories

For 1-polygraphs the relations $s_0{}^*s_1=s_0{}^*t_1$ and $t_0{}^*s_1=t_0{}^*t_1$ are verified.

For 2-polygraphs the relations $s_0{}^*s_1=s_0{}^*t_1$, $t_0{}^*s_1=t_0{}^*t_1$, $s_1{}^*s_2=s_1{}^*t_2$, $t_1{}^*s_2=t_1{}^*t_2$ are verified.

The 0-cell of a 0-polygraph is one of its elements. The elements of $G_1$ should be seen as 1-cells, the elements of $G_2$ should be seen as 2-cells and so on.

The 1-polygraph generates a 2-category, the 2-polygraph generates a 3-category.

The diagram from Fig. A6 shows an n-category.

Table A.3 illustrates the schema that may be associated to polygraphs.

The system evolves from $G_0$ an account of rules $G_0{}^*$ to $G_1$ and so on.

The rules are interconnected too but the sense of this interconnection may be reversed.

**Table A.3** Schema for polygraphs

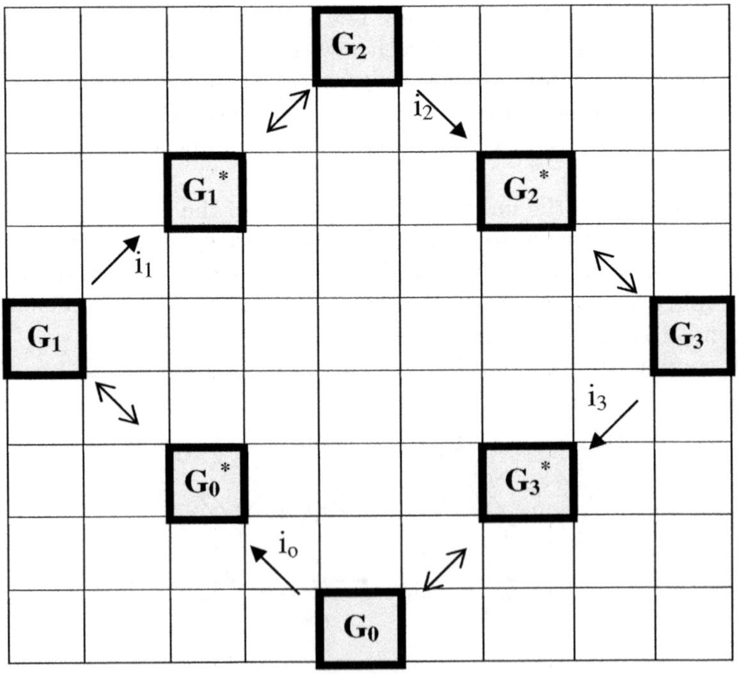

This implies an interchange of source and target roles. An integrative closure was considered in Table A.3.

The n-graphs may be considered as special cases of n-polygraphs (Burroni 1993).

Two i-cells are called parallel if they have the same source and the same target.

Any two 0-cells are considered to be parallel.

Fig. A7 shows an illustration of the n-graphs. The 0-graphs are sets and are pictured as points in the plane. The 1-graphs are directed graphs. The 2-graphs are graphs plus the 2-cells between paths of the same source and target, the 3-graphs are 2-graphs plus the 3-cells between paths of the same source and target. The 3-graphs allow calculating with graphs. It should be observed that the source and target operators have reverse directions and that the construction shown in Fig. A7 is restricted to 3-graphs.

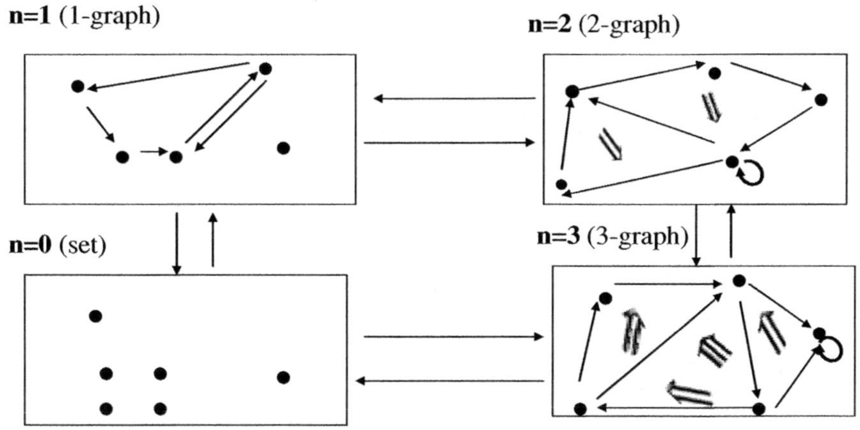


**Fig. A.7** n-graphs illustration

## A.6 Multicategories and Operads

A multicategory consists of objects, morphisms, composition operation and identities, like a standard category, the difference being that the domain of the morphism is not just a single type but a finite sequence of them (Leinster 2004). For example, vector spaces and linear maps form a category, vector spaces and multilinear maps form a multicategory. For a category an arrow, that is the morphism, has one object as its domain and one object as its codomain. For a multicategory the arrow looks like in Fig. A8 that is with a finite sequence of inputs as its domain and an output object as its codomain.

It was observed that multicategories are suitable to illustrate flow charts, separation schemes, circuit diagrams, and so forth. The unifying idea is that of information flow. Arrows can be composed when outputs are joined to

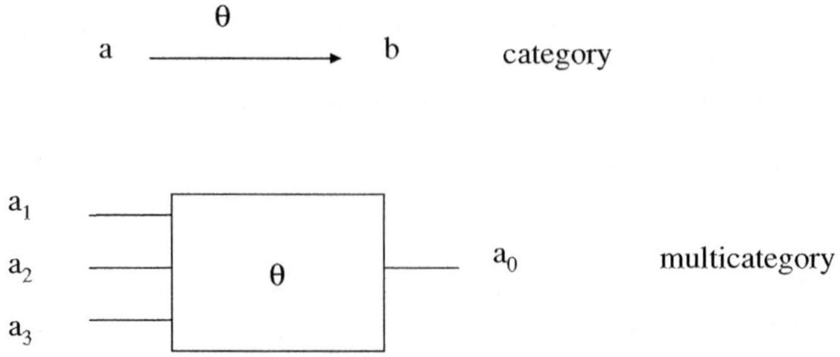


**Fig. A.8** Arrows for category and multicategory

inputs, which for categories, means that any string of arrows has a well defined composite: $a_0 \rightarrow a_1 \rightarrow a_2 \ldots \rightarrow a_n$. For multicategories, the composition means that any tree of arrows such as that in Fig. A9 has a well defined composite as an object of the type shown in Fig. A8, but with several inputs.

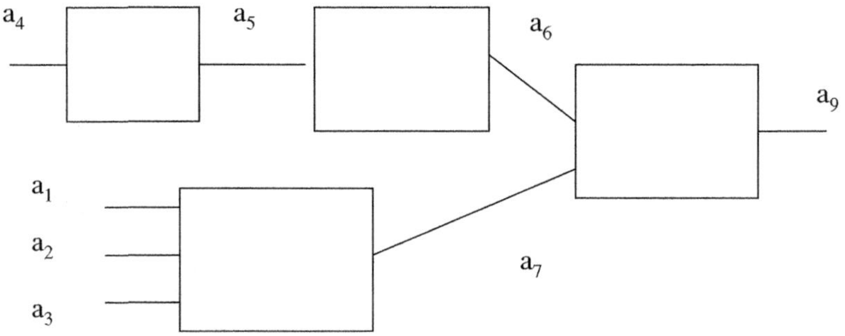


**Fig. A.9** Composable diagram of arrows in a multicategory

An operad is a multicategory with only one object. This directs to the alternative name coloured operads for multicategories (Leinster 2004).

To define an operad we need to refer to:

- A sequence of sets P (n) whose elements $\theta$ will be called n-ary operations of P and drown with n inputs
- For each n, $k_1, \ldots, k_n$ a function: $P(n) \times P(k_1) \times \ldots . \times P(k_n) = P(k_1 + \ldots \ldots + k_n)$ called composition. P(n) indicates the way to put together, while $P(k_1), \ldots ., P(k_n)$ indicates things to put together.
- An identity satisfying associativity and identity axioms

Fig. A10 shows examples of operads.

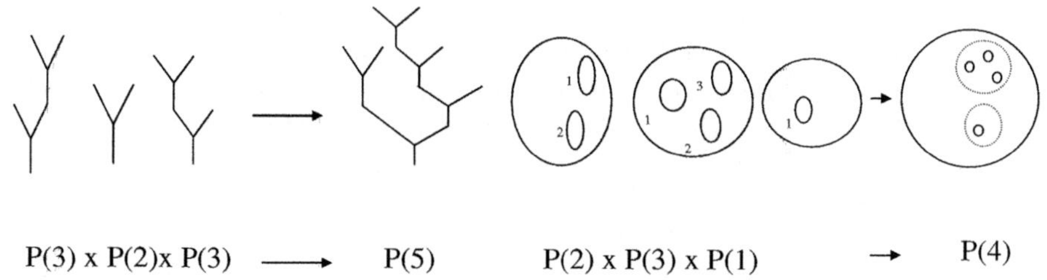


**Fig. A.10** Examples of operads

## A.7 Rewriting Systems

Rewriting systems have broad applications in general model of computation, Petri nets, and for automated theorem proving.

Rewriting rules specify the repeated replacement of sub-terms of a given formula with equivalent terms. A rewriting rule is a basic derivation that allows passing from one term in an appropriate language to another and the study of a rewriting system is the study of the compositions of such basic derivations. Higher dimensional categories naturally appear in the study of various rewriting systems.

String rewriting has been successfully applied in modelling plant development The L-system formalism (Lindenmayer 1968) is characterized by the parallel application of rewriting rules on strings representing a branching structure.

The string rewriting systems and the categorical notation for such systems are introduced following Johnson (Johnson 1991). If S is a set that is an alphabet, let us denote S* the free monoid on S. This means that the elements of S* are words in the alphabet S.

The string rewriting system consists of an alphabet S and a set R⊂ S* x S*. We write an element (s,s') ∈ R as s→s' and call it a rewrite of s to s'. An element w ∈ S* may be rewritten by finding a sub word of w witch match the left hand side of a rewrite and replacing it with the corresponding right hand side to obtain some w'. We could write w→w' and say that w and w' are → related. A sequence of such rewrites w→w'→. . . →w" is called a derivation.

Consider for example S={a,b,c,x,y,z} and R={xyz→a, ab→x}. An example of derivation beginning with w=xyzbyabc is: xyzbyzabc→abyzabc→ xyzabc→xyzxc→axc.

The application of the rewrite ab→x in the context abc→xc could have been carried out at any position in the sequence of rewrites. Thus a distinct but essentially equivalent derivation would be: xyzbyzabc→abyzabc→abyzxc →xyzxc→axc.

The rewrite taking abc→xc may be thought of as occurring in parallel with the other rewrites.

This can be made explicit in a 2-categorical perspective. The 2-categories provide a simple framework for string rewriting systems. A 2-category consists of objects indicated as points, arrows called 1-cells which go between objects and which compose in the usual way, and arrows called 2-cells which go between 1-cells and which compose horizontally or vertically. The 1-cells describe relations while the 2-cells describe relations between relations. Horizontal composition corresponds to parallel operations while vertical composition corresponds to sequential operations. In a 2-category all compositions are associative, there are identity 1-cells and 2-cells and the horizontal and vertical compositions interact according to the interchange law which say that composing first horizontally and then vertically gives the same result as composing first vertically and then horizontally. Observe that a 2-category C is a Cat-enriched category.

Given any string rewriting systems (S, R) one can freely construct a 2-category $C_{S,R}$. Then the hom category $C_{S,R}$ (∗, ∗) have as objects the elements of S* and as arrows the deviations modulo an equivalence relation. This is the category of string rewriting systems.

One of the main questions about a string rewriting system (S, R) is: when does each word in S* have a unique normal form under rewriting. This is significant because one of the principal applications of rewriting systems is the computation of congruence relations.

A rewriting system (S, R) generates a congruence ∼ R, by taking the reflexive symmetric transitive closure of the relation → on S*. If each word in S* has a unique normal form, then two words are congruent if and only if they have the same normal form. Thus to test for congruence it suffice to rewrite the words until their normal forms are found and then compare the normal forms.

Let (S, R) be a string rewriting system. A word w ∈ S* is called irreducible if w cannot be rewritten. The word w is a normal form for w' if w is irreducible and there is an arrow, that is, a derivation from w in $C_{S,R}$ (∗, ∗). The word w' has rank n if the longest derivation starting at w' contains n rewrites. The rewriting system (S, R) is terminating if each w ∈ S* has a finite rank.

The theory of rewriting focuses on terminating systems. Typically rewrites are chosen so that rewrites are complexity reducing.

A string rewriting system (S, R) is called locally confluent if for all w, x, y ∈ S*, with w → x and w → y there is some z ∈ S* and derivations from x to z and from y to z. This is also called the local confluence condition for w, x and y.

If (S, R) is locally confluent and terminating then, each word in S* has a normal form and it is unique.

A string rewriting system (S, R) is called locally confluent on overlaps if, whenever the left hand sides of two rewrite rules can be superimposed to form a single word w, then w and the two words obtained by applying the rewrites to it, say x and y satisfies the local confluence condition. In this case the two words x and y have been called a critical pair.

If (S, R) is locally confluent on overlaps and terminating then each word in S* has a normal form and it is unique.

A term rewriting system consists of:

- A set of types
- A set of typed variables
- A set of typed functions, determining a language
- A set of pairs (x, y) of terms in the language such that the variables occurring in y are a subset of the variables occurring in x. The elements of this set are called rewrite schema. A term rewriting system is called linear if for each rewrite schema (x, y) each variable occurs at most once in x and at most once in y.

The theory of term rewriting system is complicated by the interaction of substitution and rewriting.

The linear rewriting schemes may be studied as 3-categories.

The 2-categories have objects, morphisms, 2-morphisms and some axioms while the 3-categories have objects, morphisms and 3-morphisms together with some axioms.

If the structures of interest are 2-categories the computations in these structures are 3-categories. A 3-category is a 2-Cat enriched category. In this case there are arrows called 3-cells which go between 2-cells. The 3-cells are generated by the rewrite schema.

The 3-categorical formulation of term rewriting systems allows an axiomatization of rewriting and critical pair completion in terms of patterns, (certain specified 2-cells), multipliers, (2-categorical composition) and replacements, (certain specified 3-cells) (Buchberger 1987, Johnson 1991). A generalization of rewriting to categories and to actions of categories is due to Brown and Heyworth (2000).

A useful notion in the study of rewriting systems is that of Gröbner basis.

The area of computer algebra called Gröbner basis provides methods for handling the rule systems defining various types of algebraic structure. Gröbner basis have been successfully applied in theorem proving, integer programming, coding theory, signal processing and Petri nets (Adams and Loustaunau 1994, Buchberger and Winkler 1998, Chandler and Heyworth 2001). Domains of direct industrial interest as pharmaceutics, enzyme kinetics, categorized component mixture experiments, robotics, and other make use of Gröbner bases methods (Pistone et al. 2000, Piepel 1999).

Petri nets are graphical and mathematical modeling tools that may be used for information processing systems that are concurrent, asynchronous, distributed, parallel, deterministic or stochastic (Jensen and Rosenberg 1991). The Petri nets may be considered as significant illustration of graphical rewriting and polygraphs (Guiraud 2006). Petri nets are graphically useful for describing systems and tokens can simulate the dynamic and concurrent activities. The diagnostic and failure analysis is an established domain of application.

A Petri net has two types of vertices: places represented by circles and transitions represented by double lines. Edges exist only between places and transitions and are labeled with their weights. In modeling places represent conditions and transitions represent events. A transition has input and output places, which represent preconditions and post-conditions of the event.

A significant problem in Petri net theory is reachability-the problem corresponds to deciding which situations modeled by the net are possible given some sequence of events.

The reachability problems have been studied making use of Gröbner basis (Chandler and Heyworth 2001). Reachability studies are significant for evolvable technologies design and control and for failure analysis.

Multi level framework for diagnostic goal driven modeling for multi-scale chemical process systems have been studied as high level Petri nets (Németh et al. 2005).

High level Petri nets including Gröbner bases have been designed to model software interfaces to devices of evolvable platforms.

## References


Adams, W.W., Loustaunau, P.: An introduction to Gröbner bases. Graduate Studies in mathematics, vol. 3. American Math. Soc., Providence (1994)

Baez, J.: An introduction to n-categories. In: Moggi, E., Rosolini, G. (eds.) CTCS 1997. LNCS, vol. 1290, Springer, Heidelberg (1997)

Baez, J., Dolan, J.: Higher dimensional algebra and topological quantum field theory. Jour. Math. Phys 36, 6073–6105 (1995)

Baez, J., Dolan, J.: Categorification. In: Getzler, E., Kapranov, M. (eds.) Higher Category Theory, Contemp. Math., vol. 230, pp. 1–36. Amer. Math. Soc., Providence (1998)

Berners-Lee, T., Hendler, J., Lassila, O.: The semantic web. Sci. Am. 284(5), 35–43 (2001)

Brown, R., Heyworth, A.: Using rewriting systems to compute left Kan extensions and induced actions of categories. J. Symbol Comput. 29(1), 5–31 (2000)

Buchberger, B.: History and basic features for the critical pair/completion procedure. J. Symbolic Computation 3, 33–38 (1987)

Buchberger, B., Winkler, F.: Gröbner Bases and Applications, 33 Years of Gröbner Bases. Proc. London Math. Soc., 251 (1998)

Burroni, A.: Higher-dimensional word problems with applications to equational logic. Theoretical Computer Science 115, 43–62 (1993)

Chandler, A., Heyworth, A.: Gröbner bases as a tool for Petri net analysis. In: Proceedings of the 5th World Multi-Conference on Systematics, Cybernetics and Informatics (SCI 2001), Orlando, Florida (2001)

Cornell, J.: Experiments with mixtures: designs, models and the analysis of mixture data. Wiley, New York (2002)

Crans, S.: A tensor product for Gray categories. Theory and Applications of Categories 5(2), 12–69 (1999)

Crans, S.: On braidings, syllepses ans symmetries. Cahiers Topologie Geom Differentielle Categ. 41(1), 2–74 (2000)

Giry, M.: A categorical approach to probability theory. In: Banaschewski, B. (ed.) Categorical Aspects of Topology and Analysis. Lecture Notes in Mathematics, vol. 915, pp. 68–85. Springer, Berlin (1981)

Gordon, R., Power, A.J., Street, R.: Coherence for tricategories. Memoirs Amer. Math. Soc. 117, 558 (1995)

Guiraud, Y.: Two polygraphic presentations of Petri nets. Theoretical Computer Science 360(1-3), 126–146 (2006)

Jensen, K., Rosenberg, G.: High-level Petri nets: Theory and Application. Springer, Heidelberg (1991)

Johnson, M.: Linear term rewriting systems are higher dimensional string rewriting systems. In: Rattray, C.M.I., Clark, R.G. (eds.) The Unified Computation Laboratory, pp. 101–110. Oxford University Press, Oxford (1991)

Joyal, A., Street, R.H., Verity, D.: Traced monoidal categories. Math. Proc. Camb. Phil. Soc. 119, 447–468 (1996)

Kelly, G.M., Street, R.: Review of the elements of 2-categories. Lecture Notes in Math, vol. 420, pp. 75–103 (1974)

Leinster, T.: Higher Operads, Higher Categories, Cambridge (2004)

Lindenmayer, A.: Mathematical models for cellular interaction in development, I, II. J. Theor. Biol. 18, 280–315 (1968)

MacLane, S.: Categories for the Working Mathematician. Springer, New York (1971)

Marquis, J.-P.: From a Geometrical Point of View: a study in the history and philosophy of category theory. Springer, New York (2009)

Németh, E., Cameron, I.T., Hangos, K.M.: Diagnostic goal driven modelling and simulation of multiscale process systems. Computers And Chemical Engineering 29, 783–796 (2005)

Piepel, G.: Modeling methods for mixture-of-mixtures experiments applied to a tablet formulation problem. Pharmaceutical Development and Technology 4, 593–606 (1999)

Pistone, G., Riccomagno, E., Wynn, H.P.: Algebraic Statistics. Chapman & Hall/CRC, Boca Raton (2000)

Street, R.: The algebra of oriented simplexes. J. Pure Appl. Algebra 49, 283–335 (1987)

Street, R.: Categorical and combinatorial aspects of descent theory. Applied Categorical Structures 12, 537–576 (2004)

# Index

GPSR Compliance

*The European Union's (EU) General Product Safety Regulation (GPSR) is a set of rules that requires consumer products to be safe and our obligations to ensure this.*

*If you have any concerns about our products, you can contact us on ProductSafety@springernature.com*

In case Publisher is established outside the EU, the EU authorized representative is:

Springer Nature Customer Service Center GmbH
Europaplatz 3
69115 Heidelberg, Germany

**Batch number: 10755284**

Printed by Printforce, the Netherlands